Mathematical Models for Eddy Currents and Magnetostatics

Scientific Computation

For further volumes:
http://www.springer.com/series/718

Rachid Touzani • Jacques Rappaz

Mathematical Models for Eddy Currents and Magnetostatics

With Selected Applications

Springer

Rachid Touzani
Université Blaise Pascal
Clermont-Ferrand, France

Jacques Rappaz
Ecole Polytechnique Fédérale de Lausanne
Lausanne, Switzerland

ISSN 1434-8322
ISBN 978-94-017-7816-9 ISBN 978-94-007-0202-8 (eBook)
DOI 10.1007/978-94-007-0202-8
Springer Dordrecht Heidelberg New York London

Printed on acid-free paper

Springer is part of Springer Science+Business Media (www.springer.com)

Preface

Eddy current processes consist in generating currents in electrically conducting material by means of time dependent magnetic fields. The term 'eddy' originates from the fact that these induced currents create magnetic field vortices inside the conductors. Eddy current applications are widely used and exploited in industrial devices for numerous purposes such as:

- Induction heating: Temperature raise by using the Joule heating produced by eddy currents. The heated material can then be treated for hardening, forging or brasing.
- Metal melting: such as in induction furnaces, metal forming and casting, and cold crucibles.
- Non-destructive testing and evaluation to detect flaws and cracks in materials.

All these processes involve various physical phenomena (fluid flow, metal deformation, heat transfer, ...) on different space and time scales. This complexity yields a great difficulty in experimental investigations since these phenomena cannot be easily isolated. Mathematical modelling and numerical simulation constitute then a challenging alternative. This modelling clearly involves, in addition to electromagnetic phenomena, other physical fields that must be coupled to electromagnetics in order to produce reliable simulation software.

The aim of this textbook is to present a mathematical framework for the description of eddy current processes and give efficient numerical methods to solve these problems. Indeed, mathematical modelling and numerical solution of eddy current processes involve a wide variety of difficulties, especially for the fully three-dimensional case. For instance, most interesting models are formulated in unbounded domains. The numerical solution of such problems can be handled either by an artificial boundary method, which leads to inaccurate solutions, or by using integral representations which are numerically accurate but which are difficult to implement because of the presence of singularities in the involved integrals and the nonclassical matrix structure obtained after space discretization. We have here systematically adopted the latter approach. A typical difficulty in such modelling is that eddy currents are created by electrotechnical setups where a power, voltage or current are applied. This means that the derived sets of equations should not require any source term or boundary distributed data. Only a number that represents power, voltage or current is to be provided. This point of view, already used in Bossavit [32], Bossavit–Vérité [36], Hiptmair–Sterz [101], among other references, is chosen here.

This book is intended for at least two kinds of audience:

- Applied mathematicians who can find in this work a self-contained collection of mathematical results on modelling of eddy current processes with known mathematical results. The second part of the book provides also a source of mathematical and numerical problems, presented as sets of nonlinear partial differential equations. Some mathematical results are available for these problems but some open problems remain to be solved.
- Researchers and developers in the electric and electrotechnical engineering fields who are mostly interested in deriving clean mathematical models to simulate industrial applications related to eddy current processes. The numerical examples given in this book are indeed mostly borrowed from real industrial applications and have resulted in numerical software currently used in an industrial environment.

The book is divided into two parts: The first part is devoted to a mathematical presentation of various eddy current models that differ from each other by the geometry type and the second part in which various applications of eddy current processes are presented and investigated.

In the first part, we start by defining in Chap. 1 the basic functional spaces that are used for the mathematical definition of the problems. We also state the main mathematical results of vector analysis and then some results on the integral representation of harmonic fields in two and three dimensions. Most of the given results in this chapter are known and reported here for self-consistency. The notion of *radial at the infinity* is however new, as far as we know, and gives in our opinion a physical foundation to the chosen functional spaces.

In Chap. 2, the Maxwell equations in electromagnetic theory are given. We focus on eddy current or low frequency approximation. All these models are considered for time harmonic (or *quasi-static*) regimes. The last section deals with the magnetostatic case where no induction is involved. In this chapter we also describe the role of vector and scalar potentials in the derivation of the equations.

Chapter 3 presents the main two-dimensional models for eddy currents. Namely, we describe two types of models. The first one consists in considering magnetic fields that have only one non-vanishing component. This assumption leads to boundary value problem formulations; the second one considers current densities with only one non-vanishing component. Here we are led to a coupled interior/exterior problem, which implies a coupling between a partial differential equation in the conductors and an integral equation that represents an external harmonic field. Axisymmetric models are then considered. They are formulated in terms of a scalar potential.

Chapter 4 presents the fully three-dimensional case. Various formulations are given starting by the electric current density model, then the so called $\boldsymbol{H}$ (magnetic field) model due to Bossavit and Vérité [36]. This model seems to be better adapted to numerical treatment. Indeed, it gives a coupling between the magnetic field in the conductors and a scalar potential in the free space. Integral representations enable us to formulate the exterior problem on the boundary of the conductors. We end with an electric field model that is generally less used but can be envisaged for some practical applications.

In Chap. 5 we consider the models of Chap. 4 where rotational symmetry is assumed. Mathematical setting of these problems is given using cylindrical coordinates and assuming angle invariance.

In Chap. 6, we consider the derivation of mathematical models where the inductors are assumed to be thin enough. We report some mathematical results justifying the use of asymptotic approximations. Most of the given results in this chapter are proven. The three-dimensional case is however incomplete since some results are proven while the final eddy current model for a thin inductor is not mathematically justified but given as a conjecture.

Chapter 7 presents the main numerical methods to solve the given problems in Chaps. 3–5. We recall for this the finite element method for various types of problems and its coupling with the boundary element method.

The second part of the book is divided into five chapters, each one presenting a particular application of eddy current processes with some known mathematical results and numerical simulations.

Chapter 8 considers induction heating applications. We present stationary and time dependent versions of the problem and report mathematical results for these problems. We then present a numerical simulation of an induction heating problem in the 3-D case. We present next an example of optimization of an induction heating process. The presentation is based on a thixoforming problem in the aluminium industry. We formulate this in terms of an optimal control problem, then formally derive the optimality system and a numerical simulation.

Chapter 9 is devoted to magnetohydrodynamics and magnetic shaping. We start by giving the set of equations for an incompressible magnetohydrodynamic problem, and mention some mathematical results in the 2-D case. The extension to a magnetic shaping problem, formulated as a free boundary problem is then given. This problem was studied by several authors and some main mathematical results are given. We end by describing an electromagnetic casting problem considered as a typical industrial application.

In Chap. 10, we consider an application in magnetohydrodynamics using compressible flows: Inductively coupled plasma torches consist in heating gas to ionization by means of induction heating. We present the mathematical model coupling eddy current equations with compressible Navier-Stokes equations, give a numerical procedure to solve the derived set of equations and then a numerical simulation.

Chapter 11 gives a magnetostatic application in ferromagnetics. This implies a nonlinear problem where the nonlinearity is due to the dependency of the magnetic permeability on the magnetic field. We give a mathematical formulation of the derived problem, prove the existence of a solution and derive an iterative procedure for its numerical solution.

In Chap. 12, we present a complete model for the simulation of the electrolytic process in the aluminium industry. The model couples electromagnetics, incompressible hydrodynamics and a free boundary problem. We present a model with some simplifying hypotheses and give an efficient iterative procedure that has resulted in a numerical code used in industrial environment.

To summarize, Part I states the best known mathematical results and gives proofs of well–posedness of the given problems, while Part II contains a large variety of applications of eddy current processes with some mathematical results, and with numerical procedures and simulations. These two parts can be read independently in the sense that all applications of Part II assume the results of Part I as known. In addition all the chapters of Part II are independent and can then be read separately.

At the end, we mention that this book is the fruit of years of work on various projects, either academic or industrial, that have had eddy currents as a central subject. Naturally, this effort was not possible without the collaboration of many colleagues and PhD students. For this we thank:

- O. Besson, J. Bourgeois, P.-A. Chevalier, S. Clain, M. Flück, S. Gauthier, A. Masserey, C. Parietti, M. Picasso, D. Rochette, R. Rozsnyo, G. Steiner, M. Swierkosz, C. Trophime, for their contributions either by managing projects related to the presented applications in Part II, or by developing software for numerical simulations.
- M. Pierre who has provided a result in Chap. 1.
- J. Descloux who has helped us with multiple and precious advice during the execution of industrial projects related to eddy currents.
- J.-C. Nédélec, Y. Amirat, S. Clain and D. Rochette who have carefully read some parts of the manuscript and helped us improving this work.

Clermont-Ferrand, France — Rachid Touzani
Lausanne, Switzerland — Jacques Rappaz
May 2013

Contents

Part I
Eddy Current Models

Chapter 1
Mathematical Framework

1.1 Introduction

This first chapter is devoted to some basic results in functional analysis that will be used throughout the book. Most of the results quoted in this chapter are known and can be found for instance in Dautray–Lions [62]. Some sharper results can also be found in Amrouche et al. [14], Ciarlet–Sonnendrucker [51], Girault–Raviart [93], Nédélec [135, 138]. We have chosen to reproduce them here to ensure some self-consistency. Some results like in Theorems 1.3.3 and 1.4.4 are however new up to our knowledge. They constitute a contribution to establish that some mathematical hypotheses are actually originated from relevant physical considerations.

Let us start by defining some notation standards. In order to respect traditional notations in the electromagnetic literature, all fields in electromagnetics will be denoted by capital letters whereas vector fields will be distinguished with bold faces. Due to this constraint, function spaces will be invoked by using calligraphic fonts. In addition, when these spaces concern vector-valued functions then the spaces will be written with bold faces. For instance, $\mathcal{L}^2(\Omega)$ stands for the space of measurable functions which are square Lebesgue–integrable on Ω, and $\boldsymbol{\mathcal{L}}^2(\Omega) := (\mathcal{L}^2(\Omega))^d$ when $\Omega \subset \mathbb{R}^d$, $d = 2$ or 3.

A generic point of the space will be denoted by $\boldsymbol{x} \in \mathbb{R}^d$ and the vector $\boldsymbol{e}_i$, $1 \le i \le d$, will stand for the i-th vector of the canonical basis of $\mathbb{R}^d$. The euclidean norm of $\boldsymbol{x}$ is denoted by

$$|\boldsymbol{x}| = \Big(\sum_{i=1}^{d} x_i^2\Big)^{\frac{1}{2}}, \quad \text{where } \boldsymbol{x} = \sum_{i=1}^{d} x_i \boldsymbol{e}_i .$$

Finally, since we shall deal with complex valued functions, the symbol i will stand for the unit imaginary number ($\mathrm{i}^2 = -1$) and for a complex number z, then $\bar{z}$, $\mathrm{Re}(z)$, $\mathrm{Im}(z)$ and $|z|$ are respectively the complex conjugate, the real, the imaginary parts and the modulus of z.

R. Touzani and J. Rappaz, *Mathematical Models for Eddy Currents and Magnetostatics: With Selected Applications*, Scientific Computation, DOI 10.1007/978-94-007-0202-8_1,

1.2 Preliminaries

We start by giving some classical existence and uniqueness results that will be widely used in this book. For this, we assume the reader is familiar with basic notions in functional analysis of Banach and Hilbert spaces. We also assume a basic knowledge of the theory of elliptic equations.

Let $\mathcal{X}$ denote a complex Banach space equipped with the norm $\|\cdot\|_{\mathcal{X}}$, the space $\mathcal{X}'$ will stand for its antidual space, i.e. the space of continuous antilinear forms on $\mathcal{X}$, equipped with the norm

$$\|g\|_{\mathcal{X}'} := \sup_{v\in\mathcal{X},\|v\|_{\mathcal{X}}=1} < g, v >_{\mathcal{X}',\mathcal{X}},$$

where $< \cdot, \cdot >_{\mathcal{X}',\mathcal{X}}$ denotes the duality pairing between $\mathcal{X}'$ and $\mathcal{X}$.

The first result is the classical Lax–Milgram theorem (see [92] for instance).

Theorem 1.2.1 (Lax–Milgram). *Let $\mathcal{V}$ denote a complex Hilbert space and let $\mathscr{B}$ denote a sesquilinear continuous form on $\mathcal{V}$ that is, in addition, coercive, i.e. there exists a real number $\alpha > 0$ such that*

$$|\mathscr{B}(v, v)| \geq \alpha \|v\|_{\mathcal{V}}^2 \quad \forall\, v \in \mathcal{V}.$$

Then, for each antilinear continuous form $\mathscr{L} \in \mathcal{V}'$, there exists a unique $u \in \mathcal{V}$ such that

$$\mathscr{B}(u, v) = \mathscr{L}(v) \quad \forall\, v \in \mathcal{V}.$$

Moreover, there exists a constant C independent of $\mathscr{L}$ such that

$$\|u\|_{\mathcal{V}} \leq C\, \|\mathscr{L}\|_{\mathcal{V}'}.$$

The second existence and uniqueness result is the so-called Babuška–Brezzi–Ladyzhenskaya theorem, the proof of which can be found in Girault–Raviart [93] for instance.

Theorem 1.2.2. *Let $\mathcal{V}$ and $\mathcal{Q}$ denote two complex Hilbert spaces and let $\mathscr{A}$ and $\mathscr{B}$ denote two sesquilinear continuous forms on $\mathcal{V} \times \mathcal{V}$ and $\mathcal{V} \times \mathcal{Q}$ respectively. Let us define the space*

$$\mathcal{V}_0 = \{\, v \in \mathcal{V};\ \mathscr{B}(v, q) = 0\ \forall\, q \in \mathcal{Q}\,\}.$$

We assume that there exist constants $\alpha, \beta > 0$ such that

$$|\mathscr{A}(v, v)| \geq \alpha\, \|v\|_{\mathcal{V}}^2 \qquad \forall\, v \in \mathcal{V}_0, \tag{1.1}$$

$$\sup_{v \in \mathcal{V} \setminus \{0\}} \frac{|\mathscr{B}(v, q)|}{\|v\|_{\mathcal{V}}} \geq \beta \, \|q\|_{\mathcal{Q}} \qquad \forall \, q \in \mathcal{Q}. \tag{1.2}$$

Then, for each antilinear continuous forms $\mathscr{L} \in \mathcal{V}'$ and $\mathscr{N} \in \mathcal{Q}'$, there exists a unique pair $(u, p) \in \mathcal{V} \times \mathcal{Q}$ such that

$$\begin{aligned} \mathscr{A}(u, v) + \mathscr{B}(v, p) &= \mathscr{L}(v) \qquad && \forall \, v \in \mathcal{V}, \\ \mathscr{B}(u, q) &= \mathscr{N}(q) \qquad && \forall \, q \in \mathcal{Q}. \end{aligned}$$

Moreover, there exists a constant C independent of $\mathscr{L}$ and $\mathscr{N}$ such that

$$\|u\|_{\mathcal{V}} + \|p\|_{\mathcal{Q}} \leq C \, (\|\mathscr{L}\|_{\mathcal{V}'} + \|\mathscr{N}\|_{\mathcal{Q}'}).$$

For the sake of clarity, we shall in the following distinguish the two and three-dimensional cases.

1.3 The Three-Dimensional Case

We shall deal, in all the following, with a conductor device denoted by Ω. The domain Ω is firstly assumed to be open, bounded and connected for the sake of simplicity. Naturally, in view of applications, cases with multiple conductors will be mentioned. The boundary of Ω is denoted by Γ and is assumed to be smooth enough (piecewise C^1, say) with outward unit normal $\boldsymbol{n}$. We shall furthermore denote by Ω_{ext} the complement of the closure $\overline{\Omega}$ of Ω, that is $\Omega_{\text{ext}} := \mathbb{R}^3 \setminus \overline{\Omega}$. The domain Ω is connected but can be simply connected or not. This topic will be addressed when needed.

Let u denote a function defined in $\mathbb{R}^3$ such that the restriction of u to Ω (*resp.* Ω_{ext}) is continuous and admits a continuous extension on Γ. Then $u_{|\Gamma-}$, or simply u^- (*resp.* $u_{|\Gamma+}$ or u^+) will stand for the inner (*resp.* outer) restriction to Γ, with respect to Ω defined for almost every $\boldsymbol{x} \in \Gamma$ by

$$\begin{aligned} u_{|\Gamma-}(\boldsymbol{x}) &= \lim_{s \to 0,\ s>0} u(\boldsymbol{x} - s\boldsymbol{n}(\boldsymbol{x})), \\ u_{|\Gamma+}(\boldsymbol{x}) &= \lim_{s \to 0,\ s>0} u(\boldsymbol{x} + s\boldsymbol{n}(\boldsymbol{x})). \end{aligned}$$

The function $[u]_\Gamma := u_{|\Gamma+} - u_{|\Gamma-}$ will stand for the jump. Note that we shall sometimes use for convenience the notation $[u(\boldsymbol{x})]$ rather than $[u](\boldsymbol{x})$. Moreover, the subscript will be omitted when no confusion is possible.

1.3.1 Function Spaces

We now recall some function spaces and some of their elementary properties we will need in the sequel. More detailed description of these spaces can be found in [3] or [92] for instance.

We start by recalling some used differential operators. For a continuously differentiable scalar field $\phi : \mathbb{R}^3 \to \mathbb{C}$ and vector field $\boldsymbol{u} : \mathbb{R}^3 \to \mathbb{C}^3$, with $\boldsymbol{u} = u_1\boldsymbol{e}_1 + u_2\boldsymbol{e}_2 + u_3\boldsymbol{e}_3$, we define the fields and tensors:

$$\nabla\phi := \frac{\partial\phi}{\partial x_1}\boldsymbol{e}_1 + \frac{\partial\phi}{\partial x_2}\boldsymbol{e}_2 + \frac{\partial\phi}{\partial x_3}\boldsymbol{e}_3,$$

$$(\nabla\boldsymbol{u})_{ij} := \left(\frac{\partial u_i}{\partial x_j}\right), \qquad 1 \le i, j \le 3,$$

$$\operatorname{div}\boldsymbol{u} := \frac{\partial u_1}{\partial x_1} + \frac{\partial u_2}{\partial x_2} + \frac{\partial u_3}{\partial x_3},$$

$$\mathbf{curl}\,\boldsymbol{u} := \left(\frac{\partial u_3}{\partial x_2} - \frac{\partial u_2}{\partial x_3}\right)\boldsymbol{e}_1 + \left(\frac{\partial u_1}{\partial x_3} - \frac{\partial u_3}{\partial x_1}\right)\boldsymbol{e}_2 + \left(\frac{\partial u_2}{\partial x_1} - \frac{\partial u_1}{\partial x_2}\right)\boldsymbol{e}_3.$$

Let $\mathscr{D}(\Omega)$ stand for the space of indefinitely differentiable functions over Ω with a compact support in Ω and $\mathcal{L}^2(\Omega)$ the space of functions which are Lebesgue square integrable over Ω. More generally, $\mathcal{L}^p(\Omega)$ is the space of functions u defined over Ω and such that:

$$\int_\Omega |u|^p \, d\boldsymbol{x} < \infty \qquad \text{for } 1 \le p < \infty,$$

$$\operatorname*{Ess\,Sup}_{\boldsymbol{x}\in\Omega} |u(\boldsymbol{x})| < \infty \qquad \text{for } p = \infty.$$

These spaces are complex Banach spaces when endowed with the norms:

$$\|u\|_{\mathcal{L}^p(\Omega)} := \left(\int_\Omega |u|^p \, d\boldsymbol{x}\right)^{\frac{1}{p}} \qquad \text{for } 1 \le p < \infty,$$

$$\|u\|_{\mathcal{L}^\infty(\Omega)} := \operatorname*{Ess\,Sup}_{\boldsymbol{x}\in\Omega} |u(\boldsymbol{x})| \qquad \text{for } p = \infty.$$

Note that all these spaces can be defined in the same way when Ω is replaced by the unbounded domain Ω_{ext} or by $\mathbb{R}^3$. Furthermore, for a domain Ω, the subscript loc in the space name means that the property of the space holds in all compact subsets of Ω.

Let $\alpha = (\alpha_1, \alpha_2, \alpha_3)$ a multi-integer, i.e. $\alpha_1, \alpha_2, \alpha_3$ are nonnegative integers, with $|\alpha| = \alpha_1 + \alpha_2 + \alpha_3$. For $1 \le p \le \infty$, we introduce the Sobolev spaces

$$\mathcal{W}^{m,p}(\Omega) := \left\{ v \in \mathcal{L}^p(\Omega);\ \frac{\partial^{|\alpha|} v}{\partial x_1^{\alpha_1}\, \partial x_2^{\alpha_2}\, \partial x_3^{\alpha_3}} \in \mathcal{L}^p(\Omega) \text{ with } |\alpha| \le m \right\},$$

where the partial derivatives are defined in the sense of distributions. These spaces are Banach spaces when endowed with the norms

$$\|v\|_{\mathcal{W}^{m,p}(\Omega)} := \Big(\sum_{|\alpha|\le m} \Big\| \frac{\partial^{|\alpha|} v}{\partial x_1^{\alpha_1} \partial x_2^{\alpha_2} \partial x_3^{\alpha_3}} \Big\|^p_{\mathcal{L}^p(\Omega)} \Big)^{\frac{1}{p}} \qquad \text{for } 1 \le p < \infty,$$

$$\|v\|_{\mathcal{W}^{m,\infty}(\Omega)} := \sup_{|\alpha|\le m} \Big\| \frac{\partial^{|\alpha|} v}{\partial x_1^{\alpha_1} \partial x_2^{\alpha_2} \partial x_3^{\alpha_3}} \Big\|_{\mathcal{L}^\infty(\Omega)} \qquad \text{for } p = \infty.$$

The space $\mathcal{W}_0^{m,p}(\Omega)$ is defined as the closure of $\mathcal{D}(\Omega)$ in $\mathcal{W}^{m,p}(\Omega)$.

When $p = 2$, the spaces $\mathcal{W}^{m,2}(\Omega)$ or $\mathcal{W}_0^{m,2}(\Omega)$ are Hilbert spaces and we denote them by

$$\mathcal{H}^m(\Omega) = \mathcal{W}^{m,2}(\Omega),\ \mathcal{H}_0^m(\Omega) = \mathcal{W}_0^{m,2}(\Omega), \quad m \ge 1.$$

In $\mathcal{W}^{m,p}(\Omega)$ we define the quantities:

$$|v|_{\mathcal{W}^{m,p}(\Omega)} := \Big(\sum_{|\alpha|= m} \Big\| \frac{\partial^{|\alpha|} v}{\partial x_1^{\alpha_1} \partial x_2^{\alpha_2} \partial x_3^{\alpha_3}} \Big\|^p_{\mathcal{L}^p(\Omega)} \Big)^{\frac{1}{p}} \qquad \text{for } 1 \le p < \infty,$$

$$|v|_{\mathcal{W}^{m,\infty}(\Omega)} := \sup_{|\alpha|= m} \Big\| \frac{\partial^{|\alpha|} v}{\partial x_1^{\alpha_1} \partial x_2^{\alpha_2} \partial x_3^{\alpha_3}} \Big\|_{\mathcal{L}^\infty(\Omega)} \qquad \text{for } p = \infty,$$

which are semi-norms on these spaces. In addition, the semi-norm $|\cdot|_{\mathcal{H}^1(\Omega)}$ is a norm on $\mathcal{H}_0^1(\Omega)$ equivalent to the norm $\|\cdot\|_{\mathcal{H}^1(\Omega)}$.

The space of traces of functions of $\mathcal{H}^1(\Omega)$ when Ω is bounded, is the Sobolev space $\mathcal{H}^{\frac{1}{2}}(\Gamma)$, and $\mathcal{H}^{-\frac{1}{2}}(\Gamma)$ is its dual space. We shall hereafter, for the sake of simplicity, denote by an integral sign this duality product, i.e.

$$\int_\Gamma \lambda\, \bar{u}\, ds := <\lambda, u>_{\mathcal{H}^{-\frac{1}{2}}(\Gamma), \mathcal{H}^{\frac{1}{2}}(\Gamma)} \qquad \forall\, \lambda \in \mathcal{H}^{-\frac{1}{2}}(\Gamma), u \in \mathcal{H}^{\frac{1}{2}}(\Gamma).$$

We have the trace inequality (see [92]),

$$\|u\|_{\mathcal{H}^{\frac{1}{2}}(\Gamma)} \le C\, \|u\|_{\mathcal{H}^1(\Omega)} \qquad \forall\, u \in \mathcal{H}^1(\Omega). \tag{1.3}$$

Note that if $\lambda \in \mathcal{H}^{-\frac{1}{2}}(\Gamma)$ then by using the Lax-Milgram theorem (Theorem 1.2.1) on $\mathcal{H}^1(\Omega)$, we have the existence of a unique $u \in \mathcal{H}^1(\Omega)$ such that

$$\int_{\Gamma} \lambda \overline{\varphi} \, ds = \int_{\Omega} (\nabla u \cdot \nabla \overline{v} + u \overline{v}) \, d\boldsymbol{x},$$

for all $\varphi \in \mathcal{H}^{\frac{1}{2}}(\Gamma)$ and for all $v \in \mathcal{H}^1(\Omega)$ satisfying $v = \varphi$ on Γ. Let us remark that if $\lambda \in \mathcal{H}^{\frac{1}{2}}(\Gamma)$ and $u \in \mathcal{H}^2(\Omega)$ then this relation corresponds to

$$\begin{cases} -\Delta u + u = 0 & \text{in } \Omega, \\ \dfrac{\partial u}{\partial n} = \lambda & \text{on } \Gamma. \end{cases}$$

Note also that there exists a constant C such that if $g \in \mathcal{H}^{\frac{1}{2}}(\Gamma)$ there exists a function $u \in \mathcal{H}^1(\Omega)$ satisfying $u = g$ on Γ and

$$\|u\|_{\mathcal{H}^1(\Omega)} \leq C \, \|g\|_{\mathcal{H}^{\frac{1}{2}}(\Gamma)}, \tag{1.4}$$

where, in particular, the function u can be chosen such that $\Delta u = 0$ in Ω. In this case, this one is characterized by the variational equation:

$$\begin{aligned} &\int_{\Omega} \nabla u \cdot \nabla \overline{v} \, d\boldsymbol{x} = 0 && \forall \, v \in \mathcal{H}_0^1(\Omega), \\ &u = g && \text{on } \Gamma. \end{aligned}$$

Let us also mention the useful Poincaré–Friedrichs inequality (see [92] for instance):

$$\|\nabla u\|_{\mathcal{L}^2(\Omega)} + \|u\|_{\mathcal{L}^2(\Gamma)} \geq C \, \|u\|_{\mathcal{H}^1(\Omega)}. \tag{1.5}$$

Sometimes, we shall invoke other duality pairings. In this case, we shall still denote by an integral symbol these dualities because the "pivoting space" is $\mathcal{L}^2(\Gamma)$.

Another class of function spaces is involved when one deals with unbounded domains. These spaces are useful for the statement and wel–posedness of exterior problems for they assign the behaviour at the infinity. It is noteworthy that if $\psi \in \mathcal{H}^1_{\mathrm{loc}}(\mathbb{R}^3)$ and if $\psi(\boldsymbol{x})$ behaves like $|\boldsymbol{x}|^{-1}$ when $|\boldsymbol{x}| \to \infty$, then ψ is not necessarily in $\mathcal{L}^2(\mathbb{R}^3)$. For this reason, we introduce the Beppo-Levi or Nédélec space $\mathcal{W}^1(\Omega_{\mathrm{ext}})$ defined by

$$\mathcal{W}^1(\Omega_{\mathrm{ext}}) := \left\{\psi ; \ \frac{\psi}{1 + |\boldsymbol{x}|} \in \mathcal{L}^2(\Omega_{\mathrm{ext}}), \ \nabla \psi \in \boldsymbol{\mathcal{L}}^2(\Omega_{\mathrm{ext}})\right\},$$

equipped with the norm

$$\|\psi\|_{\mathcal{W}^1(\Omega_{\mathrm{ext}})} := \left(\left\| \frac{\psi}{1 + |\boldsymbol{x}|} \right\|^2_{\mathcal{L}^2(\Omega_{\mathrm{ext}})} + \|\nabla \psi\|^2_{\boldsymbol{\mathcal{L}}^2(\Omega_{\mathrm{ext}})} \right)^{\frac{1}{2}}.$$

We define the semi-norm

$$|\psi|_{\mathcal{W}^1(\Omega_{\text{ext}})} := \|\nabla\psi\|_{\mathcal{L}^2(\Omega_{\text{ext}})},$$

which is equivalent to the norm $\|\cdot\|_{\mathcal{W}^1(\Omega_{\text{ext}})}$ (see [62], Vol. 4, p. 118), i.e.

$$\|\psi\|_{\mathcal{W}^1(\Omega_{\text{ext}})} \leq C\,|\psi|_{\mathcal{W}^1(\Omega_{\text{ext}})} \qquad \forall\,\psi \in \mathcal{W}^1(\Omega_{\text{ext}}), \tag{1.6}$$

where C is independent of ψ. Note that the nonzero constant function does not belong to $\mathcal{W}^1(\Omega_{\text{ext}})$. It is also noteworthy that, in particular, if a function v belongs to $\mathcal{W}^1(\Omega_{\text{ext}})$, then the restriction of v to any bounded subset D of Ω_{ext} is a function of $\mathcal{H}^1(D)$. Consequently, we have a similar trace inequality to (1.3), i.e.

$$\|u\|_{\mathcal{H}^{\frac{1}{2}}(\Gamma)} \leq C\,\|u\|_{\mathcal{W}^1(\Omega_{\text{ext}})} \qquad \forall\,u \in \mathcal{W}^1(\Omega_{\text{ext}}). \tag{1.7}$$

A similar inequality to (1.4) can be given for the unbounded case, i.e. if $g \in \mathcal{H}^{\frac{1}{2}}(\Gamma)$, then there exists a function $u \in \mathcal{W}^1(\Omega_{\text{ext}})$ satisfying $u = g$ on Γ such that

$$\|u\|_{\mathcal{W}^1(\Omega_{\text{ext}})} \leq C\,\|g\|_{\mathcal{H}^{\frac{1}{2}}(\Gamma)}, \tag{1.8}$$

where here also, u can be chosen as a harmonic function in Ω_{ext}.

The space $\mathcal{W}^1(\mathbb{R}^3)$ is defined in an analogous way to $\mathcal{W}^1(\Omega_{\text{ext}})$. In this case, (1.6) holds and (1.7), (1.8) are meaningless.

In electromagnetism, we are often faced with the use of the div and **curl** operators. The complex spaces $\boldsymbol{\mathcal{H}}(\text{div},\Omega)$ and $\boldsymbol{\mathcal{H}}(\textbf{curl},\Omega)$ are the most appropriate tools to formulate electromagnetic problems. The space $\boldsymbol{\mathcal{H}}(\text{div},\Omega)$ is defined by

$$\boldsymbol{\mathcal{H}}(\text{div},\Omega) = \{\boldsymbol{v} \in \boldsymbol{\mathcal{L}}^2(\Omega);\ \text{div}\,\boldsymbol{v} \in \mathcal{L}^2(\Omega)\},$$

where the partial derivatives are taken in the sense of distributions. This space is a complex Hilbert space (see [62], Vol. 3, p. 204) when endowed with the inner product

$$(\boldsymbol{u},\boldsymbol{v})_{\boldsymbol{\mathcal{H}}(\text{div},\Omega)} := \int_\Omega \boldsymbol{u}\cdot\overline{\boldsymbol{v}}\,d\boldsymbol{x} + \int_\Omega \text{div}\,\boldsymbol{u}\ \text{div}\,\overline{\boldsymbol{v}}\,d\boldsymbol{x},$$

and its associated norm

$$\|\boldsymbol{v}\|_{\boldsymbol{\mathcal{H}}(\text{div},\Omega)} := \left(\|\boldsymbol{v}\|^2_{\boldsymbol{\mathcal{L}}^2(\Omega)} + \|\,\text{div}\,\boldsymbol{v}\|^2_{\mathcal{L}^2(\Omega)}\right)^{\frac{1}{2}}.$$

For functions $\boldsymbol{v} \in \boldsymbol{\mathcal{H}}(\text{div},\Omega)$, the trace of $\boldsymbol{v}\cdot\boldsymbol{n}$ on Γ can be defined as a function of $\mathcal{H}^{-\frac{1}{2}}(\Gamma)$ by the duality pairing,

$$\int_\Gamma \boldsymbol{v}\cdot\boldsymbol{n}\,\overline{\varphi}\,ds := \int_\Omega \boldsymbol{v}\cdot\nabla\overline{\varphi}\,d\boldsymbol{x} + \int_\Omega \overline{\varphi}\,\text{div}\,\boldsymbol{v}\,d\boldsymbol{x} \qquad \forall\,\varphi \in \mathcal{H}^1(\Omega). \tag{1.9}$$

Remark 1.3.1. This definition of the trace is clearly valid only for Γ as boundary of a domain Ω. It is however possible to define, in a weaker sense, traces of the normal component of a vector field in $\mathcal{H}(\mathrm{div}, \cdot)$ on any portion of Γ (see for this [14] for instance) but this is out of the scope of this book. We shall however make use of these traces with the same duality notation as (1.9).

The space $\mathcal{H}(\mathbf{curl}, \Omega)$ is defined by

$$\mathcal{H}(\mathbf{curl}, \Omega) = \{ \boldsymbol{v} \in \mathcal{L}^2(\Omega);\ \mathbf{curl}\, \boldsymbol{v} \in \mathcal{L}^2(\Omega)\}.$$

Here again, the **curl** operator is the one that involves partial derivatives in the sense of distributions. This complex space is a Hilbert space (see [62], Vol. 3, p. 204) when endowed with the inner product

$$(\boldsymbol{u}, \boldsymbol{v})_{\mathcal{H}(\mathbf{curl},\Omega)} := \int_\Omega \boldsymbol{u} \cdot \overline{\boldsymbol{v}}\, dx + \int_\Omega \mathbf{curl}\, \boldsymbol{u} \cdot \mathbf{curl}\, \overline{\boldsymbol{v}}\, dx,$$

and its associated norm

$$\|\boldsymbol{v}\|_{\mathcal{H}(\mathbf{curl},\Omega)} := (\|\boldsymbol{v}\|^2_{\mathcal{L}^2(\Omega)} + \|\,\mathbf{curl}\, \boldsymbol{v}\|^2_{\mathcal{L}^2(\Omega)})^{\frac{1}{2}}.$$

For functions $\boldsymbol{v} \in \mathcal{H}(\mathbf{curl}, \Omega)$ the trace of $\boldsymbol{v} \times \boldsymbol{n}$ on Γ can be defined as an element of $\mathcal{H}^{-\frac{1}{2}}(\Gamma)$ by the duality pairing

$$\int_\Gamma \overline{\boldsymbol{w}} \cdot (\boldsymbol{v} \times \boldsymbol{n})\, ds := \int_\Omega \boldsymbol{v} \cdot \mathbf{curl}\, \overline{\boldsymbol{w}}\, dx - \int_\Omega \overline{\boldsymbol{w}} \cdot \mathbf{curl}\, \boldsymbol{v}\, dx \qquad \forall\, \boldsymbol{w} \in \mathcal{H}^1(\Omega).$$

The same remark as 1.3.1 applies here.

If X is Ω_{ext} or $\mathbb{R}^3$, the spaces $\mathcal{H}(\mathrm{div}, X)$ and $\mathcal{H}(\mathbf{curl}, X)$ are defined in the same way as above.

The following result is a consequence to the above ones. This one will enable us to give interface conditions for fields in $\mathcal{H}(\mathrm{div}, \cdot)$ or $\mathcal{H}(\mathbf{curl}, \cdot)$.

Theorem 1.3.1. *Let* $\boldsymbol{u} : \mathbb{R}^3 \to \mathbb{C}^3$ *denote a vector field.*

1. *If* $\boldsymbol{u}_{|\Omega} \in \mathcal{H}(\mathrm{div}, \Omega)$ *and* $\boldsymbol{u}_{|\Omega_{ext}} \in \mathcal{H}(\mathrm{div}, \Omega_{ext})$, *then*

$$\boldsymbol{u} \in \mathcal{H}(\mathrm{div}, \mathbb{R}^3) \;\Leftrightarrow\; [\boldsymbol{u} \cdot \boldsymbol{n}]_\Gamma = 0.$$

2. *If* $\boldsymbol{u}_{|\Omega} \in \mathcal{H}(\mathbf{curl}, \Omega)$ *and* $\boldsymbol{u}_{|\Omega_{ext}} \in \mathcal{H}(\mathbf{curl}, \Omega_{ext})$, *then*

$$\boldsymbol{u} \in \mathcal{H}(\mathbf{curl}, \mathbb{R}^3) \;\Leftrightarrow\; [\boldsymbol{u} \times \boldsymbol{n}]_\Gamma = 0.$$

We recall that $[\cdot]_\Gamma$ *denotes the jump through* Γ.

Let us, in addition give the following useful result.

Theorem 1.3.2. *Let $v \in \mathcal{W}^1(\mathbb{R}^3)$, then we have the identity*

$$\int_{\mathbb{R}^3} |\nabla v|^2\, dx = \int_{\mathbb{R}^3} |\,\mathbf{curl}\, v|^2\, dx + \int_{\mathbb{R}^3} |\operatorname{div} v|^2\, dx,$$

where

$$|\nabla v| := \Big(\sum_{i,j=1}^{3} \Big(\frac{\partial v_i}{\partial x_j} \Big)^2 \Big)^{\frac{1}{2}}.$$

Proof. By density of the space $\mathscr{D}(\mathbb{R}^3)$ into $\mathcal{W}^1(\mathbb{R}^3)$ (see [62], Vol. 4, p. 117), it is sufficient to prove the identity for functions $v \in \mathscr{D}(\mathbb{R}^3)$. We have in each open ball B with boundary ∂B (see [62], Vol. 3, p. 210),

$$\int_B |\nabla v|^2\, dx = \int_B |\,\mathbf{curl}\, v|^2\, dx + \int_B |\operatorname{div} v|^2\, dx + I_B(v),$$

where

$$I_B(v) = \int_{\partial B} \Big(v \cdot \nabla(\overline{v} \cdot n) - v \cdot n \operatorname{div} \overline{v} \Big)\, ds.$$

Choosing B as a ball that contains the support of v, we obtain $I_B(v) = 0$. Therefore, we have for $v \in \mathscr{D}(\mathbb{R}^3)$,

$$\int_{\mathbb{R}^3} |\nabla v|^2\, dx = \int_{\mathbb{R}^3} |\,\mathbf{curl}\, v|^2\, dx + \int_{\mathbb{R}^3} |\operatorname{div} v|^2\, dx. \qquad \square$$

Let us now define some surface operators that are generally used in electromagnetism. Let v denote a smooth vector field defined on an oriented regular surface S imbedded in $\mathbb{R}^3$ with the unit normal vector n. Following ([62], Vol. 4, p. 136), we extend v as a vector field $\tilde{v}$ to a neighborhood S_δ of S of "thickness" δ,

$$S_\delta := \{ x + s n(x);\ x \in S,\ -\delta < s < \delta \}$$

with δ sufficiently small to guarantee existence and uniqueness of a local projection operator $\mathscr{P} : S_\delta \to S$. We also associate to any scalar function $\varphi : S \to \mathbb{C}$ a function $\tilde{\varphi}$ defined on S_δ by

$$\tilde{\varphi}(x) := \varphi(\mathscr{P}(x)) \qquad x \in S_\delta.$$

We can then define the surface vector field

$$\mathbf{curl}_S\, \varphi(x) = n(x) \times \nabla\tilde{\varphi}(x) \qquad x \in S. \tag{1.10}$$

For a surface vector field $\boldsymbol{v} : S \to \mathbb{C}^3$, we define

$$\tilde{\boldsymbol{v}}(\boldsymbol{x}) := \boldsymbol{v}(\mathscr{P}(\boldsymbol{x})) \qquad \boldsymbol{x} \in S.$$

The surface curl of $\boldsymbol{v}$ is then defined by

$$\operatorname{curl}_S \boldsymbol{v}(\boldsymbol{x}) = \mathbf{curl}\, \tilde{\boldsymbol{v}}(\boldsymbol{x}) \cdot \boldsymbol{n}(\boldsymbol{x}) \qquad \boldsymbol{x} \in S. \tag{1.11}$$

It is worth noting that if the scalar field φ is smooth then its extension $\tilde{\varphi}$ is a smooth scalar field in S_δ. Since $\mathbf{curl}\, \nabla\tilde{\varphi} = 0$, then by Theorem 1.3.1, we deduce that the surface vector $\mathbf{curl}_S\, \varphi = \boldsymbol{n} \times \nabla\tilde{\varphi}$ is well defined on S. In an analog way, if the vector field $\boldsymbol{v}$ is smooth then $\tilde{\boldsymbol{v}}$ is a smooth vector field in S_δ. Since $\operatorname{div} \mathbf{curl}\, \tilde{\boldsymbol{v}} = 0$, then by Theorem 1.3.1, we deduce that the surface function $\operatorname{curl}_S \boldsymbol{v} = \mathbf{curl}\, \tilde{\boldsymbol{v}} \cdot \boldsymbol{n}$ is well defined on S.

1.3.2 *Behaviour at the Infinity*

In electromagnetics, the physical fields are defined in the whole space and consequently, one has to deal with problems formulated in unbounded domains. A classical technique to numerically handle such problems is to try to represent the solution outside the conductor domains (a physical field or a potential) on the boundary of the domains. For this, we make use of integral formulations of partial differential equations and particularly of integral representation of harmonic fields since only these ones are involved in eddy currents. In addition, as far as conditions at the infinity are concerned, a physical point of view dictates that all physical fields are radial at the infinity, i.e. for $\boldsymbol{x} \in \mathbb{R}^3$ with $|\boldsymbol{x}|$ large enough, all physical fields are close to radial fields. In the following, we shall give a precise definition of this notion and then prove that this leads to appropriate behaviour at the infinity.

Definition 1.3.1. A scalar field $u : \mathbb{R}^3 \to \mathbb{C}$ (*resp.* vector field $\boldsymbol{u} : \mathbb{R}^3 \to \mathbb{C}^3$) is said to be *radial at the infinity* if there exists a function $g : (0, \infty) \to \mathbb{R}$ such that for all $\varepsilon > 0$, there is a positive constant R that satisfies

$$|u(\boldsymbol{x}) - g(|\boldsymbol{x}|)| \le \varepsilon \qquad \forall\, |\boldsymbol{x}| \ge R.$$

(*resp.* $|\boldsymbol{u}(\boldsymbol{x}) - g(|\boldsymbol{x}|)\boldsymbol{x}| \le \varepsilon$ for all $\boldsymbol{x} \in \mathbb{R}^3$, $|\boldsymbol{x}| \ge R$).

The following result is obtained in collaboration with M. Pierre (Private communication).

Theorem 1.3.3. *Let $u \in C^2(\Omega_{ext})$ be a function that is radial at the infinity, and assume that u is harmonic in Ω_{ext}. Then there exist $\alpha, \beta \in \mathbb{C}$ such that:*

$$u(\boldsymbol{x}) = \alpha + \frac{\beta}{|\boldsymbol{x}|} + \mathcal{O}\left(\frac{1}{|\boldsymbol{x}|^2}\right) \qquad \textit{for } |\boldsymbol{x}| \to \infty, \tag{1.12}$$

$$\nabla u(\boldsymbol{x}) = -\frac{\beta}{|\boldsymbol{x}|^3}\boldsymbol{x} + \mathcal{O}\left(\frac{1}{|\boldsymbol{x}|^3}\right) \qquad \textit{for } |\boldsymbol{x}| \to \infty. \tag{1.13}$$

Proof. By considering the Kelvin's transformation,

$$v(\boldsymbol{y}) = \frac{1}{|\boldsymbol{y}|} u\left(\frac{\boldsymbol{y}}{|\boldsymbol{y}|^2}\right)$$

when $\boldsymbol{y}$ is taken in a neighborhood of the origin, it is easy to prove the existence of an r_0 such that $\overline{\Omega}$ is included in the ball $\{\boldsymbol{x} \in \mathbb{R}^3;\ |\boldsymbol{x}| < 1/r_0\}$ and such that v is harmonic in $B := \{\boldsymbol{y} \in \mathbb{R}^3;\ 0 < |\boldsymbol{y}| < r_0\}$.

If $\boldsymbol{R}$ is a rotation in $\mathbb{R}^3$, then since u is radial at the infinity, we deduce by setting $w(\boldsymbol{y}) := v(\boldsymbol{y}) - v(\boldsymbol{R}\boldsymbol{y})$ that $|\boldsymbol{y}|w(\boldsymbol{y})$ is bounded when $|\boldsymbol{y}| \to 0$. By using (Proposition 16, in [62], Vol. 1, p. 259), there exists a constant α such that the function

$$\tilde{w}(\boldsymbol{y}) = w(\boldsymbol{y}) - \frac{\alpha}{|\boldsymbol{y}|}$$

can be extended to a harmonic function on the ball $\tilde{B} = \{\boldsymbol{y} \in \mathbb{R}^3;\ |\boldsymbol{y}| < r_0\}$ and we have the Poisson integral (see [62], Vol. 1, p. 249),

$$\tilde{w}(\boldsymbol{y}) = \frac{1}{2\pi r_0}\left(\frac{r_0^2}{4} - |\boldsymbol{y}|^2\right) \int_C \frac{\tilde{w}(\boldsymbol{z})}{|\boldsymbol{z} - \boldsymbol{y}|^3}\, ds(\boldsymbol{z}) \qquad \forall\ \boldsymbol{y} \text{ with } |\boldsymbol{y}| < \frac{r_0}{2},$$

where C is the sphere centered at the origin with radius $r_0/2$. It follows that for $0 < |\boldsymbol{y}| < r_0/2$,

$$\begin{aligned} v(\boldsymbol{y}) - v(\boldsymbol{R}\boldsymbol{y}) - \frac{\alpha}{|\boldsymbol{y}|} &= \frac{1}{2\pi r_0}\left(\frac{r_0^2}{4} - |\boldsymbol{y}|^2\right) \int_C \frac{v(\boldsymbol{z}) - v(\boldsymbol{R}\boldsymbol{z})}{|\boldsymbol{z} - \boldsymbol{y}|^3}\, ds(\boldsymbol{z}) \\ &\quad - \frac{\alpha}{\pi r_0^2}\left(\frac{r_0^2}{4} - |\boldsymbol{y}|^2\right) \int_C \frac{1}{|\boldsymbol{z} - \boldsymbol{y}|^3}\, ds(\boldsymbol{z}). \end{aligned}$$

Since

$$1 = \frac{1}{2\pi r_0}\left(\frac{r_0^2}{4} - |\boldsymbol{y}|^2\right) \int_C \frac{1}{|\boldsymbol{z} - \boldsymbol{y}|^3}\, ds(\boldsymbol{z}) \qquad \text{for all } |\boldsymbol{y}| < r_0/2,$$

then we obtain

$$v(\boldsymbol{y}) - v(\boldsymbol{R}\boldsymbol{y}) - \frac{\alpha}{|\boldsymbol{y}|} = \frac{1}{2\pi r_0}\left(\frac{r_0^2}{4} - |\boldsymbol{y}|^2\right) \int_C \frac{v(\boldsymbol{z}) - v(\boldsymbol{R}\boldsymbol{z})}{|\boldsymbol{z} - \boldsymbol{y}|^3}\, ds(\boldsymbol{z}) - \frac{2\alpha}{r_0}.$$

By setting

$$h(\boldsymbol{y}) = v(\boldsymbol{y}) - \frac{1}{2\pi r_0}\Big(\frac{r_0^2}{4} - |\boldsymbol{y}|^2\Big)\int_C \frac{v(z)}{|z-\boldsymbol{y}|^3}\, ds(z),$$

we obtain for $0 < |\boldsymbol{y}| < r_0/2$,

$$h(\boldsymbol{y}) = h(\boldsymbol{R}\boldsymbol{y}) + \alpha\Big(\frac{1}{|\boldsymbol{y}|} - \frac{2}{r_0}\Big).$$

Let B denote the sphere centered at the origin with radius $r_0/4$ and let n be a positive integer. For $\boldsymbol{y} \in B$, we have $\boldsymbol{R}^n\boldsymbol{y} \in B$ and

$$h(\boldsymbol{y}) = h(\boldsymbol{R}^n\boldsymbol{y}) + n\alpha\frac{2}{r_0}.$$

Clearly, we have $\alpha = 0$ since otherwise $\lim_{n\to\infty}|h(\boldsymbol{R}^n\boldsymbol{y})| = \infty$, which is in contradiction with the continuity of h. Therefore $h(\boldsymbol{y}) = h(\boldsymbol{R}\boldsymbol{y})$ for all $0 < |\boldsymbol{y}| < r_0/2$, which proves that h is radial for $0 < |\boldsymbol{y}| < r_0/2$.

Since h is harmonic for $0 < |\boldsymbol{y}| < r_0/2$, we obtain by direct computation

$$h(\boldsymbol{y}) = \frac{a}{|\boldsymbol{y}|} + b, \text{ where } a, b \in \mathbb{C}.$$

It follows that

$$v(\boldsymbol{y}) = \frac{a}{|\boldsymbol{y}|} + b + \frac{1}{2\pi r_0}\Big(\frac{r_0^2}{4} - |\boldsymbol{y}|^2\Big)\int_C \frac{v(z)}{|z-\boldsymbol{y}|^3}\, ds(z). \tag{1.14}$$

Since for $z \in \mathbb{C}$,

$$\frac{1}{|z-\boldsymbol{y}|^3} = \frac{1}{|z|^3}(1+\mathcal{O}(|\boldsymbol{y}|)) = \frac{8}{r_0^3}(1+\mathcal{O}(|\boldsymbol{y}|)) \qquad \text{when } |\boldsymbol{y}| \to 0,$$

we obtain

$$v(\boldsymbol{y}) = \frac{a}{|\boldsymbol{y}|} + \Big(b + \frac{1}{\pi r_0^2}\Big) + \mathcal{O}(|\boldsymbol{y}|) \qquad \text{when } |\boldsymbol{y}| \to 0.$$

Finally, since

$$u(\boldsymbol{x}) = \frac{1}{|\boldsymbol{x}|}\, v\Big(\frac{\boldsymbol{x}}{|\boldsymbol{x}|^2}\Big),$$

then

$$u(\boldsymbol{x}) = a + \Big(b + \frac{1}{\pi r_0^2}\Big)\frac{1}{|\boldsymbol{x}|} + \mathcal{O}(|\boldsymbol{x}|^{-2}) \qquad \text{when } |\boldsymbol{x}| \to \infty.$$

This proves (1.12) with $\alpha = a$ and $\beta = b + 1/\pi r_0^2$. To prove (1.13), we calculate ∇u from (1.14) and we obtain the result in a similar way. □

Remark 1.3.2. The same behaviour as (1.12) is obtained if u is assumed to be bounded at the infinity (see [62], Vol. 1, p. 266). This conclusion can also be drawn from the proof of Theorem 1.3.3.

1.3.3 Kernel of div and curl Operators

We present now some characterizations of kernels of the operators div and **curl**. The proofs of these theorems can be found in ([62], Vol. 3, Chap. IX or [14]) in the Sobolev space context.

Theorem 1.3.4.

1. *Let* $\boldsymbol{u}$ *denote a function in* $\mathcal{H}(\mathrm{div}, \mathbb{R}^3)$ *such that*

$$\operatorname{div} \boldsymbol{u} = 0.$$

There exists a function $\boldsymbol{v} \in \mathcal{W}^1(\mathbb{R}^3)$ *such that*

$$\boldsymbol{u} = \mathbf{curl}\, \boldsymbol{v} \qquad \text{in } \mathbb{R}^3.$$

In addition, $\boldsymbol{v}$ *can be chosen such that*

$$\operatorname{div} \boldsymbol{v} = 0 \qquad \text{in } \mathbb{R}^3. \tag{1.15}$$

In this case, $\boldsymbol{v}$ *is unique. The additional constraint* (1.15) *is usually called* Coulomb Gauge.

2. *Let* $\boldsymbol{u}$ *denote a function in* $\mathcal{H}(\mathbf{curl}, \mathbb{R}^3)$ *such that*

$$\mathbf{curl}\, \boldsymbol{u} = 0.$$

Then, there exists a unique function $\varphi \in \mathcal{W}^1(\mathbb{R}^3)$ *such that*

$$\boldsymbol{u} = \nabla \varphi.$$

We next need a characterization of the kernel of the **curl** operator in the conductor Ω as well as in the external domain. It is well known that in the case where the domain Ω is simply connected, if $\boldsymbol{u}$ is a vector field of $\mathcal{H}(\mathbf{curl}, \Omega)$ that satisfies $\mathbf{curl}\, \boldsymbol{u} = 0$ in Ω, then there exists a scalar field φ in $\mathcal{H}^1(\Omega)$ such that $\boldsymbol{u} = \nabla \varphi$. Let us now assume that Ω is not simply connected and that its boundary Γ of Ω is a surface of genus 1, i.e. a torus (cf. Fig. 1.1). Then there is a *cut* , i.e. a smooth surface S contained in Ω such that the domain $\Omega \setminus S$ is simply connected. It is clear

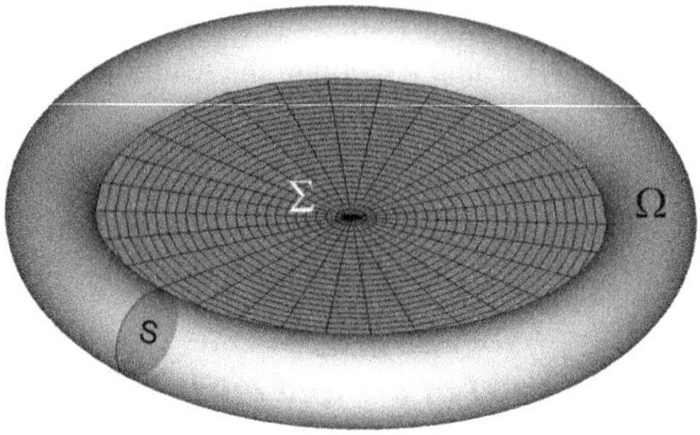

Fig. 1.1 A toroidal conductor (of genus 1) with cuts S and Σ

then that there exists a cut Σ, contained in Ω_{ext} such that the domain $\Omega_{\text{ext}} \setminus \Sigma$ is simply connected.

The following two theorems can be found in Dautray–Lions ([62], Vol. 3, pp. 219 and 230).

Theorem 1.3.5. *Assume that the boundary Γ of Ω is of genus 1, and let $\boldsymbol{u}$ denote a function in $\mathcal{H}(\mathbf{curl}, \Omega)$ such that*

$$\mathbf{curl}\, \boldsymbol{u} = 0.$$

Then there exists a function $\varphi \in \mathcal{H}^1(\Omega)$ and a complex number α such that

$$\boldsymbol{u} = \nabla\varphi + \alpha\nabla q \qquad \text{in } \Omega \setminus S, \tag{1.16}$$

where the function q is a solution of the following problem,

$$\begin{cases} \Delta q = 0 & \text{in } \Omega \setminus S, \\ [q]_S = 1, & \\ \left[\dfrac{\partial q}{\partial n}\right]_S = 0, & \\ \dfrac{\partial q}{\partial n} = 0 & \text{on } \Gamma. \end{cases} \tag{1.17}$$

In addition, there exists a positive constant C such that

$$\|\varphi\|_{\mathcal{H}^1(\Omega)} \leq C\, \|\boldsymbol{u}\|_{\mathcal{H}(\mathbf{curl},\Omega)}. \tag{1.18}$$

Note that, although the function q is determined up to an additive constant, the expansion (1.16) does not depend on this constant. In fact, the function $q \in \mathcal{H}^1(\Omega \setminus S)$ is such that

$$\int_\Omega \nabla q \cdot \nabla v \, d\boldsymbol{x} = 0 \quad \forall\, v \in \mathcal{H}^1(\Omega) \text{ and } [q]_S = 1.$$

Theorem 1.3.6. *Assume that the boundary Γ of Ω is of genus 1, and let $\boldsymbol{u}$ denote a function in $\boldsymbol{\mathcal{H}}(\mathbf{curl}, \Omega_{ext})$ such that*

$$\mathbf{curl}\, \boldsymbol{u} = 0.$$

Then there exists a function $\varphi \in \mathcal{W}^1(\Omega_{ext})$ and a complex number λ, such that

$$\boldsymbol{u} = \nabla\varphi + \lambda\nabla p \qquad \text{in } \Omega_{ext} \setminus \Sigma, \tag{1.19}$$

where the function p is a solution of the following problem,

$$\begin{cases} \Delta p = 0 & \text{in } \Omega_{ext} \setminus \Sigma, \\ [p]_\Sigma = 1, & \\ \left[\dfrac{\partial p}{\partial n}\right]_\Sigma = 0, & \\ \dfrac{\partial p}{\partial n} = 0 & \text{on } \Gamma, \\ p(\boldsymbol{x}) = \mathcal{O}(|\boldsymbol{x}|^{-1}) & \text{for } |\boldsymbol{x}| \to \infty. \end{cases} \tag{1.20}$$

Note that the function p in this case is unique. In fact, $p \in \mathcal{W}^1(\Omega_{\text{ext}} \setminus \Sigma)$ is such that

$$\int_{\Omega_{\text{ext}}} \nabla p \cdot \nabla v \, d\boldsymbol{x} = 0 \quad \forall\, v \in \mathcal{W}^1(\Omega_{\text{ext}}) \text{ and } [p]_\Sigma = 1.$$

Remark 1.3.3. If the surface Γ is of genus $N > 1$ (roughly speaking, a domain with N holes), we have to introduce N cuts $S_1, \dots, S_N$ in order to obtain a simply connected domain $\Omega \setminus \cup_{j=1}^N S_j$. If we associate to these cuts N cuts $\Sigma_1, \dots, \Sigma_N$ in order to obtain a simply connected domain $\Omega_{\text{ext}} \setminus \cup_{j=1}^N \Sigma_j$, we can derive analog results to those obtained in Theorems 1.3.5 and 1.3.6 with N functions $q_1, \dots, q_N$ and N functions $p_1, \dots, p_N$. Moreover, if Ω is made of M connected components, this process can be repeated M times.

1.3.4 Integral Representations

Let us summarize some results about integral representations of harmonic fields. Such fields appear indeed in the modelling of eddy currents in unbounded regions of the space and their representation on the boundary enable deriving stable and accurate numerical methods. The given results are all known and most of them are proved in the book series of Dautray and Lions [62] or in the lectures of J.-C. Nédélec [135, 138]. We quote them here for a sake of self consistency.

We start by introducing the Green function

$$G(\boldsymbol{x}, \boldsymbol{y}) := \frac{1}{4\pi} \frac{1}{|\boldsymbol{x} - \boldsymbol{y}|} \qquad \boldsymbol{x}, \boldsymbol{y} \in \mathbb{R}^3,\ \boldsymbol{x} \neq \boldsymbol{y}. \tag{1.21}$$

It is well known that

$$- \Delta_y G(\boldsymbol{x}, \cdot) = \delta_x \qquad \text{in } \mathbb{R}^3, \tag{1.22}$$

where δ_x is the Dirac distribution at $\boldsymbol{x}$ and Δ_y is the Laplace operator for the variable $\boldsymbol{y}$. As a consequence, it can be shown that if $f \in \mathcal{L}^2(\mathbb{R}^3)$ with a compact support, then the solution of the elliptic problem

$$\begin{cases} -\Delta u = f & \text{in } \mathbb{R}^3, \\ u(\boldsymbol{x}) = \mathcal{O}(|\boldsymbol{x}|^{-1}) & |\boldsymbol{x}| \to \infty, \end{cases}$$

is given by the expression

$$u(\boldsymbol{x}) = \int_{\mathbb{R}^3} f(\boldsymbol{y})\, G(\boldsymbol{x}, \boldsymbol{y})\, d\boldsymbol{y} \qquad \boldsymbol{x} \in \mathbb{R}^3. \tag{1.23}$$

We now give a fundamental result about integral representations.

Theorem 1.3.7. *Let u be a function of class $(\mathcal{C}^1(\overline{\Omega}) \cap \mathcal{C}^2(\Omega)) \cap (\mathcal{C}^1(\overline{\Omega}_{ext}) \cap \mathcal{C}^2(\Omega_{ext}))$ that satisfies*

$$\Delta u = 0 \qquad \text{in } \Omega \cup \Omega_{ext},$$

and assume furthermore that

$$|u(\boldsymbol{x})| = \mathcal{O}(|\boldsymbol{x}|^{-1}) \qquad \text{when } |\boldsymbol{x}| \to \infty.$$

Then we have for $\boldsymbol{x} \in \Omega \cup \Omega_{ext}$, the representation

$$u(\boldsymbol{x}) = -\int_\Gamma \left[\frac{\partial u}{\partial n}(\boldsymbol{y})\right] G(\boldsymbol{x}, \boldsymbol{y})\, ds(\boldsymbol{y}) + \int_\Gamma [u(\boldsymbol{y})]\, \frac{\partial G}{\partial n_y}(\boldsymbol{x}, \boldsymbol{y})\, ds(\boldsymbol{y}), \tag{1.24}$$

and for $\boldsymbol{x} \in \Gamma$,

$$\begin{aligned} \frac{1}{2}(u_{|\Gamma^-}(\boldsymbol{x}) + u_{|\Gamma^+}(\boldsymbol{x})) = &- \int_\Gamma \left[\frac{\partial u}{\partial n}(\boldsymbol{y})\right] G(\boldsymbol{x}, \boldsymbol{y})\, ds(\boldsymbol{y}) \\ &+ \int_\Gamma [u(\boldsymbol{y})]\, \frac{\partial G}{\partial n_y}(\boldsymbol{x}, \boldsymbol{y})\, ds(\boldsymbol{y}). \end{aligned} \tag{1.25}$$

Proof. This integral representation is well known (see Nédélec [135]). However, for convenience of the reader we establish its proof for $\boldsymbol{x} \in \Omega$.

Let $R > 0$ be such that the open ball B_R centered at the origin with radius R satisfies $\overline{\Omega} \subset B^R$. We denote by ∂B_R the boundary of this ball and $\Omega_{\text{ext}}^R = B_R \setminus \overline{\Omega}$. For $\boldsymbol{x} \in \Omega$, $G(\boldsymbol{x}, \cdot)$ is harmonic in Ω_{ext} and we have by using Green's formula,

$$\begin{aligned} 0 &= \int_{\Omega_{\text{ext}}^R} \Delta u(\boldsymbol{y})\, G(\boldsymbol{x}, \boldsymbol{y})\, ds(\boldsymbol{y}) - \int_{\Omega_{\text{ext}}^R} u(\boldsymbol{y})\, \Delta_{\boldsymbol{y}} G(\boldsymbol{x}, \boldsymbol{y})\, ds(\boldsymbol{y}) \\ &= -\int_\Gamma \frac{\partial u^+}{\partial n}(\boldsymbol{y})\, G(\boldsymbol{x}, \boldsymbol{y})\, ds(\boldsymbol{y}) + \int_\Gamma u^+(\boldsymbol{y}) \frac{\partial G}{\partial n_{\boldsymbol{y}}}(\boldsymbol{x}, \boldsymbol{y})\, ds(\boldsymbol{y}) \\ &\quad + \int_{\partial B_R} \frac{\partial u}{\partial n}(\boldsymbol{y})\, G(\boldsymbol{x}, \boldsymbol{y})\, ds(\boldsymbol{y}) - \int_{\partial B_R} u(\boldsymbol{y}) \frac{\partial G}{\partial n_{\boldsymbol{y}}}(\boldsymbol{x}, \boldsymbol{y})\, ds(\boldsymbol{y}), \end{aligned}$$

where the unit normal on ∂B_R is oriented outside B_R and $u^+ = u_{|\Omega_{\text{ext}}}$. By using the fact that $u(\boldsymbol{x}) = \mathcal{O}(|\boldsymbol{x}|^{-1})$ when $|\boldsymbol{x}| \to \infty$, Theorem 1.3.3 and the expression of G, we easily prove that

$$\lim_{R \to \infty} \left(\int_{\partial B_R} \frac{\partial u}{\partial n}(\boldsymbol{y})\, G(\boldsymbol{x}, \boldsymbol{y})\, ds(\boldsymbol{y}) - \int_{\partial B_R} u(\boldsymbol{y}) \frac{\partial G}{\partial n_{\boldsymbol{y}}}(\boldsymbol{x}, \boldsymbol{y})\, ds(\boldsymbol{y}) \right) = 0.$$

It follows that

$$\int_\Gamma \frac{\partial u^+}{\partial n}(\boldsymbol{y})\, G(\boldsymbol{x}, \boldsymbol{y})\, ds(\boldsymbol{y}) - \int_\Gamma u^+(\boldsymbol{y}) \frac{\partial G}{\partial n_{\boldsymbol{y}}}(\boldsymbol{x}, \boldsymbol{y})\, ds(\boldsymbol{y}) = 0. \tag{1.26}$$

Let $\varepsilon > 0$ be such that the open ball $B_\varepsilon(\boldsymbol{x})$ centered at $\boldsymbol{x}$ with radius ε satisfies $\overline{B}_\varepsilon(\boldsymbol{x}) \subset \Omega$. We denote by $\partial B_\varepsilon(\boldsymbol{x})$ the boundary of this ball and $\Omega_\varepsilon := \Omega \setminus \overline{B}_\varepsilon(\boldsymbol{x})$. Since $G(\boldsymbol{x}, \cdot)$ and u are harmonic in Ω_ε, we have by using the Green's formula in Ω_ε,

$$\begin{aligned} 0 &= -\int_{\Omega_\varepsilon} u(\boldsymbol{y})\, \Delta_{\boldsymbol{y}} G(\boldsymbol{x}, \boldsymbol{y})\, d\boldsymbol{y} + \int_{\Omega_\varepsilon} G(\boldsymbol{x}, \boldsymbol{y})\, \Delta u(\boldsymbol{y})\, d\boldsymbol{y} \\ &= -\int_\Gamma u^-(\boldsymbol{y}) \frac{\partial G}{\partial n_{\boldsymbol{y}}}(\boldsymbol{x}, \boldsymbol{y})\, ds(\boldsymbol{y}) + \int_\Gamma \frac{\partial u^-}{\partial n}(\boldsymbol{y})\, G(\boldsymbol{x}, \boldsymbol{y})\, ds(\boldsymbol{y}) \\ &\quad - \int_{\partial B_\varepsilon(\boldsymbol{x})} u^-(\boldsymbol{y}) \frac{\partial G}{\partial n_{\boldsymbol{y}}}(\boldsymbol{x}, \boldsymbol{y})\, ds(\boldsymbol{y}) + \int_{\partial B_\varepsilon(\boldsymbol{x})} \frac{\partial u^-}{\partial n}\, G(\boldsymbol{x}, \boldsymbol{y})\, ds(\boldsymbol{y}), \end{aligned}$$

where the unit normal vector to $\partial B_\varepsilon(\boldsymbol{x})$ points to the center of $B_\varepsilon(\boldsymbol{x})$. It is easy to prove that

$$\lim_{\varepsilon \to 0} \int_{\partial B_\varepsilon(\boldsymbol{x})} u^-(\boldsymbol{y}) \frac{\partial G}{\partial n_{\boldsymbol{y}}}(\boldsymbol{x}, \boldsymbol{y})\, ds(\boldsymbol{y}) = u(\boldsymbol{x}),$$

and

$$\lim_{\varepsilon \to 0} \int_{\partial B_\varepsilon(x)} \frac{\partial u^-}{\partial n}(y)\, G(x, y)\, ds(y) = 0.$$

It follows that

$$u(x) = -\int_\Gamma u^-(y) \frac{\partial G}{\partial n_y}(x, y)\, ds(y) + \int_\Gamma \frac{\partial u^-}{\partial n}(y)\, G(x, y)\, ds(y). \qquad (1.27)$$

We obtain by subtracting (1.26) from (1.27),

$$u(x) = \int_\Gamma [u(y)]\, \frac{\partial G}{\partial n_y}(x, y)\, ds(y) - \int_\Gamma \left[\frac{\partial u}{\partial n}(y)\right] G(x, y)\, ds(y). \qquad \square$$

The results of Theorem 1.3.7 can now be used to derive integral representations of solutions of various problems in unbounded domains. Consider the Dirichlet problem

$$\begin{cases} \Delta u = 0 & \text{in } \Omega \cup \Omega_{\text{ext}}, \\ u = g & \text{on } \Gamma, \\ u(x) = \mathcal{O}(|x|^{-1}) & |x| \to \infty, \end{cases} \qquad (1.28)$$

where $g \in \mathcal{H}^{\frac{1}{2}}(\Gamma)$. Denoting by p the jump $[\frac{\partial u}{\partial n}]_\Gamma$, we have, according to Theorem 1.3.7, the integral representation

$$u(x) = -\int_\Gamma p(y)\, G(x, y)\, ds(y) \qquad x \in \mathbb{R}^3.$$

In particular, for $x \in \Gamma$, the function p satisfies the integral equation

$$g(x) = -\int_\Gamma p(y)\, G(x, y)\, ds(y) \qquad x \in \Gamma. \qquad (1.29)$$

As a problem with the unknown p, (1.29) admits the variational formulation:

$$\begin{cases} \text{Find } p \in \mathcal{H}^{-\frac{1}{2}}(\Gamma) \text{ such that} \\ \mathscr{B}(p, q) = -\int_\Gamma g\, \overline{q}\, ds \quad \forall\, q \in \mathcal{H}^{-\frac{1}{2}}(\Gamma), \end{cases} \qquad (1.30)$$

where $\mathscr{B}$ is the sesquilinear form on $\mathcal{H}^{-\frac{1}{2}}(\Gamma) \times \mathcal{H}^{-\frac{1}{2}}(\Gamma)$ given by

$$\mathscr{B}(p,q) := \int_{\Gamma} \int_{\Gamma} G(\boldsymbol{x},\boldsymbol{y})\, p(\boldsymbol{y})\, \overline{q}(\boldsymbol{x})\, ds(\boldsymbol{y})\, ds(\boldsymbol{x}).$$

Theorem 1.3.8 ([62], Vol. 4, p. 123). *The sequilinear form $\mathscr{B}$ is continuous on $\mathcal{H}^{-\frac{1}{2}}(\Gamma) \times \mathcal{H}^{-\frac{1}{2}}(\Gamma)$ and there is a constant C such that*

$$|\mathscr{B}(q,q)| \geq C\, \|q\|^2_{\mathcal{H}^{-\frac{1}{2}}(\Gamma)} \qquad \forall\, q \in \mathcal{H}^{-\frac{1}{2}}(\Gamma).$$

As a consequence of Theorem 1.3.8 and of the Lax-Milgram theorem (Theorem 1.2.1), there exists an isomorphism $T : \mathcal{H}^{\frac{1}{2}}(\Gamma) \to \mathcal{H}^{-\frac{1}{2}}(\Gamma)$ satisfying

$$\mathscr{B}(Tf,q) = \int_{\Gamma} f \overline{q}\, ds \qquad \forall\, q \in \mathcal{H}^{-\frac{1}{2}}(\Gamma),\ f \in \mathcal{H}^{\frac{1}{2}}(\Gamma).$$

By setting $K = T^{-1} : \mathcal{H}^{-\frac{1}{2}}(\Gamma) \to \mathcal{H}^{\frac{1}{2}}(\Gamma)$ we have the identity

$$\int_{\Gamma} (Kp)\, \overline{q}\, ds := \mathscr{B}(p,q) \qquad \forall\, p,q \in \mathcal{H}^{-\frac{1}{2}}(\Gamma). \tag{1.31}$$

So we obtain $u = -K\left[\frac{\partial u}{\partial n}\right]_{\Gamma}$ on Γ when u satisfies (1.32).

We have the regularity result.

Theorem 1.3.9 ([62], Vol. 4, p. 124). *Assume that the boundary Γ is smooth enough (C^{∞}, say). Then, the mapping K defines an isomorphism of $\mathcal{H}^{s-1}(\Gamma)$ onto $\mathcal{H}^{s}(\Gamma)$ for all $s \in \mathbb{R}$.*

It is then possible to calculate the normal derivative of u^+ by the following result:

Theorem 1.3.10 ([62], Vol. 4, p. 131). *Let u denote the solution of* (1.28) *and let $p = [\partial u/\partial n]_{\Gamma}$. Then the exterior normal derivative of u is given by*

$$\frac{\partial u^+}{\partial n}(\boldsymbol{x}) = \frac{1}{2} p(\boldsymbol{x}) - \int_{\Gamma} p(\boldsymbol{y}) \frac{\partial G}{\partial \boldsymbol{n}_{\boldsymbol{x}}}(\boldsymbol{x},\boldsymbol{y})\, ds(\boldsymbol{y}) \qquad \boldsymbol{x} \in \Gamma. \qquad \square$$

A second type of integral equations appears in integral representation of harmonic fields. Consider, for this, the exterior Dirichlet problem:

$$\begin{cases} \Delta u = 0 & \text{in } \Omega_{\text{ext}}, \\ u = g & \text{on } \Gamma, \\ u(\boldsymbol{x}) = \mathcal{O}(|\boldsymbol{x}|^{-1}) & \text{for } |\boldsymbol{x}| \to \infty, \end{cases} \tag{1.32}$$

where $g \in \mathcal{H}^{\frac{1}{2}}(\Gamma)$. If

$$\int_\Gamma \frac{\partial u}{\partial n}\, ds = 0, \tag{1.33}$$

we can define a harmonic function u in Ω such that $[\partial u/\partial n] = 0$ on Γ and we set $\varphi := [u]$ on Γ. Remark that the extension of u on Ω is determined up to an additive constant. This is the reason for which we consider φ in $\mathcal{H}^{\frac{1}{2}}(\Gamma)/\mathbb{C}$ in the following. We obtain from (1.25) the representation

$$\frac{1}{2}\varphi(\boldsymbol{x}) + \int_\Gamma \varphi(\boldsymbol{y}) \frac{\partial G}{\partial n_{\boldsymbol{y}}}(\boldsymbol{x}, \boldsymbol{y})\, ds(\boldsymbol{y}) = g(\boldsymbol{x}). \qquad \boldsymbol{x} \in \Gamma, \tag{1.34}$$

Equation (1.34) can be written in operator form as

$$(\frac{1}{2}I + R)\varphi = g, \tag{1.35}$$

where I is the identity operator in $\mathcal{H}^{\frac{1}{2}}(\Gamma)/\mathbb{C}$ and $R : \mathcal{H}^{\frac{1}{2}}(\Gamma)/\mathbb{C} \to \mathcal{H}^{\frac{1}{2}}(\Gamma)/\mathbb{C}$ is the mapping defined by

$$R\varphi := \int_\Gamma \varphi(\boldsymbol{y}) \frac{\partial G}{\partial n_{\boldsymbol{y}}}(\cdot, \boldsymbol{y})\, ds(\boldsymbol{y}). \tag{1.36}$$

We have the following result (cf. [62], Vol. 4, p. 128).

Lemma 1.3.1. *Under the same hypotheses as in Theorem 1.3.9, the mapping R is linear and continuous from $\mathcal{H}^s(\Gamma)$ into $\mathcal{H}^{s+1}(\Gamma)$ for all real s.*

In order to formulate (1.35) in $\mathcal{H}^{\frac{1}{2}}(\Gamma)$, we shall add an extra condition to g to ensure (1.33). Let us define the exterior problem:

$$\begin{cases} \Delta v = 0 & \text{in } \Omega_{\text{ext}}, \\ v = 1 & \text{on } \Gamma, \\ v(\boldsymbol{x}) = \mathcal{O}(|\boldsymbol{x}|^{-1}) & \text{for } |\boldsymbol{x}| \to \infty. \end{cases}$$

To apply the Green formula in Ω_{ext}, we denote by B_R the ball with center 0 and radius $R > 0$, large enough so that $B_R \supset \overline{\Omega}$, and by Γ_R the boundary of B_R. Owing to the properties of u and v and (1.13) we have

$$\lim_{R\to\infty} \left(\int_{\Gamma_R} u \frac{\partial v}{\partial n}\, ds - \int_{\Gamma_R} v \frac{\partial u}{\partial n}\, ds \right) = 0.$$

Then the Green formula in Ω_{ext} yields

$$\int_\Gamma u \frac{\partial v}{\partial n}\, ds = \int_\Gamma v \frac{\partial u}{\partial n}\, ds,$$

which implies

$$\int_\Gamma g \frac{\partial v}{\partial n} \, ds = \int_\Gamma \frac{\partial u}{\partial n} \, ds = 0.$$

We have the following result.

Theorem 1.3.11 ([62], Vol. 4, p. 126). *Assume that $g \in \mathcal{H}^{\frac{1}{2}}(\Gamma)$ and satisfies*

$$\int_\Gamma g \frac{\partial v}{\partial n} \, ds = 0. \tag{1.37}$$

Then (1.35) *admits a unique solution $\varphi \in \mathcal{H}^{\frac{1}{2}}(\Gamma)/\mathbb{C}$. Moreover, the operator $\frac{1}{2}I + R$ defines an isomorphism of $\mathcal{H}^{\frac{1}{2}}(\Gamma)/\mathbb{C}$ onto the space of functions $g \in \mathcal{H}^{\frac{1}{2}}(\Gamma)$ that satisfy* (1.37).

1.3.5 The Exterior Steklov–Poincaré Operator

A major tool that will be used in coupled exterior–interior problems is the so-called *Exterior Steklov–Poincaré operator*. This one is defined by (The normal vector $\boldsymbol{n}$ points outward of Ω):

$$P : \varphi \in \mathcal{H}^{\frac{1}{2}}(\Gamma) \mapsto -\frac{\partial u}{\partial n}\Big|_\Gamma \in \mathcal{H}^{-\frac{1}{2}}(\Gamma), \tag{1.38}$$

where u is the unique solution, in $\mathcal{W}^1(\Omega_{\text{ext}})$, of the exterior problem

$$\begin{cases} \Delta u = 0 & \text{in } \Omega_{\text{ext}}, \\ u = \varphi & \text{on } \Gamma, \\ u(\boldsymbol{x}) = \mathcal{O}(|\boldsymbol{x}|^{-1}) & \text{for } |\boldsymbol{x}| \to \infty. \end{cases} \tag{1.39}$$

We shall in the sequel give an integral representation of this operator and show some of its properties. For this, let us denote, for $\varphi \in \mathcal{H}^{\frac{1}{2}}(\Gamma)$ by u_φ its corresponding solution of (1.39). We have by definition of $\mathcal{W}^1(\Omega_{\text{ext}})$ and (1.13),

$$\int_{\Omega_{\text{ext}}} \nabla u_\varphi \cdot \nabla \bar{v} \, d\boldsymbol{x} = 0 \qquad \forall \, v \in \mathcal{W}^1(\Omega_{\text{ext}}) \text{ with } v = 0 \text{ on } \Gamma.$$

Then by the Green formula applied to

$$0 = \int_{\Omega_{\text{ext}}} \Delta u_\varphi u_\psi \, d\boldsymbol{x}$$

we obtain

$$\int_\Gamma (P\varphi)\overline{\psi}\,ds = \int_{\Omega_{\text{ext}}} \nabla u_\varphi \cdot \nabla \overline{u}_\psi \,d\boldsymbol{x} \qquad \forall\, \varphi, \psi \in \mathcal{H}^{\frac{1}{2}}(\Gamma).$$

Theorem 1.3.12. *The operator P is continuous, selfadjoint and coercive.*

Proof. Let $\varphi, \psi \in \mathcal{H}^{\frac{1}{2}}(\Gamma)$. We have from (1.8) and the definition of $\mathcal{H}^{-\frac{1}{2}}(\Gamma)$ as the dual space of $\mathcal{H}^{\frac{1}{2}}(\Gamma)$, for all $\psi \in \mathcal{H}^{\frac{1}{2}}(\Gamma)$,

$$\left| \int_\Gamma (P\varphi)\overline{\psi}\,ds \right| \le \|u_\varphi\|_{\mathcal{W}^1(\Omega_{\text{ext}})} \|u_\psi\|_{\mathcal{W}^1(\Omega_{\text{ext}})} \le C\, \|\varphi\|_{\mathcal{H}^{\frac{1}{2}}(\Gamma)} \|\psi\|_{\mathcal{H}^{\frac{1}{2}}(\Gamma)}.$$

It results then that

$$\|P\varphi\|_{\mathcal{H}^{-\frac{1}{2}}(\Gamma)} \le C\, \|\varphi\|_{\mathcal{H}^{\frac{1}{2}}(\Gamma)} \qquad \forall\, \varphi \in \mathcal{H}^{\frac{1}{2}}(\Gamma),$$

which implies the continuity of P. Furthermore, P is selfadjoint by

$$\int_\Gamma (P\varphi)\overline{\psi}\,ds = \int_{\Omega_{\text{ext}}} \nabla u_\varphi \cdot \nabla \overline{u}_\psi \,d\boldsymbol{x} = \int_\Gamma (P\psi)\overline{\varphi}\,ds.$$

Finally, we have from the trace inequality (1.7) and from (1.6),

$$\int_\Gamma (P\psi)\overline{\psi}\,ds = \int_{\Omega_{\text{ext}}} |\nabla u_\psi|^2 \,d\boldsymbol{x} \ge C\, \|\psi\|^2_{\mathcal{H}^{\frac{1}{2}}(\Gamma)}. \qquad \square$$

In order to obtain an integral representation for the operator P, we extend the solution u of (1.39) to $\mathbb{R}^3$ with a continuous trace, i.e. such that $\Delta u = 0$ in Ω and $[u]_\Gamma = 0$. We have then from (1.24) and (1.25) the identity

$$u(\boldsymbol{x}) = -\int_\Gamma p(\boldsymbol{y})\, G(\boldsymbol{x}, \boldsymbol{y})\, ds(\boldsymbol{y}) \qquad \boldsymbol{x} \in \mathbb{R}^3,$$

where $p = \left[\dfrac{\partial u}{\partial n}\right]_\Gamma$. Moreover we have from Theorem 1.3.10,

$$\frac{\partial u^+}{\partial n}(\boldsymbol{x}) = \frac{1}{2} p(\boldsymbol{x}) - \int_\Gamma p(\boldsymbol{y}) \frac{\partial G}{\partial n_x}(\boldsymbol{x}, \boldsymbol{y})\, ds(\boldsymbol{y}) \qquad \boldsymbol{x} \in \Gamma. \tag{1.40}$$

Let us consider the two operators K and R defined in (1.31) and (1.36) respectively and define the adjoint operator R' of R by

$$\int_\Gamma R' p(\boldsymbol{x}) \psi(\boldsymbol{x})\, ds(\boldsymbol{x}) := \int_\Gamma p(\boldsymbol{x})\, R\psi(\boldsymbol{x})\, ds(\boldsymbol{x})$$
$$\forall\, p \in \mathcal{H}^{-\frac{1}{2}}(\Gamma),\ \psi \in \mathcal{H}^{\frac{1}{2}}(\Gamma). \tag{1.41}$$

Note that the operator R' is actually defined on $\mathcal{H}^{-\frac{1}{2}}(\Gamma)$ and that its range is included in the space

$$\{\chi \in \mathcal{H}^{-\frac{1}{2}}(\Gamma);\ \int_\Gamma \chi\, ds = 0\}.$$

Using the fact that the mapping $K : \mathcal{H}^{-\frac{1}{2}}(\Gamma) \to \mathcal{H}^{\frac{1}{2}}(\Gamma)$ is an isomorphism, we have from (1.40), for $\varphi, \psi \in \mathcal{H}^{\frac{1}{2}}(\Gamma)$,

$$\begin{aligned}\int_\Gamma (P\varphi)\, \psi\, ds &= -\int_\Gamma \frac{\partial u_\varphi}{\partial n}\, \psi\, ds \\ &= -\frac{1}{2}\int_\Gamma p\psi\, ds + \int_\Gamma p\,(R\psi)\, ds \\ &= -\int_\Gamma (\frac{1}{2}I - R')p\, \psi\, ds.\end{aligned}$$

Using the relation $Kp = \varphi$, we obtain the expression

$$P = (-\frac{1}{2}I + R')\, K^{-1}. \tag{1.42}$$

An alternative to the previous representation of the exterior Steklov–Poincaré operator consists in extending (1.39) to the interior domain and then using an integral representation to calculate the jump of the normal derivative.

Let $\varphi \in \mathcal{H}^{\frac{1}{2}}(\Gamma)$ and let us denote by $\tilde{u}$ the unique solution in $\mathcal{H}^1(\Omega)$ of the Dirichlet problem:

$$\begin{cases} \Delta\tilde{u} = 0 & \text{in } \Omega, \\ \tilde{u} = \varphi & \text{on } \Gamma. \end{cases} \tag{1.43}$$

Using (1.25), we have, in addition for the jump of the normal derivative on Γ, the identity

$$\varphi(\boldsymbol{x}) + \int_\Gamma \Big(\frac{\partial u}{\partial n}(\boldsymbol{y}) - \frac{\partial \tilde{u}}{\partial n}(\boldsymbol{y})\Big)\, G(\boldsymbol{x},\boldsymbol{y})\, ds(\boldsymbol{y}) = 0, \qquad \boldsymbol{x} \in \Gamma.$$

By setting $\lambda = \dfrac{\partial u}{\partial n}$, multiplying this equation by $\overline{\mu} \in \mathcal{H}^{-\frac{1}{2}}(\Gamma)$ and integrating over Γ, we obtain

$$\begin{aligned}\int_\Gamma\int_\Gamma \lambda(\boldsymbol{y})\, G(\boldsymbol{x},\boldsymbol{y})\,\overline{\mu}(\boldsymbol{x})\, ds(\boldsymbol{y})\, ds(\boldsymbol{x}) &= \int_\Gamma\int_\Gamma \frac{\partial \tilde{u}}{\partial n}(\boldsymbol{y})\, G(\boldsymbol{x},\boldsymbol{y})\,\overline{\mu}(\boldsymbol{x})\, ds(\boldsymbol{y})\, ds(\boldsymbol{x}) \\ &\quad - \int_\Gamma \varphi(\boldsymbol{x})\,\overline{\mu}(\boldsymbol{x})\, ds(\boldsymbol{x}).\end{aligned}$$

We can then define the problem:

$$\begin{cases} \text{Find } \lambda \in \mathcal{H}^{-\frac{1}{2}}(\Gamma) \text{ such that} \\ \mathscr{B}(\lambda, \mu) = \mathscr{L}(\mu) \qquad \forall\, \mu \in \mathcal{H}^{-\frac{1}{2}}(\Gamma), \end{cases} \tag{1.44}$$

where

$$\mathscr{B}(\lambda, \mu) := \int_\Gamma \int_\Gamma \lambda(y)\overline{\mu}(x)\, G(x, y)\, ds(y)\, ds(x),$$

$$\mathscr{L}(\mu) := \int_\Gamma \int_\Gamma \frac{\partial \tilde{u}}{\partial n}(y)\, \overline{\mu}(x)\, G(x, y)\, ds(y)\, ds(x) - \int_\Gamma \varphi\, \overline{\mu}\, ds.$$

Theorem 1.3.13. *Problem* (1.44) *has a unique solution.*

Proof. We use the Lax-Milgram theorem (Theorem 1.2.1), the coercivity of the form $\mathscr{B}$ will be guaranteed by Theorem 1.3.8. The continuity of the antilinear form $\mathscr{L}$ is also a known result in integral representation theory (see [62], Vol. 4, p. 123). □

Remark 1.3.4. Since $\lambda = \dfrac{\partial u}{\partial n}$, then we have $\lambda = -P\varphi$. Consequently when $\varphi \in \mathcal{H}^{\frac{1}{2}}(\Gamma)$ is given, a procedure to compute $P\varphi$ would consist in solving (1.43) on the bounded domain Ω and then solving the boundary integral problem (1.44).

1.4 The Two-Dimensional Case

Two-dimensional configurations are obtained by considering a conductor domain Ω given by the cylinder $\Lambda \times \mathbb{R}$ where Λ is an open, bounded and connected set. Here again, the connectivity hypothesis is assumed to simplify problem statements. More general configurations will be considered when needed. The boundary of Λ is denoted by γ. In the sequel, we assume that this one is smooth enough (C^1, say). Its connected components are assumed to be closed curves of class C^1, with outward unit normal $\boldsymbol{n}$. The (unbounded) domain Λ_{ext} is defined as the complement of the closure $\overline{\Lambda}$ of Λ, i.e. $\Lambda_{\text{ext}} = \mathbb{R}^2 \setminus \overline{\Lambda}$. A generic point of $\mathbb{R}^2$ is also denoted here by $\boldsymbol{x} = x_1\, \boldsymbol{e}_1 + x_2\, \boldsymbol{e}_2$.

In the sequel, we shall limit ourselves to the statement of results that actually differ from the three-dimensional case.

Let u denote a function defined in $\mathbb{R}^2$, then $u_{|\gamma^-}$, or simply u^- (*resp.* $u_{|\gamma^+}$ or u^+) will stand for the inner (*resp.* outer) restriction to γ. The function $[u]_\gamma := u_{|\gamma^+} - u_{|\gamma^-}$ will stand for the jump.

1.4.1 Function Spaces

The gradient and divergence operators are obviously defined in a similar way than for the three-dimensional case. For the curl operator, scalar and vector versions are defined, for functions $\phi : \mathbb{R}^2 \to \mathbb{C}$ and $\boldsymbol{u} : \mathbb{R}^2 \to \mathbb{C}^2$ with $\boldsymbol{u} = u_1 \boldsymbol{e}_1 + u_2 \boldsymbol{e}_2$, by

$$\operatorname{curl} \boldsymbol{u} := \frac{\partial u_2}{\partial x_1} - \frac{\partial u_1}{\partial x_2},$$

$$\mathbf{curl}\, \phi := \frac{\partial \phi}{\partial x_2} \boldsymbol{e}_1 - \frac{\partial \phi}{\partial x_1} \boldsymbol{e}_2.$$

It is easy to verify that if $\boldsymbol{w}$ is a three-dimensional vector field given by $\boldsymbol{w} = \boldsymbol{u} + \phi\, \boldsymbol{e}_3$, where $\boldsymbol{u} = u_1 \boldsymbol{e}_1 + u_2 \boldsymbol{e}_2$ and that does not depend on x_3, then $\mathbf{curl}\, \boldsymbol{w} = \mathbf{curl}\, \phi + (\operatorname{curl} \boldsymbol{u})\, \boldsymbol{e}_3$. Function spaces $\mathscr{D}(\Lambda)$, $\mathcal{L}^p(\Lambda)$, $\mathcal{W}^{m,p}(\Lambda)$, $\mathcal{H}^1(\Lambda)$, $\mathcal{H}^{\frac{1}{2}}(\gamma)$ and $\mathcal{H}^{-\frac{1}{2}}(\gamma)$ are defined in the same way as for the three-dimensional case. Moreover, properties (1.4) and (1.5) are also valid in the present case.

We define the Beppo-Levi or Nédélec space $\mathcal{W}^1(\Lambda_{\text{ext}})$ by

$$\mathcal{W}^1(\Lambda_{\text{ext}}) := \left\{\psi;\ \frac{\psi}{(1 + |\boldsymbol{x}|) \ln(2 + |\boldsymbol{x}|)} \in \mathcal{L}^2(\Lambda_{\text{ext}}),\ \nabla\psi \in \boldsymbol{\mathcal{L}}^2(\Lambda_{\text{ext}})\right\},$$

equipped with the norm

$$\|\psi\|_{\mathcal{W}^1(\Lambda_{\text{ext}})} := \left(\left\|\frac{\psi}{(1 + |\boldsymbol{x}|) \ln(2 + |\boldsymbol{x}|)}\right\|^2_{\mathcal{L}^2(\Lambda_{\text{ext}})} + \|\nabla\psi\|^2_{\boldsymbol{\mathcal{L}}^2(\Lambda_{\text{ext}})}\right)^{\frac{1}{2}}.$$

We also define $\mathcal{W}^1_0(\Lambda_{\text{ext}})$ as the closure of $\mathscr{D}(\Lambda_{\text{ext}})$ in $\mathcal{W}^1(\Lambda_{\text{ext}})$ and we also define the semi-norm

$$|\psi|_{\mathcal{W}^1(\Lambda_{\text{ext}})} := \|\nabla\psi\|_{\boldsymbol{\mathcal{L}}^2(\Lambda_{\text{ext}})}, \qquad \psi \in \mathcal{W}^1(\Lambda_{\text{ext}}).$$

In particular, it is known (see [135]) that we have the inequality

$$\|\psi\|_{\mathcal{W}^1(\Lambda_{\text{ext}})} \le C\, |\psi|_{\mathcal{W}^1(\Lambda_{\text{ext}})} \qquad \forall\ \psi \in \mathcal{W}^1_0(\Lambda_{\text{ext}}), \tag{1.45}$$

which implies that the above semi-norm is equivalent to the norm on $\mathcal{W}^1_0(\Lambda_{\text{ext}})$. The space $\mathcal{W}^1(\mathbb{R}^2)$ is defined in a similar way to $\mathcal{W}^1(\Lambda_{\text{ext}})$. Note that in opposition to the three-dimensional case, the constant function belongs to $\mathcal{W}^1(\Lambda_{\text{ext}})$. This explains why (1.6) does not hold in the two-dimensional case and we have to add in (1.45) the fact that the functions vanish on Γ. In addition, it can be proved that the semi-norm $|\cdot|_{\mathcal{W}^1(\Lambda_{\text{ext}})}$ and the norm $\|\cdot\|_{\mathcal{W}^1(\Lambda_{\text{ext}})}$ are equivalent on the space $\mathcal{W}^1(\Lambda_{\text{ext}})/\mathbb{C}$, i.e. we have

$$\|\psi\|_{\mathcal{W}^1(\Lambda_{\text{ext}})} \le C\, |\psi|_{\mathcal{W}^1(\Lambda_{\text{ext}})} \qquad \forall\ \psi \in \mathcal{W}^1(\Lambda_{\text{ext}})/\mathbb{C}. \tag{1.46}$$

While the space $\mathcal{H}(\text{div}, \Lambda)$ is defined similarly as for the three-dimensional case, the two-dimensional analog of $\mathcal{H}(\text{curl}, \Lambda)$ is actually defined for the curl operator by

$$\mathcal{H}(\text{curl}, \Lambda) := \{ \boldsymbol{v} \in \mathcal{L}^2(\Lambda);\ \text{curl}\, \boldsymbol{v} \in \mathcal{L}^2(\Lambda)\}.$$

This space is a complex Hilbert space when endowed with the inner product

$$(\boldsymbol{u}, \boldsymbol{v})_{\mathcal{H}(\text{curl},\Lambda)} := \int_\Lambda \boldsymbol{u} \cdot \overline{\boldsymbol{v}}\, d\boldsymbol{x} + \int_\Lambda \text{curl}\, \boldsymbol{u}\ \text{curl}\, \overline{\boldsymbol{v}}\, d\boldsymbol{x},$$

and its associated norm

$$\|\boldsymbol{v}\|_{\mathcal{H}(\text{curl},\Lambda)} := (\|\boldsymbol{v}\|^2_{\mathcal{L}^2(\Lambda)} + \|\,\text{curl}\, \boldsymbol{v}\|^2_{\mathcal{L}^2(\Lambda)})^{\frac{1}{2}}.$$

To define traces of functions of this space, it is convenient to introduce a unit tangent to γ by $\boldsymbol{\tau} = -n_2\, \boldsymbol{e}_1 + n_1\, \boldsymbol{e}_2$. It can then be shown that the tangential component can be defined for a field $\boldsymbol{v} \in \mathcal{H}(\text{curl}, \Lambda)$ by the duality pairing

$$\int_\gamma \phi\, (\boldsymbol{v} \cdot \boldsymbol{\tau})\, ds := \int_\Lambda \phi\ \text{curl}\, \boldsymbol{v}\, d\boldsymbol{x} - \int_\Lambda \boldsymbol{v} \cdot \mathbf{curl}\, \phi\, d\boldsymbol{x} \qquad \forall\, \phi \in \mathcal{H}^1(\Lambda).$$

The spaces $\mathcal{H}(\text{div}, \cdot)$ and $\mathcal{H}(\text{curl}, \cdot)$ are defined in the same way on Λ_{ext} and $\mathbb{R}^2$.

A two-dimensional analog to Theorem 1.3.1 can be stated.

Theorem 1.4.1. *Let* $\boldsymbol{u} : \mathbb{R}^2 \to \mathbb{C}^2$ *denote a vector field.*

1. *If* $\boldsymbol{u}_{|\Lambda} \in \mathcal{H}(\text{div}, \Lambda)$ *and* $\boldsymbol{u}_{|\Lambda_{ext}} \in \mathcal{H}(\text{div}, \Lambda_{ext})$*, then*

$$\boldsymbol{u} \in \mathcal{H}(\text{div}, \mathbb{R}^2) \;\Leftrightarrow\; [\boldsymbol{u} \cdot \boldsymbol{n}]_\gamma = 0.$$

2. *If* $\boldsymbol{u}_{|\Lambda} \in \mathcal{H}(\text{curl}, \Lambda)$ *and* $\boldsymbol{u}_{|\Lambda_{ext}} \in \mathcal{H}(\text{curl}, \Lambda_{ext})$*, then*

$$\boldsymbol{u} \in \mathcal{H}(\text{curl}, \mathbb{R}^2) \;\Leftrightarrow\; [\boldsymbol{u} \cdot \boldsymbol{\tau}]_\gamma = 0.$$

Remark 1.4.1. It is noteworthy that if $\boldsymbol{u} \in \mathcal{H}(\text{curl}, \Lambda)$ and if $\boldsymbol{u}^\perp$ is the vector field given by

$$\boldsymbol{u}^\perp := u_2\, \boldsymbol{e}_1 - u_1\, \boldsymbol{e}_2$$

then we have $\text{div}\, \boldsymbol{u}^\perp = \text{curl}\, \boldsymbol{u}$. For this reason, the space $\mathcal{H}(\text{curl}, \Lambda)$ is rarely used.

The kernels of operators div and curl can be characterized in the two-dimensional case in a similar way to the three-dimensional one. We shall however consider the kernel of the divergence operator since this one slightly differs from the three-dimensional case. For the proof of this result, we refer to Girault–Raviart [93] for instance.

Theorem 1.4.2. *Let $\boldsymbol{u}$ denote a vector field in* $\mathcal{H}(\mathrm{div}, \mathbb{R}^2)$ *such that* $\mathrm{div}\, \boldsymbol{u} = 0$. *Then there exists a scalar field* $\phi \in \mathcal{W}^1(\mathbb{R}^2)$ *such that*

$$\boldsymbol{u} = \mathbf{curl}\, \phi \qquad \text{in } \mathbb{R}^2.$$

Another result of the same type will be helpful in the sequel.

Theorem 1.4.3. *Let f denote a scalar field in $\mathcal{L}^2(\Lambda)$. Then there exists a unique vector field* $\boldsymbol{v} \in \mathcal{H}^1(\Lambda)^2$ *such that*

$$\begin{aligned} f &= \mathrm{curl}\, \boldsymbol{v} && \text{in } \Lambda, \\ \mathrm{div}\, \boldsymbol{v} &= 0 && \text{in } \Lambda, \\ \boldsymbol{v} \cdot \boldsymbol{n} &= 0 && \text{on } \gamma. \end{aligned}$$

Proof. Let ϕ denote a function in $\mathcal{H}_0^1(\Lambda)$ such that

$$-\Delta\phi = f \qquad \text{in } \Lambda.$$

Since the boundary of Λ is smooth, we have $\phi \in \mathcal{H}^2(\Lambda)$. Let us define $\boldsymbol{v} = \mathbf{curl}\, \phi \in \mathcal{H}^1(\Lambda)$. We have

$$\mathrm{div}\, \boldsymbol{v} = 0 \qquad \text{in } \Lambda.$$

Moreover, since $f = -\mathrm{div}\, \nabla\phi = \mathrm{curl}\, \mathbf{curl}\, \phi$, we deduce

$$f = -\Delta\phi = \mathrm{curl}\, \boldsymbol{v}.$$

Finally

$$\boldsymbol{v} \cdot \boldsymbol{n} = \nabla\phi \cdot \boldsymbol{\tau} = 0 \qquad \text{on } \gamma.$$

The uniqueness of $\boldsymbol{v}$ results from the uniqueness of ϕ. □

1.4.2 Behaviour at the Infinity

As far as the behaviour of physical fields at the infinity is involved, things do substantially differ from the three-dimensional case. Introducing the same notion of *radial field at the infinity*, we will show here that a harmonic field can have a logarithmic behaviour at the infinity, which means in particular that it cannot, in principle, vanish at the infinity and is not even regular enough to be in $\mathcal{W}^1(\Lambda_{\mathrm{ext}})$. Here again we speak of a scalar field $u : \mathbb{R}^2 \to \mathbb{C}$ that is *radial at the infinity* using Definition 1.3.1. The following result is obtained in collaboration with M. Pierre (Private communication).

Theorem 1.4.4. *Let $u \in \mathcal{C}^2(\Lambda_{ext})$ be a radial at the infinity and harmonic function. Then there exist constants $\alpha, \beta \in \mathbb{R}$ such that*

$$u(\boldsymbol{x}) = \alpha \ln |\boldsymbol{x}| + \beta + \mathcal{O}(|\boldsymbol{x}|^{-1}) \qquad \textit{for } |\boldsymbol{x}| \to \infty, \tag{1.47}$$

$$\nabla u(\boldsymbol{x}) = \alpha \frac{\boldsymbol{x}}{|\boldsymbol{x}|^2} + \mathcal{O}(|\boldsymbol{x}|^{-2}) \qquad \textit{for } |\boldsymbol{x}| \to \infty. \tag{1.48}$$

Proof. Consider the new variable $\boldsymbol{y} = \boldsymbol{x}/|\boldsymbol{x}|^2$ and let v be the function defined by

$$v(\boldsymbol{y}) = u(\boldsymbol{x}) = u\Big(\frac{\boldsymbol{y}}{|\boldsymbol{y}|^2}\Big) \qquad \text{when } |\boldsymbol{y}| \to 0.$$

We verify that there exists $r_0 > 0$ such that v is harmonic in the set $B = \{\boldsymbol{y} \in \mathbb{R}^2;\ 0 < |\boldsymbol{y}| < r_0\}$. Moreover, since u is radial at the infinity, the function $w(\boldsymbol{y}) := v(\boldsymbol{y}) - v(\boldsymbol{R}\,\boldsymbol{y})$ is harmonic on B and bounded when $\boldsymbol{y}$ tends to zero for all rotations $\boldsymbol{R}$ in $\mathbb{R}^2$. Therefore, w can be extended to a harmonic function on the ball $\tilde{B} = \{\boldsymbol{y} \in \mathbb{R}^2;\ |\boldsymbol{y}| < r_0\}$ and we have the Poisson integral (see [62], Vol. 1, p. 249):

$$w(\boldsymbol{y}) = \frac{1}{\pi r_0}\Big(\frac{r_0^2}{4} - |\boldsymbol{y}|^2\Big) \int_C \frac{w(z)}{|z - \boldsymbol{y}|^2}\, ds(z) \qquad \forall\ \boldsymbol{y} \text{ with } |\boldsymbol{y}| < \frac{r_0}{2},$$

where C is the circle centered at the origin with radius $\frac{r_0}{2}$. We have for $0 < |\boldsymbol{y}| < r_0/2$:

$$\begin{aligned} v(\boldsymbol{y}) - \frac{1}{\pi r_0}\Big(\frac{r_0^2}{4} - |\boldsymbol{y}|^2\Big) \int_C \frac{v(z)}{|z - \boldsymbol{y}|^2}\, ds(z) \\ = v(\boldsymbol{R}\,\boldsymbol{y}) - \frac{1}{\pi r_0}\Big(\frac{r_0^2}{4} - |\boldsymbol{R}\,\boldsymbol{y}|^2\Big) \int_C \frac{v(z)}{|z - \boldsymbol{R}\,\boldsymbol{y}|^2}\, ds(z). \end{aligned}$$

It follows that the function

$$h(\boldsymbol{y}) := v(\boldsymbol{y}) - \frac{1}{\pi r_0}\Big(\frac{r_0^2}{4} - |\boldsymbol{y}|^2\Big) \int_C \frac{v(z)}{|z - \boldsymbol{y}|^2}\, ds(z)$$

is harmonic for $0 < |\boldsymbol{y}| < r_0/2$ and it is a radial function. By direct computation we obtain $h(\boldsymbol{y}) = a \ln |\boldsymbol{y}| + b$ where $a, b \in \mathbb{R}$, so that

$$v(\boldsymbol{y}) = a \ln |\boldsymbol{y}| + b + \frac{1}{\pi r_0}\Big(\frac{r_0^2}{4} - |\boldsymbol{y}|^2\Big) \int_C \frac{v(z)}{|z - \boldsymbol{y}|^2}\, ds(z) \tag{1.49}$$

for $0 < |\boldsymbol{y}| < r_0/2$. Since

$$|z - \boldsymbol{y}|^{-2} = |z|^{-2}(1 + \mathcal{O}(|\boldsymbol{y}|)) \qquad \text{when } |\boldsymbol{y}| \to 0,$$

we obtain

$$v(y) = a \ln |y| + b + \frac{1}{\pi r_0} \int_C v(z)\, ds(z) + \mathcal{O}(|y|) \qquad |x| \to 0.$$

It follows that

$$u(x) = -a \ln |x| + b + \frac{1}{\pi r_0} \int_C u\Big(\frac{z}{|z|^2}\Big)\, ds(z) + \mathcal{O}(|x|^{-1})$$

$$\text{when } |x| \to \infty.$$

We then obtain (1.47) by setting

$$\alpha = -a, \quad \beta = b + \frac{1}{\pi r_0} \int_C u\Big(\frac{z}{|z|^2}\Big)\, ds(z).$$

To obtain (1.48), we have from (1.49)

$$u(x) = -a \ln |x| + b + \frac{1}{\pi r_0}\Big(\frac{r_0^2}{4} - \frac{1}{|x|^2}\Big) \int_C v\Big(\frac{z}{|z|^2}\Big)\, ds(z).$$

Taking the gradient of this expression and proceeding in a similar way, we retrieve (1.48). □

1.4.3 Integral Representations

We introduce the Green function in two dimensions,

$$G(x, y) := -\frac{1}{2\pi} \ln |x - y|, \qquad x \in \mathbb{R}^2,\ x \neq y. \tag{1.50}$$

We check indeed that

$$-\Delta_y G(x, \cdot) = \delta_x \qquad \text{in } \mathbb{R}^2. \tag{1.51}$$

Let us now consider the problem of finding a radial at the infinity function u such that

$$-\Delta u = f \qquad \text{in } \mathbb{R}^2, \tag{1.52}$$

where f is a function in $\mathcal{L}^2(\mathbb{R}^2)$ with a compact support.

Theorem 1.4.5. *The solution u of* (1.52) *is given by*

$$u(\boldsymbol{x}) = \int_{\mathbb{R}^2} f(\boldsymbol{y})\, G(\boldsymbol{x},\boldsymbol{y})\, d\boldsymbol{y} + \beta \qquad \boldsymbol{x} \in \mathbb{R}^2, \tag{1.53}$$

where β is the complex constant given by (1.47)*. Moreover, if f satisfies*

$$\int_{\mathbb{R}^2} f(\boldsymbol{x})\, d\boldsymbol{x} = 0, \tag{1.54}$$

then, we have $u \in \mathcal{W}^1(\mathbb{R}^2)$, i.e., we have $\alpha = 0$ in (1.47)*.*

Proof. Let B_r denote the ball centered at the origin with radius $r > 0$ and let ∂B_r denote its boundary. The radius r is chosen such that B_r contains the support of f. We have from the Green formula for all $\boldsymbol{x} \in \mathbb{R}^2$,

$$\begin{aligned}
&\int_{B_r} u(\boldsymbol{y})\, \Delta_y G(\boldsymbol{x},\boldsymbol{y})\, d\boldsymbol{y} - \int_{B_r} \Delta u(\boldsymbol{y})\, G(\boldsymbol{x},\boldsymbol{y})\, d\boldsymbol{y} \\
&\quad = \int_{\partial B_r} u(\boldsymbol{y}) \frac{\partial}{\partial \boldsymbol{n}_y} G(\boldsymbol{x},\boldsymbol{y})\, ds(\boldsymbol{y}) - \int_{\partial B_r} \frac{\partial u}{\partial n}(\boldsymbol{y})\, G(\boldsymbol{x},\boldsymbol{y})\, ds(\boldsymbol{y}).
\end{aligned}$$

From (1.52) and (1.51), we obtain

$$\begin{aligned}
&- u(\boldsymbol{x}) + \int_{B_r} f(\boldsymbol{y}) G(\boldsymbol{x},\boldsymbol{y})\, d\boldsymbol{y} \\
&\quad = \int_{\partial B_r} u(\boldsymbol{y}) \frac{\partial G}{\partial \boldsymbol{n}_y}(\boldsymbol{x},\boldsymbol{y})\, ds(\boldsymbol{y}) - \int_{\partial B_r} \frac{\partial u}{\partial n}(\boldsymbol{y})\, G(\boldsymbol{x},\boldsymbol{y})\, ds(\boldsymbol{y}),
\end{aligned}$$

for all $\boldsymbol{x} \in \mathbb{R}^2$. Let us evaluate the boundary integrals for large r. We note first that when $|\boldsymbol{y}| \to \infty$, we have

$$\begin{aligned}
|\boldsymbol{x} - \boldsymbol{y}| &= |\boldsymbol{y}| + \mathcal{O}(1), \\
\ln |\boldsymbol{x} - \boldsymbol{y}| &= \ln |\boldsymbol{y}| + \mathcal{O}(|\boldsymbol{y}|^{-1}).
\end{aligned}$$

Using Theorem 1.4.4 and the identity $\boldsymbol{n}(\boldsymbol{y}) = \boldsymbol{y}/|\boldsymbol{y}|$ on ∂B_r, we have with $\boldsymbol{y} = r(\cos\theta \boldsymbol{e}_1 + r \sin\theta \boldsymbol{e}_2)$,

$$\begin{aligned}
&\int_{\partial B_r} u(\boldsymbol{y}) \frac{\partial G}{\partial \boldsymbol{n}_y}(\boldsymbol{x},\boldsymbol{y})\, ds(\boldsymbol{y}) - \int_{\partial B_r} \frac{\partial u}{\partial n}(\boldsymbol{y})\, G(\boldsymbol{x},\boldsymbol{y})\, ds(\boldsymbol{y}) \\
&= -\frac{1}{2\pi} \int_0^{2\pi} (\alpha \ln r + \beta + \mathcal{O}(r^{-1}))\Big(-\frac{1}{r} + \frac{(\boldsymbol{x} - \boldsymbol{y}) \cdot \boldsymbol{x}}{|\boldsymbol{x} - \boldsymbol{y}|^2\, r} \Big) r\, d\theta \\
&\quad - \frac{1}{2\pi} \int_0^{2\pi} \Big(\frac{\alpha}{r} + \mathcal{O}(r^{-2})\Big)\big(\ln r + \mathcal{O}(r^{-1})\big) r\, d\theta \\
&= \beta + \mathcal{O}(r^{-1} \ln r).
\end{aligned}$$

Taking the limit $r \to \infty$, we eventually obtain

$$u(\boldsymbol{x}) = \int_{\mathbb{R}^2} f(\boldsymbol{y}) G(\boldsymbol{x}, \boldsymbol{y})\, d\boldsymbol{y} + \beta.$$

Let us now assume (1.54). By integrating (1.52) on B_r, we obtain

$$0 = -\int_{B_r} \Delta u \, d\boldsymbol{x} = -\int_{\partial B_r} \frac{\partial u}{\partial n}\, ds.$$

Using again Theorem 1.4.4, we have

$$0 = -\alpha + \mathcal{O}(r^{-1}) \qquad \text{when } r \to \infty,$$

and consequently $\alpha = 0$. □

Remark 1.4.2. The problem of finding a function $u \in \mathcal{W}^1(\mathbb{R}^2)/\mathbb{C}$ such that

$$\int_{\mathbb{R}^2} \nabla u \cdot \nabla \overline{v}\, d\boldsymbol{x} = \int_{\mathbb{R}^2} f \overline{v}\, d\boldsymbol{x} \qquad \forall\, v \in \mathcal{W}^1(\mathbb{R}^2)/\mathbb{C},$$

has a unique solution. This one is a solution of (1.52) if and only if $\int_{\mathbb{R}^2} f\, d\boldsymbol{x} = 0$. This is a consequence of the Lax-Milgram theorem and the fact that the norm on $\mathcal{W}^1(\mathbb{R}^2)/\mathbb{C}$ and the semi-norm are equivalent thanks to (1.46).

By using Theorem 1.4.4, we can now quote an analog to Theorem 1.3.7 for the two-dimensional case. The proof of this result can be found in [135].

Theorem 1.4.6. *Let u be a function of class $(C^2(\Lambda) \cap C^1(\overline{\Lambda})) \cap (C^2(\Lambda_{ext}) \cap C^1(\overline{\Lambda}_{ext}))$, that is radial at the infinity and that satisfies*

$$\Delta u = 0 \qquad \text{in } \Lambda \cup \Lambda_{ext}.$$

Then we have, for $\boldsymbol{x} \in \Lambda \cup \Lambda_{ext}$, the representation

$$u(\boldsymbol{x}) = -\int_{\gamma} \left[\frac{\partial u}{\partial n}(\boldsymbol{y})\right] G(\boldsymbol{x}, \boldsymbol{y})\, ds(\boldsymbol{y}) + \int_{\gamma} [u(\boldsymbol{y})]\, \frac{\partial G}{\partial n_{\boldsymbol{y}}}(\boldsymbol{x}, \boldsymbol{y})\, ds(\boldsymbol{y}) + \xi, \tag{1.55}$$

and for $\boldsymbol{x} \in \gamma$,

$$\begin{aligned} \frac{1}{2}(u_{|\gamma^-}(\boldsymbol{x}) + u_{|\gamma^+}(\boldsymbol{x})) = &-\int_{\gamma} \left[\frac{\partial u}{\partial n}(\boldsymbol{y})\right] G(\boldsymbol{x}, \boldsymbol{y})\, ds(\boldsymbol{y}) \\ &+ \int_{\gamma} [u(\boldsymbol{y})]\, \frac{\partial G}{\partial n_{\boldsymbol{y}}}(\boldsymbol{x}, \boldsymbol{y})\, ds(\boldsymbol{y}) + \xi, \end{aligned} \tag{1.56}$$

where ξ is a constant.

Remark 1.4.3. If u is a harmonic function in Λ_{ext} that satisfies

$$\lim_{|\boldsymbol{x}|\to\infty} |u(\boldsymbol{x})| = 0,$$

then $\xi = 0$.

We can now proceed as for the three-dimensional case to derive some useful integral representation results. For this we consider the exterior problem:

$$\begin{cases} \Delta u = 0 & \text{in } \Lambda_{\text{ext}}, \\ u = g & \text{on } \gamma, \\ u(\boldsymbol{x}) = \mathcal{O}(1) & \text{for } |\boldsymbol{x}| \to \infty, \end{cases} \tag{1.57}$$

where $g \in \mathcal{H}^{\frac{1}{2}}(\gamma)$.

Lemma 1.4.1. *Problem* (1.57) *has a unique solution* $u \in \mathcal{W}^1(\Lambda_{ext})$.

Proof. Let B_r denote the ball centered at the origin, with radius $r > 0$ such that $\overline{\Lambda} \subset B_r$. There exists $\varphi \in H^1(B_r \setminus \overline{\Lambda})$ such that $\varphi = g$ on γ and $\varphi = 0$ on ∂B_r. Extending φ by zero to the complement of B_r, we obtain

$$\varphi \in \mathcal{W}^1(\Lambda_{\text{ext}}) \text{ and } \varphi = g \text{ on } \gamma.$$

Let $w \in \mathcal{W}_0^1(\Lambda_{\text{ext}})$ be such that

$$\int_{\Lambda_{\text{ext}}} \nabla w \cdot \nabla \overline{\psi} \, d\boldsymbol{x} = -\int_{\Lambda_{\text{ext}}} \nabla \varphi \cdot \nabla \overline{\psi} \, d\boldsymbol{x} \qquad \forall\, \psi \in \mathcal{W}_0^1(\Lambda_{\text{ext}}).$$

By the Lax-Milgram theorem, such a w exists. Moreover, if $u = w + \varphi$, then $u \in \mathcal{W}^1(\Lambda_{\text{ext}})$ and $u = g$ on γ. We have

$$\int_{\Lambda_{\text{ext}}} \nabla u \cdot \nabla \overline{\psi} \, d\boldsymbol{x} = 0 \qquad \forall\, \psi \in \mathcal{W}_0^1(\Lambda_{\text{ext}}),$$

and then u is harmonic in Λ_{ext}. Since $u \in \mathcal{W}^1(\Lambda_{\text{ext}})$, we have $u(\boldsymbol{x}) = \alpha + \mathcal{O}(|\boldsymbol{x}|^{-1})$ when $|\boldsymbol{x}| \to \infty$. We conclude by noting that the solution of (1.57) is unique since for $g = 0$ the unique solution is 0. □

Let us define the interior problem

$$\begin{cases} \Delta u = 0 & \text{in } \Lambda, \\ u = g & \text{on } \gamma. \end{cases}$$

If p stands for the jump $[\frac{\partial u}{\partial n}]_\gamma$, we have, from Theorem 1.4.6, the integral representation

$$u(\boldsymbol{x}) = -\int_\gamma p(\boldsymbol{y})\, G(\boldsymbol{x},\boldsymbol{y})\, ds(\boldsymbol{y}) + \xi \qquad \boldsymbol{x} \in \mathbb{R}^2,$$

where $\xi \in \mathbb{C}$, and for $\boldsymbol{x} \in \gamma$,

$$g(\boldsymbol{x}) = -\int_\gamma p(\boldsymbol{y})\, G(\boldsymbol{x},\boldsymbol{y})\, ds(\boldsymbol{y}) + \xi \qquad \boldsymbol{x} \in \gamma. \tag{1.58}$$

Let $\tilde{\mathcal{H}}^{-\frac{1}{2}}(\gamma)$ denote the space

$$\tilde{\mathcal{H}}^{-\frac{1}{2}}(\gamma) := \{ q \in \mathcal{H}^{-\frac{1}{2}}(\gamma);\ \int_\gamma q\, ds = 0 \}.$$

Problem (1.58) admits the variational formulation,

$$\begin{cases} \text{Find } p \in \tilde{\mathcal{H}}^{-\frac{1}{2}}(\gamma) \text{ such that} \\ \mathscr{B}(p,q) = \displaystyle\int_\gamma g\overline{q}\, ds \qquad \forall\, q \in \tilde{\mathcal{H}}^{-\frac{1}{2}}(\gamma), \end{cases} \tag{1.59}$$

where $\mathscr{B}$ is the sesquilinear form given by

$$\mathscr{B}(p,q) := \int_\gamma \int_\gamma G(\boldsymbol{x},\boldsymbol{y})\, p(\boldsymbol{y})\, \overline{q}(\boldsymbol{x})\, ds(\boldsymbol{y})\, ds(\boldsymbol{x}).$$

Theorem 1.4.7. *Problem* (1.59) *has a unique solution* $p \in \tilde{\mathcal{H}}^{-\frac{1}{2}}(\gamma)$. *Moreover, the exterior normal derivative of the solution u of* (1.57) *is given by*

$$\frac{\partial u^+}{\partial n}(\boldsymbol{x}) = \frac{1}{2}\, p(\boldsymbol{x}) - \int_\gamma p(\boldsymbol{y})\, \frac{\partial G}{\partial n_{\boldsymbol{x}}}(\boldsymbol{x},\boldsymbol{y})\, ds(\boldsymbol{y}) \qquad \boldsymbol{x} \in \gamma.$$

Proof. From [135], we deduce that the sesquilinear form $\mathscr{B}$ is continuous on $\tilde{\mathcal{H}}^{-\frac{1}{2}}(\gamma)$ and that there is a constant C such that

$$|\mathscr{B}(q,q)| \geq C\, \|q\|^2_{\mathcal{H}^{-\frac{1}{2}}(\gamma)} \qquad \forall\, q \in \tilde{\mathcal{H}}^{-\frac{1}{2}}(\gamma).$$

It follows from the Lax-Milgram theorem (Theorem 1.2.1) that (1.59) has a unique solution $p \in \tilde{\mathcal{H}}^{-\frac{1}{2}}(\gamma)$. □

As a consequence, (1.59) defines an isomorphism $K : \tilde{\mathcal{H}}^{-\frac{1}{2}}(\gamma) \to \mathcal{H}^{\frac{1}{2}}(\gamma)/\mathbb{C}$ by the identity

$$\int_\gamma (Kp)\overline{q}\, ds := \mathscr{B}(p,q) \qquad p,q \in \tilde{\mathcal{H}}^{-\frac{1}{2}}(\gamma). \tag{1.60}$$

We have the regularity result.

Theorem 1.4.8 ([135]). *Assume that the boundary γ is smooth enough. Then, the mapping K^{-1} defines an isomorphism of $\mathcal{H}^s(\gamma)/\mathbb{C}$ onto the space*

$$\{q \in \mathcal{H}^{s-1}(\gamma);\ \int_\gamma q\, ds = 0\}$$

for all $s \in \mathbb{R}$.

Similarly to the three-dimensional case, we consider the Dirichlet Problem (1.57).

This problem has a unique solution and we deduce from Theorem 1.4.4 that

$$u(\boldsymbol{x}) = \beta + \mathcal{O}(|\boldsymbol{x}|^{-1}) \qquad |\boldsymbol{x}| \to \infty, \tag{1.61}$$

$$|\nabla u(\boldsymbol{x})| = \mathcal{O}(|\boldsymbol{x}|^{-2}) \qquad |\boldsymbol{x}| \to \infty. \tag{1.62}$$

In addition, since (1.62) implies that

$$\lim_{r\to\infty} \int_{\partial B_r} \frac{\partial u}{\partial n}\, ds = 0,$$

where ∂B_r is the circle centered at the origin with radius r, then we have

$$\int_\gamma \frac{\partial u}{\partial n}\, ds = 0.$$

Therefore, u can be defined up to a complex constant, in Λ as a harmonic function with $[\frac{\partial u}{\partial n}]_\gamma = 0$. Note that this contrasts with the 3-D case where the extra-condition (1.37), imposed by (1.33) has to be added.

Let $\varphi = [u]_\gamma$, which is defined up to a complex constant. We have for φ from Theorem 1.4.6, the integral representation

$$g(\boldsymbol{x}) = \frac{1}{2}\varphi(\boldsymbol{x}) + \int_\gamma \varphi(\boldsymbol{y})\, \frac{\partial G}{\partial n_{\boldsymbol{y}}}(\boldsymbol{x},\boldsymbol{y})\, ds(\boldsymbol{y}) + \xi \qquad \boldsymbol{x} \in \gamma, \tag{1.63}$$

where $\xi \in \mathbb{C}$. Further, it is noteworthy that by setting $u = 1$ in Λ and $u = 0$ in Λ_{ext} in (1.56), one has $\xi = 0$ and we obtain

$$\int_\gamma \frac{\partial G}{\partial n_y}(x, y)\, ds(y) = -\frac{1}{2} \qquad \text{for } x \in \gamma,$$

and consequently

$$g(x)-\xi = \frac{1}{2}(\varphi(x)+\eta)+\int_\gamma (\varphi(y)+\eta)\,\frac{\partial G}{\partial n_y}(x, y)\, ds(y)+\xi \quad \forall\, x \in \gamma,\ \forall\, \xi, \eta \in \mathbb{C}.$$

Now, we can write (1.63) in operator form as

$$(\frac{1}{2}I + R)\varphi = g, \tag{1.64}$$

where R is the mapping defined by

$$R\varphi := \int_\gamma \varphi(y)\frac{\partial G}{\partial n_y}(\cdot, y)\, ds(y). \tag{1.65}$$

We have the following property on R (cf. [108]):

Lemma 1.4.2. *The mapping R is linear and continuous from $\mathcal{H}^s(\gamma)$ into $\mathcal{H}^{s+1}(\gamma)$ for all real s.*

We have the following result.

Theorem 1.4.9 ([135]). *Assume that $g \in \mathcal{H}^{\frac{1}{2}}(\gamma)$. Then* (1.64) *admits a unique solution $\varphi \in \mathcal{H}^{\frac{1}{2}}(\gamma)$ up to a complex constant. Moreover, the operator $\frac{1}{2}I + R$ defines an isomorphism of $\mathcal{H}^{\frac{1}{2}}(\gamma)/\mathbb{C}$ onto $\mathcal{H}^{\frac{1}{2}}(\gamma)/\mathbb{C}$.*

1.4.4 The Exterior Steklov–Poincaré Operator

The two-dimensional version of the exterior Steklov–Poincaré operator is defined exactly in the same way as previously. To summarize, we define

$$P : g \in \mathcal{H}^{\frac{1}{2}}(\gamma) \mapsto -\frac{\partial u}{\partial n}\Big|_\gamma \in \mathcal{H}^{-\frac{1}{2}}(\gamma), \tag{1.66}$$

where u is the unique solution, in $\mathcal{W}^1(\Lambda_{\text{ext}})$, of the exterior problem

$$\begin{cases} \Delta u = 0 & \text{in } \Lambda_{\text{ext}}, \\ u = g & \text{on } \gamma, \\ u(\boldsymbol{x}) = \mathcal{O}(1) & |\boldsymbol{x}| \to \infty. \end{cases} \tag{1.67}$$

Then, we have

$$P = (-\frac{1}{2}I + R')\, K^{-1},$$

where the operators K and R are defined in (1.60) and (1.65) respectively, and R' is the adjoint operator of R defined as in (1.41). Using the same techniques as in Sect. 1.3.5, we can prove the following

Theorem 1.4.10. *The operator P is continuous, selfadjoint and coercive.*

Here also, like for the three-dimensional case, we can resort to an alternative formulation that combines solving an interior problem with the use of a simple layer integral representation. We skip this description since this one is identical to the one given in Sect. 1.3.5.

Chapter 2
Maxwell and Eddy Current Equations

2.1 Introduction

Maxwell equations stand for the set of partial differential equations that describe electric and magnetic phenomena. Maxwell equations were derived in several steps successively by Coulomb, Faraday, Ampère and Maxwell. Their treatment contains numerous difficulties either from the mathematical or numerical point of view. In particular, the presence of a large number of unknown fields, conditions at the infinity, high frequency require specific techniques to handle them. The literature in Mathematics and Physics is rather plentiful in this field and the reader is referred to most popular textbooks in electromagnetic theory (e.g. Feynman [74], Jackson [107], Landau and Lifshitz [116], Robinson [155]) and to Nédélec [138], Monk [131] and many others for the numerical solution of these equations.

Our purpose throughout this textbook is to study Maxwell equations in the particular situation where the source current in an electromagnetic setup has a low frequency. The term "low" means here that the characteristic length of the considered conducting bodies is small when compared to the wavelength of the inflowing current. Dimensional analysis considerations show that, in this case, propagation (hyperbolic) terms can be neglected beside all other terms. In other words, wave propagation phenomena are neglected and we have the creation of the so-called eddy currents inside the conductors. Such configurations are present in some specific industrial setups when induction properties of electromagnetic phenomena have to be exploited. For example, electric conduction generates heat by dissipation (Joule effect) and this feature can be used to raise conductor temperature for many purposes (e.g. forging, welding, surface processing). Another typical situation is the one where Lorentz forces can be used to stir liquid metals (e.g. cold crucibles, solidification). Many other examples can be found in metallurgy and other fields of application. In all these situations, the use of low frequency currents helps creating eddy currents with a negligible effect of displacement currents.

In this chapter, we start by presenting the general setting of Maxwell equations. Through these equations, we shall show the existence of a *potential vector* which

R. Touzani and J. Rappaz, *Mathematical Models for Eddy Currents and Magnetostatics: With Selected Applications*, Scientific Computation, DOI 10.1007/978-94-007-0202-8_2,

will play a central role in the analysis and modeling of electromagnetic phenomena. We shall then restrict ourselves to the main object of the present monograph: Study of low frequency regimes. We show the validity of such an approximation and consider static (time independent) cases for which we derive models in electrostatics and magnetostatics. Then, we consider time harmonic regimes that are useful to study time periodic currents.

2.2 Maxwell Equations

In all the sequel, we shall denote as usual by $\boldsymbol{B}$, $\boldsymbol{H}$, $\boldsymbol{D}$, $\boldsymbol{E}$ and $\boldsymbol{J}$ respectively magnetic induction field, magnetic field, electric displacement current field, electric field and electric current density field.

2.2.1 *General Setting*

Maxwell–Ampère and Faraday equations are respectively given by:

$$\frac{\partial \boldsymbol{D}}{\partial t} - \mathbf{curl}\, \boldsymbol{H} + \boldsymbol{J} = 0, \tag{2.1}$$

$$\frac{\partial \boldsymbol{B}}{\partial t} + \mathbf{curl}\, \boldsymbol{E} = 0. \tag{2.2}$$

Sometimes, (2.1) is replaced by an equation to take into account a *source current* $\boldsymbol{J}_S$, that is

$$\frac{\partial \boldsymbol{D}}{\partial t} - \mathbf{curl}\, \boldsymbol{H} + \boldsymbol{J} + \boldsymbol{J}_S = 0. \tag{2.3}$$

We have in addition the magnetic flux conservation equation

$$\operatorname{div} \boldsymbol{B} = 0. \tag{2.4}$$

Remark that taking the divergence of (2.2) yields

$$\frac{\partial}{\partial t} \operatorname{div} \boldsymbol{B} = 0.$$

This means that (2.4) can be interpreted as an initial condition to (2.2) since its validity for the initial time $t = 0$ guarantees it for all times thanks to (2.2).

Equations (2.1)–(2.4) are valid in the whole space $\mathbb{R}^3$ and for all times $t > 0$.

The related constitutive equations for this system are:

$$\boldsymbol{B} = \mu \boldsymbol{H}, \tag{2.5}$$

$$\boldsymbol{D} = \varepsilon \boldsymbol{E}, \tag{2.6}$$

in $\mathbb{R}^3$. The functions μ and ε stand for *magnetic permeability* and *electric permittivity* respectively. Relation (2.5) is called *Magnetic induction law* and (2.6) is the *Electric induction law*. In the sequel, we shall assume, for obvious physical reasons that the functions ε and μ fulfill the following conditions:

$$0 < \mu_m \le \mu \le \mu_M, \tag{2.7}$$

$$0 < \varepsilon_m \le \varepsilon \le \varepsilon_M, \tag{2.8}$$

where $\mu_m, \mu_M, \varepsilon_m, \varepsilon_M$ are defined lower and upper bounds for μ and ε. In addition, μ and ε are constant equal to μ_0 and ε_0, called respectively *Magnetic permeability* and *electric permittivity of the vacuum.*

Remark 2.2.1. The charge density ϱ_q can be deduced by a charge conservation equation that is

$$\operatorname{div} \boldsymbol{D} = \varrho_q. \tag{2.9}$$

2.2.2 Presence of Conductors

In the presence of conductors Ω moving with velocity $\boldsymbol{v}$, we adopt the *Ohm's law*:

$$\boldsymbol{J} = \sigma\,(\boldsymbol{E} + \boldsymbol{v} \times \boldsymbol{B}) \qquad \text{in } \mathbb{R}^3, \tag{2.10}$$

where σ is the *electric conductivity* of the given conductor occupying Ω, and $\sigma = 0$ outside the conductors. The function σ is assumed to satisfy the hypothesis:

$$0 < \sigma_m \le \sigma \le \sigma_M. \tag{2.11}$$

Outside the conductors we have $\boldsymbol{J} = 0$ and (2.10) can be considered with $\sigma = 0$. In most situations, we will consider eddy currents in non moving conductors for which we have $\boldsymbol{J} = \sigma \boldsymbol{E}$.

2.2.3 Wave Propagation

Equations (2.1)–(2.6) are of hyperbolic type. To see this, let us consider, for instance, the simple case of a non moving homogeneous isotropic medium, i.e. the case where ε, μ and σ are constant and $\boldsymbol{v} = 0$. We have from (2.1)–(2.6),

$$\varepsilon \frac{\partial \boldsymbol{E}}{\partial t} - \mathbf{curl}\, \boldsymbol{H} + \boldsymbol{J} = 0,$$

$$\mu \frac{\partial \boldsymbol{H}}{\partial t} + \mathbf{curl}\, \boldsymbol{E} = 0.$$

Taking the **curl** of the first equation and the time derivative of the second one, we obtain by subtracting, the equation

$$\varepsilon\mu \frac{\partial^2 \boldsymbol{H}}{\partial t^2} + \mathbf{curl}\,\mathbf{curl}\, \boldsymbol{H} = \mathbf{curl}\, \boldsymbol{J}.$$

Since $\operatorname{div} \boldsymbol{H} = \mu \operatorname{div} \boldsymbol{B} = 0$, we deduce

$$\varepsilon\mu \frac{\partial^2 \boldsymbol{H}}{\partial t^2} - \Delta \boldsymbol{H} = \mathbf{curl}\, \boldsymbol{J}.$$

If the current density $\boldsymbol{J} = \boldsymbol{J}_S$ is given, we obtain a hyperbolic problem that describes propagation of electromagnetic waves in the space. If the Ohm's law (2.10) is assumed in Ω, we obtain the equations:

$$\varepsilon\mu \frac{\partial^2 \boldsymbol{H}}{\partial t^2} + \mu\sigma \frac{\partial \boldsymbol{H}}{\partial t} - \Delta \boldsymbol{H} = 0 \qquad \text{in } \Omega,$$

$$\varepsilon\mu \frac{\partial^2 \boldsymbol{H}}{\partial t^2} - \Delta \boldsymbol{H} = 0 \qquad \text{in } \Omega_{\text{ext}},$$

with appropriate interface conditions and condition at the infinity.

Here also we have wave propagation but, waves are damped in the conductors, the damping being proportional to the electric conductivity σ. We have the same conclusion if the coefficients ε, μ and σ are not constant but the equations are slightly more complex.

2.2.4 The Vector Potential

One of the principal ingredients in electromagnetism is the use of a vector potential. This one is defined in the following way: Using (2.4), we deduce from Theorem 1.3.4 the existence of a vector valued function $\boldsymbol{A} : \mathbb{R}^3 \to \mathbb{C}^3$, called *vector potential*, such that

$$\boldsymbol{B} = \mathbf{curl}\, \boldsymbol{A} \quad \text{in } \mathbb{R}^3. \tag{2.12}$$

Such a vector field is in addition unique if we impose the gauge condition

$$\operatorname{div} \boldsymbol{A} = 0. \tag{2.13}$$

As far as the regularity of the vector potential $\boldsymbol{A}$ is involved, we see that if the unknowns of the Maxwell equations are sought in the space $\mathcal{L}^2(\mathbb{R}^3)$, Theorem 1.3.4 says that necessarily $\boldsymbol{A} \in \mathcal{W}^1(\mathbb{R}^3)$.

We shall characterize later this vector in more specific situations.

2.3 Low Frequency Approximation

In many situations, like in alternating current configurations, low frequencies enable neglecting the displacement current term $\partial \boldsymbol{D}/\partial t$ in the Maxwell equations. This leads to the set of equations:

$$\textbf{curl}\, \boldsymbol{H} = \boldsymbol{J}, \tag{2.14}$$

$$\frac{\partial \boldsymbol{B}}{\partial t} + \textbf{curl}\, \boldsymbol{E} = 0, \tag{2.15}$$

$$\operatorname{div} \boldsymbol{B} = 0, \tag{2.16}$$

$$\boldsymbol{B} = \mu \boldsymbol{H}. \tag{2.17}$$

Some authors have rigourously proved the validity of such an approximation. We mention here a result of Ammari, Buffa and Nédélec [13] where the authors use a formulation with source currents. They show that the external magnetic and electric fields (outside the conductors) are approximated at the first order with respect to the frequency by the system (2.14)–(2.17).

Our study deals mainly with this set of equations, and especially in the presence of the so-called *Eddy Currents* in the conductors. It is also to be specified that, in this section, we consider the magnetic permeability μ as a known function of the position $\boldsymbol{x}$. This allows later to describe nonlinear problems in which μ depends on $\boldsymbol{H}$. However, when we address the numerical solution of this kind of problems, we use an iterative method in which μ can be taken variable but given.

2.3.1 A Vector Potential Formulation

Let us now see how the vector potential $\boldsymbol{A}$ can be characterized using the system of equations (2.14)–(2.17). Looking for solutions of (2.14)–(2.17) in the space $\mathcal{L}^2(\mathbb{R}^3)$, we deduce from Theorem 1.3.4 that $\boldsymbol{A} \in \mathcal{W}^1(\mathbb{R}^3)$. From (2.14) and (2.17), we deduce

$$\textbf{curl}\,(\mu^{-1}\, \textbf{curl}\, \boldsymbol{A}) = \boldsymbol{J} \qquad \text{in } \mathbb{R}^3. \tag{2.18}$$

Let us now assume that the current density $\boldsymbol{J}$ is a function of $\boldsymbol{\mathcal{L}}^2(\mathbb{R}^3)$ with a compact support contained in a domain Ω. We have the system of equations:

$$\mathbf{curl}\,(\mu^{-1}\,\mathbf{curl}\,\boldsymbol{A}) = \boldsymbol{J} \qquad \text{in } \mathbb{R}^3, \tag{2.19}$$

$$\operatorname{div}\boldsymbol{A} = 0 \qquad \text{in } \mathbb{R}^3, \tag{2.20}$$

$$|\boldsymbol{A}(\boldsymbol{x})| = \mathcal{O}(|\boldsymbol{x}|^{-1}) \qquad \text{for } |\boldsymbol{x}| \to \infty, \tag{2.21}$$

the condition at the infinity being a consequence of $\boldsymbol{A} \in \boldsymbol{\mathcal{W}}^1(\mathbb{R}^3)$.

To prove existence and uniqueness of a solution of (2.19)–(2.21) for given $\boldsymbol{J}$, we derive a variational formulation of it. Let us take a function $\boldsymbol{w} \in \mathscr{D}(\mathbb{R}^3)$. If $\boldsymbol{J}$ is smooth enough, we have by the Green formula,

$$\int_{\mathbb{R}^3} \mathbf{curl}\,(\mu^{-1}\,\mathbf{curl}\,\boldsymbol{A}) \cdot \overline{\boldsymbol{w}}\,d\boldsymbol{x} = \int_{\mathbb{R}^3} \mu^{-1}\,\mathbf{curl}\,\boldsymbol{A} \cdot \mathbf{curl}\,\overline{\boldsymbol{w}}\,d\boldsymbol{x}.$$

This leads to the variational formulation of (2.19)–(2.21):

$$\text{Find } \boldsymbol{A} \in \boldsymbol{\mathcal{V}} \quad \text{such that} \quad \mathscr{B}(\boldsymbol{A},\boldsymbol{w}) = \mathscr{L}(\boldsymbol{w}) \qquad \forall\, \boldsymbol{w} \in \boldsymbol{\mathcal{V}}, \tag{2.22}$$

where $\boldsymbol{\mathcal{V}}$ is the space

$$\boldsymbol{\mathcal{V}} := \{\boldsymbol{w} \in \boldsymbol{\mathcal{W}}^1(\mathbb{R}^3);\ \operatorname{div}\boldsymbol{w} = 0\},$$

equipped with the semi-norm $|\cdot|_{\boldsymbol{\mathcal{W}}^1(\mathbb{R}^3)}$, which is a norm on $\boldsymbol{\mathcal{W}}^1(\mathbb{R}^3)$ (see [62], Vol. 4, p. 118), and

$$\mathscr{B}(\boldsymbol{A},\boldsymbol{w}) := \int_{\mathbb{R}^3} \mu^{-1}\,\mathbf{curl}\,\boldsymbol{A} \cdot \mathbf{curl}\,\overline{\boldsymbol{w}}\,d\boldsymbol{x},$$

$$\mathscr{L}(\boldsymbol{w}) := \int_{\Omega} \boldsymbol{J} \cdot \overline{\boldsymbol{w}}\,d\boldsymbol{x}.$$

We have the following result.

Theorem 2.3.1. *Assume that* μ *satisfies* (2.7). *Then* (2.22) *has a unique solution* $\boldsymbol{A} \in \boldsymbol{\mathcal{V}}$. *Moreover, there is a constant* C *such that*

$$\|\boldsymbol{A}\|_{\boldsymbol{\mathcal{W}}^1(\mathbb{R}^3)} \le C\,\|\boldsymbol{J}\|_{\boldsymbol{\mathcal{L}}^2(\mathbb{R}^3)}. \tag{2.23}$$

Proof. Using Theorem 1.3.2, we deduce that the quantity

$$|||\boldsymbol{w}||| := (\|\mathbf{curl}\,\boldsymbol{w}\|^2_{\boldsymbol{\mathcal{L}}^2(\mathbb{R}^3)} + \|\operatorname{div}\boldsymbol{w}\|^2_{\mathcal{L}^2(\mathbb{R}^3)})^{\frac{1}{2}} = \|\mathbf{curl}\,\boldsymbol{w}\|_{\boldsymbol{\mathcal{L}}^2(\mathbb{R}^3)}$$

defines a norm on the space $\boldsymbol{\mathcal{V}}$ that is equivalent to the norm of $\boldsymbol{\mathcal{W}}^1(\mathbb{R}^3)$. This implies that the sesquilinear form $\mathscr{B}$ is continuous and coercive on $\boldsymbol{\mathcal{V}} \times \boldsymbol{\mathcal{V}}$. The antilinear form $\mathscr{L}$ is also continuous on $\boldsymbol{\mathcal{V}}$. The Lax–Milgram Theorem 1.2.1 gives then the conclusion. □

In the particular case where the magnetic permeability μ is constant (equal to μ_0), we have an integral formula for the vector potential $\boldsymbol{A}$ and the magnetic induction $\boldsymbol{B}$.

Theorem 2.3.2. *Let $\boldsymbol{J}$ be a given vector field in the space $\boldsymbol{\mathcal{L}}^2(\mathbb{R}^3)$ with a compact support and assume that $\mu = \mu_0$ in $\mathbb{R}^3$. Then the potential $\boldsymbol{A}$ and the magnetic induction $\boldsymbol{B}$ are respectively given by:*

$$\boldsymbol{A}(\boldsymbol{x}) = \mu_0 \int_{\mathbb{R}^3} G(\boldsymbol{x}, \boldsymbol{y})\, \boldsymbol{J}(\boldsymbol{y})\, d\boldsymbol{y}, \tag{2.24}$$

$$\boldsymbol{B}(\boldsymbol{x}) = \mu_0 \int_{\mathbb{R}^3} \nabla_x G(\boldsymbol{x}, \boldsymbol{y}) \times \boldsymbol{J}(\boldsymbol{y})\, d\boldsymbol{y}, \tag{2.25}$$

for $\boldsymbol{x} \in \mathbb{R}^3$, where G is the Green kernel in dimension 3, defined by (1.21).

Proof. Let us first note that when $\mu = \mu_0$ is constant, Eqs. (2.19)–(2.21) become:

$$\mathbf{curl}\,\mathbf{curl}\,\boldsymbol{A} = \mu_0\,\boldsymbol{J} \quad \text{in } \mathbb{R}^3, \tag{2.26}$$

$$\operatorname{div} \boldsymbol{A} = 0 \quad \text{in } \mathbb{R}^3, \tag{2.27}$$

$$|\boldsymbol{A}(\boldsymbol{x})| = \mathcal{O}(|\boldsymbol{x}|^{-1}) \quad \text{for } |\boldsymbol{x}| \to \infty. \tag{2.28}$$

Using the vector identity

$$-\Delta \boldsymbol{A} = \mathbf{curl}\,\mathbf{curl}\,\boldsymbol{A} - \nabla \operatorname{div} \boldsymbol{A},$$

we deduce

$$-\Delta \boldsymbol{A} = \mu_0\,\boldsymbol{J} \qquad \text{in } \mathbb{R}^3.$$

A vector field that satisfies the above identity and (2.28), can be written, thanks to (1.23),

$$\boldsymbol{A}(\boldsymbol{x}) = \mu_0 \int_{\mathbb{R}^3} \boldsymbol{J}(\boldsymbol{y})\, G(\boldsymbol{x}, \boldsymbol{y})\, d\boldsymbol{y}.$$

Remark that this solution is unique and since the support of $\boldsymbol{J}$ is compact we have (2.21).

The proof of (2.25) is simply obtained by applying the **curl** operator to (2.24), the integrand being an integrable function. □

Relation (2.25) enables calculating the magnetic induction generated by a conductor Ω where an electric current of density $\boldsymbol{J}$ flows.

2.3.2 A Scalar Potential Problem

The material developed in the previous subsection shows that when $\mu = \mu_0$, the magnetic induction can be directly calculated from the current density $\boldsymbol{J}$ by the formula (2.25). When this is not the case (2.19)–(2.21) is not well adapted to numerical solution. For this, we can proceed as the following.

Let us consider that the electric current density $\boldsymbol{J}$ is a given vector field and has a compact support Ω with boundary Γ. We assume that $\mu = \mu_0$ outside $\overline{\Omega}$. Let us introduce a magnetic field $\boldsymbol{H}_0$ defined by

$$\boldsymbol{H}_0(\boldsymbol{x}) = \int_{\mathbb{R}^3} \nabla_x G(\boldsymbol{x}, \boldsymbol{y}) \times \boldsymbol{J}(\boldsymbol{y})\, ds(\boldsymbol{y}) \qquad \boldsymbol{x} \in \mathbb{R}^3. \tag{2.29}$$

This field is due to $\boldsymbol{J}$ when $\mu = \mu_0$ in Ω. According to (2.25), we have the equations:

$$\mathbf{curl}\, \boldsymbol{H}_0 = \boldsymbol{J}, \tag{2.30}$$

$$\operatorname{div} \boldsymbol{H}_0 = 0 \tag{2.31}$$

in $\mathbb{R}^3$. Subtracting (2.30) from (2.14), we find

$$\mathbf{curl}\, (\boldsymbol{H} - \boldsymbol{H}_0) = 0 \qquad \text{in } \mathbb{R}^3.$$

This implies (see Theorem 1.3.4) the existence of a scalar field $\psi : \mathbb{R}^3 \to \mathbb{C}$ such that

$$\boldsymbol{H} - \boldsymbol{H}_0 = -\nabla\psi \qquad \text{in } \mathbb{R}^3. \tag{2.32}$$

Multiplying this equation by μ and using (2.16), (2.17), we obtain

$$\operatorname{div}(\mu(\nabla\psi - \boldsymbol{H}_0)) = 0 \qquad \text{in } \mathbb{R}^3,$$

and consequently

$$\operatorname{div}(\mu\nabla\psi) = \operatorname{div}(\mu\boldsymbol{H}_0) \qquad \text{in } \Omega \cup \Omega_{\text{ext}}, \tag{2.33}$$

where Ω denotes the conductor, $\Omega_{\text{ext}} = \mathbb{R}^3 \setminus \overline{\Omega}$ and in addition

$$\left[\mu\frac{\partial\psi}{\partial n}\right]_\Gamma = [\mu\boldsymbol{H}_0 \cdot \boldsymbol{n}]_\Gamma.$$

By using (2.31), (2.16), (2.17), we successively obtain

$$\left[\mu\frac{\partial\psi}{\partial n}\right]_\Gamma = [\mu \boldsymbol{H}_0\cdot\boldsymbol{n}]_\Gamma = [(\mu-\mu_0)\boldsymbol{H}_0\cdot\boldsymbol{n}]_\Gamma = (\mu_0-\mu)\boldsymbol{H}_0\cdot\boldsymbol{n} \qquad \text{on } \Gamma.$$

We then obtain the problem:

$$-\operatorname{div}(\mu\nabla\psi) = -\operatorname{div}(\mu\boldsymbol{H}_0) \qquad \text{in } \Omega, \tag{2.34}$$

$$\Delta\psi = 0 \qquad \text{in } \Omega_{\text{ext}}, \tag{2.35}$$

$$[\psi]_\Gamma = 0, \tag{2.36}$$

$$\left[\mu\frac{\partial\psi}{\partial n}\right]_\Gamma = (\mu_0-\mu)\,\boldsymbol{H}_0\cdot\boldsymbol{n}, \tag{2.37}$$

$$\psi(\boldsymbol{x}) = \mathcal{O}(|\boldsymbol{x}|^{-1}) \qquad \text{for } |\boldsymbol{x}|\to\infty. \tag{2.38}$$

Let us now show that (2.34)–(2.38) is well posed if we seek $\psi \in \mathcal{W}^1(\mathbb{R}^3)$. Using the exterior Steklov–Poincaré operator P defined in Sect. 1.3.5, we can formulate equations (2.35)–(2.38) as

$$-\mu\frac{\partial\psi^-}{\partial n} = \mu_0\,P\psi - (\mu-\mu_0)\,\boldsymbol{H}_0\cdot\boldsymbol{n}. \tag{2.39}$$

Multiplying (2.34) by a function $\overline{\theta} \in \mathcal{H}^1(\Omega)$ and using the Green formula with $\operatorname{div}(\mu_0\boldsymbol{H}_0) = 0$, we obtain then

$$\int_\Omega \mu\nabla\psi\cdot\nabla\overline{\theta}\,d\boldsymbol{x} - \int_\Gamma \mu\frac{\partial\psi^-}{\partial n}\,\overline{\theta}\,ds = \int_\Omega (\mu-\mu_0)\boldsymbol{H}_0\cdot\nabla\overline{\theta}\,d\boldsymbol{x} - \int_\Gamma (\mu-\mu_0)\,\boldsymbol{H}_0\cdot\boldsymbol{n}\,\overline{\theta}\,ds.$$

By using (2.39), this leads to the variational problem:

$$\begin{cases} \text{Find } \psi \in \mathcal{H}^1(\Omega) \text{ such that} \\ \displaystyle\int_\Omega \mu\,\nabla\psi\cdot\nabla\overline{\theta}\,d\boldsymbol{x} + \mu_0\int_\Gamma P(\psi)\,\overline{\theta}\,ds = \int_\Omega (\mu-\mu_0)\,\boldsymbol{H}_0\cdot\nabla\overline{\theta}\,d\boldsymbol{x} \\ \hfill \forall\,\theta \in \mathcal{H}^1(\Omega). \end{cases} \tag{2.40}$$

Let us prove the following result.

Theorem 2.3.3. *Assume that μ is given in $\mathcal{L}^\infty(\Omega)$ and satisfies* (2.7). *Assume in addition that the restriction of $\boldsymbol{H}_0$ to Ω is given by* (2.29). *Then* (2.40) *admits a unique solution.*

Proof. We define the sesquilinear and antilinear forms:

$$\mathscr{B}(\psi,\theta) := \int_\Omega \mu\, \nabla\psi \cdot \nabla\overline{\theta}\, d\boldsymbol{x} + \mu_0 \int_\Gamma P(\psi)\,\overline{\theta}\, ds,$$

$$\mathscr{L}(\theta) := \int_\Omega (\mu - \mu_0)\, \boldsymbol{H}_0 \cdot \nabla\overline{\theta}\, d\boldsymbol{x}.$$

Since $\boldsymbol{H}_0 \in \boldsymbol{\mathcal{L}}^2(\Omega)$, then we have by using (2.7),

$$\begin{aligned} |\mathscr{L}(\theta)| &\le (\mu_0 + \mu_M)\, \|\boldsymbol{H}_0\|_{\boldsymbol{\mathcal{L}}^2(\Omega)}\, \|\nabla\theta\|_{\boldsymbol{\mathcal{L}}^2(\Omega)} \\ &\le C\, \|\theta\|_{\mathcal{H}^1(\Omega)}. \end{aligned}$$

The form $\mathscr{L}$ is hence continuous on $\mathcal{H}^1(\Omega)$. The sesquilinear form $\mathscr{B}$ is also continuous since we have from Theorem 1.3.12, the trace theorem [92] and (2.7),

$$|\mathscr{B}(\psi,\theta)| \le C\, \|\psi\|_{\mathcal{H}^1(\Omega)}\, \|\theta\|_{\mathcal{H}^1(\Omega)}.$$

The coercivity of $\mathscr{B}$ is obtained thanks to Theorem 1.3.12, (2.7) and (1.5),

$$\begin{aligned} \mathscr{B}(\theta,\theta) &= \int_\Omega \mu\, |\nabla\theta|^2\, d\boldsymbol{x} + \mu_0 \int_\Gamma P(\theta)\,\overline{\theta}\, ds \\ &\ge \mu_m\, \|\nabla\theta\|^2_{\boldsymbol{\mathcal{L}}^2(\Omega)} + C_1\, \|\theta\|^2_{\mathcal{H}^{\frac{1}{2}}(\Gamma)} \\ &\ge C_2\, \|\theta\|^2_{\mathcal{H}^1(\Omega)}. \end{aligned}$$

Existence and uniqueness of a solution is then a consequence of the Lax–Milgram theorem (Theorem 1.2.1). □

2.4 Static Cases

Static cases stand for configurations where all fields are time independent. We obtain from (2.14)–(2.17) after dropping time derivatives:

$$\mathbf{curl}\, \boldsymbol{H} = \boldsymbol{J}, \tag{2.41}$$

$$\mathbf{curl}\, \boldsymbol{E} = 0, \tag{2.42}$$

$$\operatorname{div}(\mu \boldsymbol{H}) = 0. \tag{2.43}$$

This situation enables decoupling electricity and magnetism in the following way.

2.4.1 Electrostatics

From (2.42) and Theorem 1.3.5, we deduce the existence of a scalar field $\phi : \mathbb{R}^3 \to \mathbb{C}$ such that

$$\boldsymbol{E} = -\nabla\phi \qquad \text{in } \mathbb{R}^3. \tag{2.44}$$

Assuming that Ohm's law is satisfied in the static conductor Ω, i.e.

$$\boldsymbol{J} = \sigma \boldsymbol{E} \qquad \text{in } \Omega, \tag{2.45}$$

where σ is assumed to satisfy (2.11), we obtain from (2.41)–(2.43),

$$\operatorname{div}(\sigma\nabla\phi) = 0 \qquad \text{in } \Omega. \tag{2.46}$$

Equation (2.46) is an elliptic equation that requires appropriate boundary conditions. In many situations, the boundary Γ of Ω is split into parts where Dirichlet or Neumann conditions can be enforced.

- If a part of the boundary is electrically isolated, we prescribe

$$\frac{\partial\phi}{\partial n} = 0$$

  on this part (Homogeneous Neumann boundary condition).
- If a part of the boundary is connected to an electricity generator, we prescribe the potential ϕ when we have a voltage generator (Dirichlet condition) or the normal derivative of ϕ by

$$\boldsymbol{J} \cdot \boldsymbol{n} = \sigma \frac{\partial\phi}{\partial n},$$

  when we have a current generator (Neumann condition).

Let us assume, for instance, that Γ is divided into three parts Γ_1, Γ_2 and Γ_3 such that

$$\inf_{\boldsymbol{x}\in\Gamma_1,\ \boldsymbol{y}\in\Gamma_3} |\boldsymbol{x} - \boldsymbol{y}| > 0,$$

i.e. Γ_1 and Γ_3 are not connected. Assume furthermore that the potential ϕ satisfies the conditions:

$$\phi = V \qquad \text{on } \Gamma_1,$$

$$\phi = 0 \qquad \text{on } \Gamma_3,$$

$$\frac{\partial \phi}{\partial n} = 0 \qquad \text{on } \Gamma_2,$$

where V is given. By defining the space

$$\mathcal{X} := \{ \psi \in \mathcal{H}^1(\Omega);\ \psi = 0 \text{ on } \Gamma_1 \cup \Gamma_3 \},$$

multiplying (2.46) by $\theta \in \mathcal{X}$ and using the Green formula, we obtain

$$\int_\Omega \sigma \nabla \phi \cdot \nabla \theta \, d\boldsymbol{x} = 0. \tag{2.47}$$

Hence, the mathematical problem consists in seeking a function $\phi \in \mathcal{H}^1(\Omega)$ such that $\phi = V$ on Γ_1, $\phi = 0$ on Γ_3 satisfying (2.47) for all $\theta \in \mathcal{X}$. Let ϕ_0 denote a function in $\mathcal{H}^1(\Omega)$ such that $\phi_0 = V$ on Γ_1 and $\phi = 0$ on Γ_3 and let $\psi = \phi - \phi_0$. We easily check that $\psi \in \mathcal{X}$ and

$$\int_\Omega \sigma \nabla \psi \cdot \nabla \overline{\theta} \, d\boldsymbol{x} = \int_\Omega \sigma \nabla \phi_0 \cdot \nabla \overline{\theta} \, d\boldsymbol{x} \qquad \forall\, \theta \in \mathcal{X}.$$

By the Lax–Milgram theorem (Theorem 1.2.1) in $\mathcal{X}$, this problem possesses a unique solution. It follows that (2.47) is well posed and we have $\phi = \psi + \phi_0$.

2.4.2 *Magnetostatics*

Let us assume we are in presence of a conductor Ω and a given current of density $\boldsymbol{J}_0$ and let us assume, as usual, that $\mu = \mu_0$ in $\Omega_{\text{ext}} = \mathbb{R}^3 \setminus \overline{\Omega}$. We define the vector field

$$\boldsymbol{M} = (\mu - \mu_0)\, \boldsymbol{H},$$

called *Magnetization*. Here above, $\boldsymbol{H}$ is assumed to be the magnetic field generated by $\boldsymbol{J}_0$, i.e. $\textbf{curl}\, \boldsymbol{H} = \boldsymbol{J}_0$.

In general, materials for which the magnetic permeability μ is not constant are called *Ferromagnetic materials*. By definition, $\boldsymbol{M} = 0$ for nonferromagnetic conductors.

In ferromagnetic materials, μ depends generally on the magnetic field $\boldsymbol{H}$. When the function $\mu = \mu(|\boldsymbol{H}|)$ is known, it suffices to compute $\boldsymbol{H}$ in order to deduce $\boldsymbol{M}$. To do this, we define $\boldsymbol{H}_0$ like in (2.30)–(2.31), i.e., $\boldsymbol{H}_0$ is the magnetic field without ferromagnetic conductors. It follows by using Sect. 2.3.2 that

$$\boldsymbol{H} = \boldsymbol{H}_0 - \nabla \psi$$

where ψ satisfies:

$$-\operatorname{div}(\mu\nabla\psi) = -\operatorname{div}(\mu \boldsymbol{H}_0) \qquad \text{in } \Omega, \tag{2.48}$$

$$\Delta\psi = 0 \qquad \text{in } \Omega_{\text{ext}}, \tag{2.49}$$

$$[\psi]_\Gamma = 0 \qquad \text{on } \Gamma, \tag{2.50}$$

$$\left[\mu\frac{\partial\psi}{\partial n}\right]_\Gamma = (\mu_0 - \mu)\,\boldsymbol{H}_0\cdot\boldsymbol{n} \qquad \text{on } \Gamma, \tag{2.51}$$

$$\psi(\boldsymbol{x}) = \mathcal{O}(|\boldsymbol{x}|^{-1}) \qquad \text{for } |\boldsymbol{x}| \to \infty. \tag{2.52}$$

Problem (2.48)–(2.52) is a nonlinear elliptic problem when we replace $\boldsymbol{H}$ in $\mu(|\boldsymbol{H}|)$ by $\boldsymbol{H}_0 - \nabla\psi$. Note that the nonlinearity appears as well in the partial differential equation (2.48) as in the boundary condition (2.51). We shall consider such problems in view of applications (Chap. 11).

2.5 Time–Harmonic Regime

We are frequently faced with the case where data are periodic functions of time. This corresponds to the case where a source alternating (AC) current is given. To handle this situation, a time–harmonic solution can be sought. This one is considered by developing the solution in Fourier series in time. We then seek solutions of (2.1)–(2.6) of the form:

$$\boldsymbol{H}(\boldsymbol{x},t) = \operatorname{Re}(e^{\mathrm{i}\omega t}\boldsymbol{H}(\boldsymbol{x})),$$

$$\boldsymbol{D}(\boldsymbol{x},t) = \operatorname{Re}(e^{\mathrm{i}\omega t}\boldsymbol{D}(\boldsymbol{x})),$$

$$\boldsymbol{J}(\boldsymbol{x},t) = \operatorname{Re}(e^{\mathrm{i}\omega t}\boldsymbol{J}(\boldsymbol{x})),$$

$$\boldsymbol{E}(\boldsymbol{x},t) = \operatorname{Re}(e^{\mathrm{i}\omega t}\boldsymbol{E}(\boldsymbol{x})),$$

for $\boldsymbol{x} \in \mathbb{R}^3$, where $\omega \in \mathbb{R}$ is the angular frequency that we choose positive for convenience. Relations (2.1)–(2.2), (2.5)–(2.6) lead to

$$\mathrm{i}\omega\boldsymbol{D} - \mathbf{curl}\,\boldsymbol{H} + \boldsymbol{J} = 0, \tag{2.53}$$

$$\mathrm{i}\omega\boldsymbol{B} + \mathbf{curl}\,\boldsymbol{E} = 0, \tag{2.54}$$

$$\boldsymbol{B} = \mu\boldsymbol{H}, \tag{2.55}$$

$$\boldsymbol{D} = \varepsilon\boldsymbol{E}. \tag{2.56}$$

Note that we have, for the sake of simplicity, kept the same notations for the involved fields although we are now concerned with time–independent complex functions. Note also that, since div **curl** = 0, (2.4) is a consequence of (2.54) if $\omega \neq 0$.

Remark 2.5.1. An analog to Sect. 2.2.3 can be made for (2.53)–(2.54). We have, when ε and μ are constant,

$$\mathbf{curl}\,\mathbf{curl}\,\boldsymbol{H} - \omega^2\varepsilon\mu\,\boldsymbol{H} = \mathbf{curl}\,\boldsymbol{J}.$$

Using the relation div $\boldsymbol{H} = 0$, we obtain the Helmholtz equation

$$-\Delta\boldsymbol{H} - \omega^2\varepsilon\mu\boldsymbol{H} = \mathbf{curl}\,\boldsymbol{J}.$$

2.6 Eddy Current Equations

The remaining chapters are devoted to the derivation and analysis of eddy current models. We consider, in the sequel, a low frequency approximation of the system of equations (2.53)–(2.56) with appropriate behaviour at the infinity. In this case, we can neglect the term $\mathrm{i}\omega\boldsymbol{D}$ in (2.53).

As far as problem data are concerned we are faced with two types of approaches:

1. A first approach consists in assuming that a *source current* $\boldsymbol{J}_0$ is given with a support contained in one (or many) conductor(s). The current density is then written as $\boldsymbol{J} = \tilde{\boldsymbol{J}} + \boldsymbol{J}_0$ where $\tilde{\boldsymbol{J}}$ is the induced current density that is supposed to obey to Ohm's law (2.10) with null velocity ($\boldsymbol{v} = 0$). We obtain then the system of equations:

$$\mathbf{curl}\,\boldsymbol{H} - \tilde{\boldsymbol{J}} = \boldsymbol{J}_0 \qquad \text{in } \mathbb{R}^3, \tag{2.57}$$

$$\mathrm{i}\omega\mu\boldsymbol{H} + \mathbf{curl}\,\boldsymbol{E} = 0 \qquad \text{in } \mathbb{R}^3, \tag{2.58}$$

$$\tilde{\boldsymbol{J}} = \sigma\boldsymbol{E} \qquad \text{in } \mathbb{R}^3, \tag{2.59}$$

$$|\boldsymbol{H}(\boldsymbol{x})| = \mathcal{O}(|\boldsymbol{x}|^{-1}) \qquad \text{for } |\boldsymbol{x}| \to \infty, \tag{2.60}$$

$$|\boldsymbol{E}(\boldsymbol{x})| = \mathcal{O}(|\boldsymbol{x}|^{-1}) \qquad \text{for } |\boldsymbol{x}| \to \infty, \tag{2.61}$$

   with σ extended by 0 outside Ω. Note here that the actual current density $\tilde{\boldsymbol{J}}$ satisfies div $\tilde{\boldsymbol{J}} = 0$ only if the source current $\boldsymbol{J}_0$ is divergence free. This condition is furthermore necessary to ensure that the eddy current problem is a good approximation of the Maxwell equations when ω is small enough (See [13]).
2. An alternative method consists in assuming that we are given either voltage or total current intensity that can be directly prescribed by a power generator. The difficulty relies here on the obtention of an adapted formulation that has the

voltage or the current as unique data. A variant consists in supplying current power. This corresponds more to realistic and industrial setups.

The first method is the most used one in the literature. Actually, the inductors are supplied with currents and it is not necessary to prescribe the electric source in the system. This method is simpler to formulate but does not correspond to realistic situations unless the conductors supporting source currents are thin enough so one can approximate a current density with its average.

The second procedure corresponds to an idealization of the real setup in the sense that voltage is given by prescribing a *cut* in the inductor represented by a non simply connected domain. This cut (see Fig. 1.1) stands for a virtual link of the inductor to the power generator and problem data are the constants given in Theorems 1.3.5 and 1.3.6. In this case, we are constrained to assume that (2.54) is valid in Ω and in Ω_{ext} but not in the whole space in order to introduce a source current. As in [34,36], we have chosen to treat in most applications this second category of formulations.

Time harmonic eddy current equations are given by the set of partial differential equations:

$$\mathbf{curl}\, \boldsymbol{H} - \boldsymbol{J} = 0 \qquad \text{in } \mathbb{R}^3, \tag{2.62}$$

$$\mathrm{i}\omega \boldsymbol{B} + \mathbf{curl}\, \boldsymbol{E} = 0 \qquad \text{in } \Omega \cup \Omega_{\text{ext}}, \tag{2.63}$$

$$\operatorname{div} \boldsymbol{B} = 0 \qquad \text{in } \mathbb{R}^3, \tag{2.64}$$

$$\boldsymbol{B} = \mu \boldsymbol{H} \qquad \text{in } \mathbb{R}^3, \tag{2.65}$$

$$\boldsymbol{J} = \sigma \boldsymbol{E} \qquad \text{in } \mathbb{R}^3, \tag{2.66}$$

$$|\boldsymbol{H}(\boldsymbol{x})| = \mathcal{O}(|\boldsymbol{x}|^{-1}) \qquad \text{for } |\boldsymbol{x}| \to \infty, \tag{2.67}$$

$$|\boldsymbol{E}(\boldsymbol{x})| = \mathcal{O}(|\boldsymbol{x}|^{-1}) \qquad \text{for } |\boldsymbol{x}| \to \infty. \tag{2.68}$$

This set of equations has to be supplemented with appropriate boundary and interface conditions on Γ. For this, let us remark that:

1. As said before, in the case where a source current density is prescribed, (2.62) is to be replaced by

$$\mathbf{curl}\, \boldsymbol{H} - \boldsymbol{J} = \boldsymbol{J}_S \qquad \text{in } \mathbb{R}^3. \tag{2.69}$$

2. Equation (2.64) is necessary, since from the previous remark, this one is no more a consequence of (2.63). However, if (2.63) is satisfied with $\omega \neq 0$, then (2.64) is equivalent to assuming that the jump $[\boldsymbol{B} \cdot \boldsymbol{n}]$ is null.
3. Equation (2.62) implies

$$\operatorname{div} \boldsymbol{J} = 0 \quad \text{in } \mathbb{R}^3. \tag{2.70}$$

4. From the set of equations (2.62)–(2.68) we can derive interface conditions (involving continuities and jumps) at boundaries of the conductors. For this, if we formally obtain by using relations

$$\operatorname{div} \boldsymbol{J} = 0 \quad \text{in } \mathbb{R}^3, \qquad \boldsymbol{J} = 0 \quad \text{in } \Omega_{\text{ext}},$$

   that

$$\boldsymbol{J} \cdot \boldsymbol{n} = 0 \quad \text{on } \Gamma. \tag{2.71}$$

   In addition, (2.62) implies that

$$[\boldsymbol{H} \times \boldsymbol{n}]_\Gamma = 0, \tag{2.72}$$

   when $\boldsymbol{J}$ has no Dirac masses on Γ, i.e., no surface currents flow on Γ.
5. Equation (2.63) is assumed to be valid only in the conductors and the free space. This is necessary when specific data are to be prescribed like voltage and total current. Depending on the models this restriction is to be relaxed by a prescription of the equation on $\mathbb{R}^3 \setminus S$, or even $\mathbb{R}^3 \setminus \partial S$, where S is a cut (or union of cuts) in the conductors. This restriction is however not necessary ((2.63) is valid in $\mathbb{R}^3$) if a source current density is specified.

Chapter 3
Two-Dimensional Models

3.1 Introduction

As it is explained in Chap. 1, two-dimensional models are obtained by assuming that any conductor domain $\Omega \subset \mathbb{R}^3$ is cylindrical, i.e. of the form $\Omega = \Lambda \times \mathbb{R}$, where Λ is a bounded open set of $\mathbb{R}^2$. The invariance direction is parallel to the x_3-axis. The two-dimensional feature can be expressed in two ways that lead to significantly different models: either the magnetic field $\boldsymbol{H}$ or the current density $\boldsymbol{J}$ are aligned with the x_3 direction. Another feature of two-dimensional models is that conditions (2.67), (2.68) are no longer valid since the fields $\boldsymbol{H}$ and $\boldsymbol{E}$ are assumed x_3-independent and can then not vanish at the infinity except in the trivial case where they are all identically equal to zero. Let us also mention that we deal in this chapter with linear models, i.e. where the function μ does not depend on the magnetic field (*nonferromagnetic materials*). However, we consider μ as a function of $\boldsymbol{x} \in \Lambda$ in view of a possible treatment of nonlinear problems in which we have $\mu = \mu(|\boldsymbol{H}(\boldsymbol{x})|)$.

Let us assume we are given a collection of N cylindrical conductors $\Omega_k = \Lambda_k \times \mathbb{R}$ and set

$$\Lambda = \bigcup_{k=1}^{N} \Lambda_k, \ \Lambda_{\text{ext}} = \mathbb{R}^2 \setminus \overline{\Lambda}.$$

In practice we shall develop simple models for a limited number of conductors, the generalizations to many conductors being straightforward. All two-dimensional models will be obtained from (2.62)–(2.66) assuming x_3–invariance to which we add a relaxed version of conditions (2.67)–(2.68). To set this, we write

$$\boldsymbol{H}(\boldsymbol{x}) = \tilde{\boldsymbol{H}}(\boldsymbol{x}) + H(\boldsymbol{x})\,\boldsymbol{e}_3,$$

$$\boldsymbol{E}(\boldsymbol{x}) = \tilde{\boldsymbol{E}}(\boldsymbol{x}) + E(\boldsymbol{x})\,\boldsymbol{e}_3,$$

$$\boldsymbol{B}(\boldsymbol{x}) = \tilde{\boldsymbol{B}}(\boldsymbol{x}) + B(\boldsymbol{x})\,\boldsymbol{e}_3,$$

R. Touzani and J. Rappaz, *Mathematical Models for Eddy Currents and Magnetostatics: With Selected Applications*, Scientific Computation, DOI 10.1007/978-94-007-0202-8_3,

$$J(x) = \tilde{J}(x) + J(x)\, e_3,$$

for $x = x_1 e_1 + x_2 e_2 \in \mathbb{R}^2$, with $\tilde{H}(x) = H_1(x) e_1 + H_2(x) e_2$, etc. We then define the set of equations:

$$\mathbf{curl}\, H = \tilde{J} \qquad \text{in } \mathbb{R}^2, \tag{3.1}$$

$$\text{curl}\, \tilde{H} = J \qquad \text{in } \mathbb{R}^2, \tag{3.2}$$

$$\mathrm{i}\omega \tilde{B} + \mathbf{curl}\, E = 0 \qquad \text{in } \Lambda \cup \Lambda_{\text{ext}}, \tag{3.3}$$

$$\mathrm{i}\omega B + \text{curl}\, \tilde{E} = 0 \qquad \text{in } \Lambda \cup \Lambda_{\text{ext}}, \tag{3.4}$$

$$\text{div}\, \tilde{B} = 0 \qquad \text{in } \mathbb{R}^2, \tag{3.5}$$

$$\tilde{B} = \mu \tilde{H},\ B = \mu H \qquad \text{in } \mathbb{R}^2, \tag{3.6}$$

$$\tilde{J} = \sigma \tilde{E},\ J = \sigma E \qquad \text{in } \mathbb{R}^2, \tag{3.7}$$

$$|\tilde{H}(x)| + |\tilde{E}(x)| = \mathcal{O}(|x|^{-1}) \qquad \text{for } |x| \to \infty. \tag{3.8}$$

$$|H(x)| + |E(x)| = \mathcal{O}(|x|^{-1}) \qquad \text{for } |x| \to \infty. \tag{3.9}$$

We recall that the scalar and vector curl operators in 2-D are defined in Sect. 1.4.1 and div is the two-dimensional divergence operator. Furthermore, we recall that we have extended the conductivity σ to $\mathbb{R}^2$ by zero in order to obtain (3.1), (3.2) and (3.7) in $\mathbb{R}^2$.

As far as interface conditions are concerned, (3.1)–(3.5) yield the following conditions:

$$[H]_\gamma = [\tilde{H} \cdot t]_\gamma = 0, \tag{3.10}$$

$$[B]_\gamma = [\tilde{B} \cdot n]_\gamma = 0, \tag{3.11}$$

$$[E]_\gamma = [\tilde{E} \cdot t]_\gamma = 0, \tag{3.12}$$

$$[J]_\gamma = [\tilde{J} \cdot n]_\gamma = 0, \tag{3.13}$$

where γ is the boundary of Λ, $n = n_1 e_1 + n_2 e_2$ is the outward unit normal vector and t is the unit tangent vector to γ defined by

$$t := -n_2 e_1 + n_1 e_2.$$

In the sequel, the development of two-dimensional models will sometimes require the use of a scalar potential. More precisely, using (3.5), Theorem 1.4.2 implies the existence of a scalar potential $A \in \mathcal{W}^1(\mathbb{R}^2)$ such that

$$\tilde{\mathbf{B}} = \mathbf{curl}\, A \qquad \text{in } \mathbb{R}^2. \tag{3.14}$$

The potential A is clearly defined up to an additive constant.

3.2 A Solenoidal Two-Dimensional Model

Let us consider the simplest configuration consisting in two conductors $\Omega_1 = \Lambda_1 \times \mathbb{R}$ and $\Omega_2 = \Lambda_2 \times \mathbb{R}$ where Λ_1 is an inductor, i.e. Λ_1 is an annulus that "encloses" Λ_2, (cf. Fig. 3.1). Let Λ_{ext} denote the bounded domain located between Λ_1 and Λ_2 and the domains Λ, $\hat{\Lambda}$, and $\tilde{\Lambda}$ be respectively defined by

$$\Lambda := \Lambda_1 \cup \Lambda_2, \ \hat{\Lambda} := \Lambda_1 \cup \overline{\Lambda}_{\text{ext}} \cup \Lambda_2, \ \tilde{\Lambda} := \mathbb{R}^2 \setminus \overline{\hat{\Lambda}}.$$

Let also γ_1, γ_2 stand for the respective boundaries of Λ_1 and Λ_2 and let $\gamma = \gamma_1 \cup \gamma_2$. Clearly, the outward unit normal to the boundaries γ_1 and γ_2 is given by the vector $\boldsymbol{n} = n_1\boldsymbol{e}_1 + n_2\boldsymbol{e}_2$ as shown on Fig. 3.1. This figure shows also the partition of the boundary $\gamma_1 = \gamma_1^- \cup \gamma_1^+$.

From a physical point of view, the inductor Λ_1 is assumed to be connected to a generator of an alternating current and can be a cut of an infinite coil in the x_3–direction with turns in the plane Ox_1x_2, and Λ_{ext} is the free space (or vacuum) between conductors.

The main issue in deriving the present model is to state that for geometries like in Fig. 3.1, the current density is sought in the form

$$\boldsymbol{J}(x_1, x_2, x_3) = J_1(x_1, x_2)\,\boldsymbol{e}_1 + J_2(x_1, x_2)\,\boldsymbol{e}_2. \tag{3.15}$$

In other words, $\boldsymbol{J}$ does not depend on x_3 and $J = 0$.

Theorem 3.2.1. *Let* $(\boldsymbol{H}, \boldsymbol{J}, \boldsymbol{B}, \boldsymbol{E})$ *denote smooth vector fields that satisfy* (3.1)–(3.9). *Then the magnetic field* $\boldsymbol{H}$ *has the form*

$$\boldsymbol{H} = H\boldsymbol{e}_3, \tag{3.16}$$

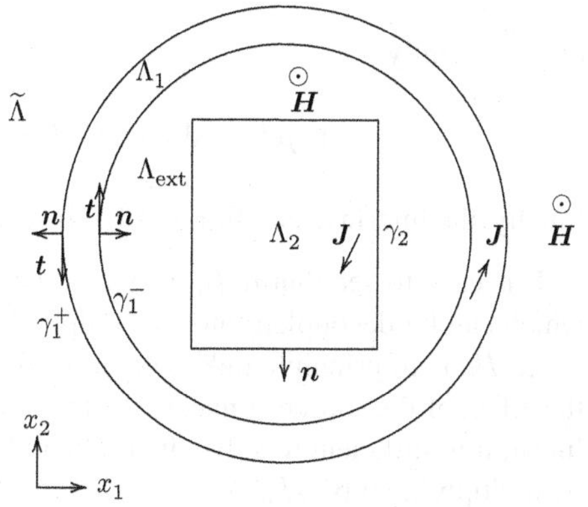

Fig. 3.1 A typical configuration of the conductors for the solenoidal model

i.e. $\tilde{\boldsymbol{H}} = 0$. *Moreover, the function* H *satisfies the set of equations:*

$$\mathrm{i}\omega\mu H - \mathrm{div}(\sigma^{-1}\nabla H) = 0 \qquad \text{in } \Lambda, \tag{3.17}$$

$$H = H_0 \qquad \text{in } \Lambda_{ext}, \tag{3.18}$$

$$H = 0 \qquad \text{in } \tilde{\Lambda}, \tag{3.19}$$

$$[H]_\gamma = 0, \tag{3.20}$$

where $H_0 \in \mathbb{C}$ *is a constant to determine.*

Proof. Using (3.2) with $J = 0$, (3.6), and (3.14) we obtain

$$\mathrm{curl}\,(\mu^{-1}\,\mathbf{curl}\,A) = -\,\mathrm{div}(\mu^{-1}\nabla A) = 0 \qquad \text{in } \mathbb{R}^2.$$

Since from Theorem 1.4.2, $A \in \mathcal{W}^1(\mathbb{R}^2)$, we deduce

$$\int_{\mathbb{R}^2} \mu^{-1}|\nabla A|^2\,d\boldsymbol{x} = 0.$$

This implies that $\nabla A = \mathbf{0}$ in $\mathbb{R}^2$ and consequently $\tilde{\boldsymbol{B}} = \mathbf{0}$ and $\tilde{\boldsymbol{H}} = \mathbf{0}$. Expression (3.16) is then obtained.

We now derive equations in the vacuum $\Lambda_{\mathrm{ext}} \cup \tilde{\Lambda}$: We have from (3.1), since $\tilde{\boldsymbol{J}} = 0$ in $\Lambda_{\mathrm{ext}} \cup \tilde{\Lambda}$, $\mathbf{curl}\,H = 0$ and then $\nabla H = 0$. This implies that H is constant in Λ_{ext} and in $\tilde{\Lambda}$, and then from (3.9),

$$H = \text{Const.} \qquad \text{in } \Lambda_{\mathrm{ext}},$$

$$H = 0 \qquad \text{in } \tilde{\Lambda}.$$

In $\Lambda_1 \cup \Lambda_2$, we have from (3.4), (3.6), (3.7) and (3.1),

$$\mathrm{i}\omega\mu H + \mathrm{curl}(\sigma^{-1}\,\mathbf{curl}\,H) = 0 \qquad \text{in } \Lambda_1 \cup \Lambda_2,$$

or equivalently

$$\mathrm{i}\omega\mu H - \mathrm{div}(\sigma^{-1}\nabla H) = 0 \qquad \text{in } \Lambda_1 \cup \Lambda_2,$$

Finally, the interface condition (3.10) is rewritten as (3.20). □

It is easy to see that if H_0 is given, then (3.17)–(3.20) is a well posed problem which can be decoupled into two elliptic boundary value problems in Λ_1 and Λ_2. Since H_0 is in principle unknown, it is convenient to deduce an equation to model the effect of the source connected to the inductor Λ_1. The mapping $H_0 \mapsto H$ being linear, it is sufficient to solve the problem for $H_0 = 1$ and then rescale the solution by multiplying it by H_0.

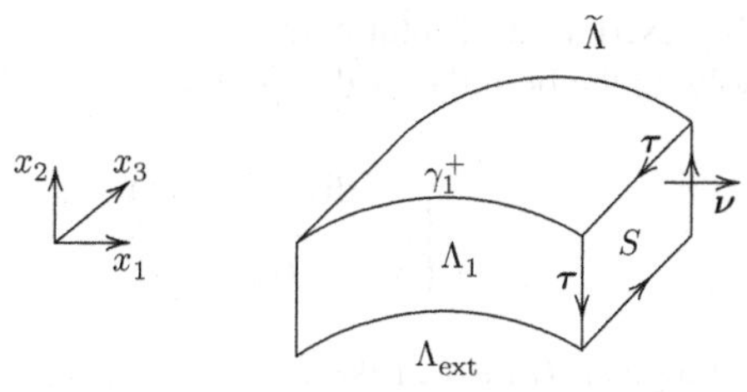

Fig. 3.2 A piece of the inductor

In practical applications, according to electrotechnical devices, one may distinguish three types of data:

(i) The total current in the inductor Λ_1 is given. We shall see that this is the case where H_0 is given.
(ii) The current voltage is given.
(iii) The current power is given.

We note that this topic is thoroughly studied in Hiptmair-Sterz [101].

3.2.1 *Total Current Data*

Consider Fig. 3.2 and denote by S a rectangle parallel to Ox_3, which is a unit section of the inductor delimited by $\tilde{\Lambda}$ and Λ_{ext} (see Fig. 3.2). Let in addition $\boldsymbol{\nu}$ denote the unit normal to the surface S oriented once for all and let $\boldsymbol{\tau}$ denote the unit tangent vector to the boundary ∂S of S, oriented according to Ampère rule (as depicted on Fig. 3.2).

Integrating (3.1) over S and using the Stokes theorem, we get

$$\begin{aligned}\int_S \tilde{\boldsymbol{J}} \cdot \boldsymbol{\nu}\, dS &= \int_S \left(\frac{\partial H}{\partial x_2}\nu_1 - \frac{\partial H}{\partial x_1}\nu_2\right) dS \\ &= \int_{\partial S} H\tau_3\, ds \\ &= H_{|\Lambda_{\text{ext}}} - H_{|\tilde{\Lambda}} \\ &= H_0. \end{aligned} \tag{3.21}$$

The integral of $\tilde{\boldsymbol{J}} \cdot \boldsymbol{\nu}$ is nothing else but the total current flowing in a portion of the inductor of unit thickness.[1] Expression (3.21) show that this value is H_0.

[1]In most applications, H_0/N is the total current flowing in a turn of an infinite coil when N is the total number of turns by unit coil length.

A prescription of total current (when H_0 is given by $\int_S \tilde{\boldsymbol{J}} \cdot \boldsymbol{\nu}\, ds$) consists then in solving the boundary value problem:

$$\begin{cases} -\operatorname{div}(\sigma^{-1}\nabla H) + \mathrm{i}\omega\mu H = 0 & \text{in } \Lambda_2, \\ H = H_0 & \text{on } \gamma_2, \end{cases} \tag{3.22}$$

for a given $H_0 \in \mathbb{C}$. Likewise, we have in Λ_1 the problem:

$$\begin{cases} -\operatorname{div}(\sigma^{-1}\nabla H) + \mathrm{i}\omega\mu H = 0 & \text{in } \Lambda_1, \\ H = H_0 & \text{on } \gamma_1^-, \\ H = 0 & \text{on } \gamma_1^+. \end{cases} \tag{3.23}$$

We have the following result:

Theorem 3.2.2. *Problems* (3.22) *and* (3.23) *have unique solutions in* $\mathcal{H}^1(\Lambda_2)$ *and* $\mathcal{H}^1(\Lambda_1)$ *respectively.*

Proof. Problem (3.22) can be easily transformed into a homogeneous Dirichlet problem by setting $u = H - H_0$. We thus define

$$\mathscr{B}(u, v) := \int_{\Lambda_2} \sigma^{-1}\nabla u \cdot \nabla \overline{v}\, d\boldsymbol{x} + \mathrm{i}\omega \int_{\Lambda_2} \mu u \overline{v}\, d\boldsymbol{x},$$

$$\mathscr{L}(v) := -\mathrm{i}\omega H_0 \int_{\Lambda_2} \mu \overline{v}\, d\boldsymbol{x}.$$

A variational formulation of (3.22) is then

$$\text{Find } u \in \mathcal{H}_0^1(\Lambda_2) \text{ such that } \quad \mathscr{B}(u, v) = \mathscr{L}(v) \quad \forall\, v \in \mathcal{H}_0^1(\Lambda_2).$$

The Lax–Milgram theorem (Theorem 1.2.1) can then be applied.

In a same way, standard results for elliptic problems (see Dautray–Lions [62] for instance) show that (3.23) possesses a unique solution. □

3.2.2 *Voltage Data*

In three-dimensional configurations, the current voltage in an electric set-up can be defined by the following relation (see [126, 139]):

$$V = \int_{\xi} (\mathrm{i}\omega \boldsymbol{A} + \boldsymbol{E}) \cdot \boldsymbol{t}\, ds, \tag{3.24}$$

where ξ is any closed curve in the conductor that cannot be retracted (with a continuous homotopy) to a point, $\boldsymbol{t}$ is a unit tangent vector to ξ and $\boldsymbol{A}$ is the vector potential obtained by (2.12) and (2.13). For the present two-dimensional case, we deduce from Theorem 1.4.3 the existence of a vector potential $\tilde{\boldsymbol{A}}$ such that

$$B = \operatorname{curl} \tilde{\boldsymbol{A}}, \quad \operatorname{div} \tilde{\boldsymbol{A}} = 0 \qquad \text{in } \hat{\Lambda}. \tag{3.25}$$

As a two-dimensional analog to (3.24), we define the circuit voltage here by the expression

$$V = \int_{\gamma_1^-} \big((\mathrm{i}\omega \tilde{\boldsymbol{A}} + \tilde{\boldsymbol{E}}) \cdot \boldsymbol{t}\big)^- \, ds, \tag{3.26}$$

where we recall that the tangent vector $\boldsymbol{t}$ is given by $\boldsymbol{t} = -n_2 \boldsymbol{e}_1 + n_1 \boldsymbol{e}_2$ and the superscript '-' means that the trace of $(\mathrm{i}\omega \tilde{\boldsymbol{A}} + \tilde{\boldsymbol{E}}) \cdot \boldsymbol{t}$ is taken inside Λ_1. Likewise, we shall denote hereafter by f^- the trace of a function f on γ_1^- taken inside Λ_{ext}. Note that although the orientation of the tangent vector in (3.26) determines the sign of V and then the one of H, this one has no influence on more relevant quantities like $|H|$ which is involved in the computation of the power as we shall see it later.

Remark 3.2.1. From (3.4), (3.6) we deduce in particular

$$\mathrm{i}\omega\mu H + \operatorname{curl} \tilde{\boldsymbol{E}} = 0 \qquad \text{in } \Lambda_1. \tag{3.27}$$

By using the Stokes theorem in Λ_1 we should obtain then, using (3.25) and (3.6),

$$\begin{aligned}
\int_{\gamma_1^-} (\mathrm{i}\omega \tilde{\boldsymbol{A}} + \tilde{\boldsymbol{E}})^- \cdot \boldsymbol{t} \, ds + \int_{\gamma_1^+} (\mathrm{i}\omega \tilde{\boldsymbol{A}} + \tilde{\boldsymbol{E}})^- \cdot \boldsymbol{t} \, ds &= \int_{\Lambda_1} (\mathrm{i}\omega \operatorname{curl} \tilde{\boldsymbol{A}} + \operatorname{curl} \tilde{\boldsymbol{E}}) \, d\boldsymbol{x} \\
&= \int_{\Lambda_1} (\mathrm{i}\omega\mu H - \mathrm{i}\omega\mu H) \, d\boldsymbol{x} \\
&= 0.
\end{aligned}$$

It follows that

$$\int_{\gamma_1^-} (\mathrm{i}\omega \tilde{\boldsymbol{A}} + \boldsymbol{E})^- \cdot \boldsymbol{t} \, ds = -\int_{\gamma_1^+} (\mathrm{i}\omega \tilde{\boldsymbol{A}} + \boldsymbol{E})^- \cdot \boldsymbol{t} \, ds.$$

However, we cannot assume that the Faraday equation

$$\mathrm{i}\omega\mu H + \operatorname{curl} \tilde{\boldsymbol{E}} = 0$$

is valid in $\mathbb{R}^2$ when $V \neq 0$. In fact, if this was true, we would obtain from the Stokes theorem on $\overline{\Lambda}_2 \cup \Lambda_{\text{ext}}$,

$$V = \int_{\gamma_1^-} (\mathrm{i}\omega \tilde{\boldsymbol{A}} + \tilde{\boldsymbol{E}}) \cdot \boldsymbol{t} \, ds = \int_{\overline{\Lambda}_2 \cup \Lambda_{\text{ext}}} (\mathrm{i}\omega\mu H + \operatorname{curl} \tilde{\boldsymbol{E}}) \, d\boldsymbol{x} = 0.$$

As a direct consequence of the previous remark, one must add to the Faraday equation a singularity supported by γ_1, γ_2 or both, on which we admit a jump of $\tilde{\boldsymbol{E}} \cdot \boldsymbol{t}$. This singularity is clearly related to the voltage.

Let us define the following Hilbert space

$$\mathcal{H} := \{\phi \in \mathcal{H}_0^1(\hat{\Lambda});\ \phi_{|\Lambda_{\text{ext}}} = \text{Const.}\}.$$

Since we have for $\phi \in \mathcal{H}$,

$$\int_{\Lambda} |\nabla \phi|^2 \, d\boldsymbol{x} = \int_{\hat{\Lambda}} |\nabla \phi|^2 \, d\boldsymbol{x},$$

then, owing to the Poincaré–Friedrichs inequality (1.5), the expression

$$|\phi|_{\mathcal{H}} := \Big(\int_{\Lambda} |\nabla \phi|^2 \, d\boldsymbol{x} \Big)^{\frac{1}{2}}.$$

defines a norm on $\mathcal{H}$.

Theorem 3.2.3. *We have*

$$\int_{\hat{\Lambda}} (\mathrm{i}\omega\mu H \overline{\phi} + \tilde{\boldsymbol{E}} \cdot \mathbf{curl}\, \overline{\phi}) \, d\boldsymbol{x} = V \overline{\phi}_{|\Lambda_{ext}} \qquad \forall\, \phi \in \mathcal{H}. \tag{3.28}$$

Proof. Let $\tilde{\boldsymbol{E}}^1 = \tilde{\boldsymbol{E}}_{|\Lambda_1}$, $\tilde{\boldsymbol{E}}' = \tilde{\boldsymbol{E}}_{|\Lambda_{\text{ext}}}$. By the Green formula, we check that for $\phi \in \mathcal{H}$,

$$\begin{aligned}\int_{\hat{\Lambda}} \tilde{\boldsymbol{E}} \cdot \mathbf{curl}\, \overline{\phi} \, d\boldsymbol{x} &= \int_{\Lambda \cup \Lambda_{\text{ext}}} \operatorname{curl} \tilde{\boldsymbol{E}} \, \overline{\phi} \, d\boldsymbol{x} \\ &\quad - \Big(\int_{\gamma_1^-} [\tilde{\boldsymbol{E}} \cdot \boldsymbol{t}] \, ds + \int_{\gamma_2} [\tilde{\boldsymbol{E}} \cdot \boldsymbol{t}] \, ds \Big) \overline{\phi}_{|\Lambda_{\text{ext}}}.\end{aligned}$$

Therefore using (3.4) and (3.6), we have

$$\int_{\hat{\Lambda}} (\mathrm{i}\omega\mu H \overline{\phi} + \tilde{\boldsymbol{E}} \cdot \mathbf{curl}\, \overline{\phi}) \, d\boldsymbol{x} = -\Big(\int_{\gamma_1^-} [\tilde{\boldsymbol{E}} \cdot \boldsymbol{t}] \, ds + \int_{\gamma_2} [\tilde{\boldsymbol{E}} \cdot \boldsymbol{t}] \, ds \Big) \overline{\phi}_{|\Lambda_{\text{ext}}}. \tag{3.29}$$

On the other hand, we have by the Stokes theorem and (3.4),

$$\begin{aligned}\mathrm{i}\omega \int_{\gamma_1^-} \tilde{\boldsymbol{A}} \cdot \boldsymbol{t} \, ds &= \mathrm{i}\omega \int_{\Lambda_{\text{ext}} \cup \Lambda_2} \operatorname{curl} \tilde{\boldsymbol{A}} \, d\boldsymbol{x} \\ &= - \int_{\Lambda_{\text{ext}} \cup \Lambda_2} \operatorname{curl} \tilde{\boldsymbol{E}} \, d\boldsymbol{x} \\ &= - \int_{\gamma_1^-} \tilde{\boldsymbol{E}}' \cdot \boldsymbol{t} \, ds - \int_{\gamma_2} [\tilde{\boldsymbol{E}} \cdot \boldsymbol{t}] \, ds.\end{aligned}$$

Then using (3.26):

$$V = \int_{\gamma_1^-} \left((\mathrm{i}\omega \tilde{\boldsymbol{A}} + \tilde{\boldsymbol{E}}^1) \cdot \boldsymbol{t} \right)^- ds = -\int_{\gamma_1^-} [\tilde{\boldsymbol{E}} \cdot \boldsymbol{t}]\, ds - \int_{\gamma_2} [\tilde{\boldsymbol{E}} \cdot \boldsymbol{t}]\, ds. \tag{3.30}$$

From (3.29) and (3.30) we deduce

$$V \overline{\phi}_{|\Lambda_{\mathrm{ext}}} = \int_{\hat{\Lambda}} (\mathrm{i}\omega\mu H \overline{\phi} + \tilde{\boldsymbol{E}} \cdot \mathbf{curl}\, \overline{\phi})\, d\boldsymbol{x}$$

for all $\phi \in \mathcal{H}$. □

Remark 3.2.2. The proof of the previous result shows in particular, that the voltage V can also be defined by the expression

$$V = -\int_{\gamma_1^-} [\tilde{\boldsymbol{E}} \cdot \boldsymbol{t}]\, ds - \int_{\gamma_2} [\tilde{\boldsymbol{E}} \cdot \boldsymbol{t}]\, ds.$$

Equation (3.28) enables giving a variational formulation for the eddy current problem with prescribed voltage. For this we note that by using (3.1), (3.7) and the fact that ϕ is constant in Λ_{ext}, we obtain for all $\phi \in \mathcal{H}$:

$$\begin{aligned} \int_{\hat{\Lambda}} (\mathrm{i}\omega\mu H\, \overline{\phi} + \tilde{\boldsymbol{E}} \cdot \mathbf{curl}\, \overline{\phi})\, d\boldsymbol{x} &= \int_{\hat{\Lambda}} \mathrm{i}\omega\mu H\, \overline{\phi}\, d\boldsymbol{x} + \int_{\Lambda} \sigma^{-1}\, \mathbf{curl}\, H \cdot \mathbf{curl}\, \overline{\phi}\, d\boldsymbol{x} \\ &= \int_{\hat{\Lambda}} \mathrm{i}\omega\mu H\, \overline{\phi}\, d\boldsymbol{x} + \int_{\Lambda} \sigma^{-1} \nabla H \cdot \nabla \overline{\phi}\, d\boldsymbol{x}. \end{aligned}$$

This suggests the variational formulation:

$$\begin{cases} \text{Find } H \in \mathcal{H} \text{ such that} \\ \displaystyle\int_{\Lambda} \sigma^{-1} \nabla H \cdot \nabla \overline{\phi}\, d\boldsymbol{x} + \mathrm{i}\omega \int_{\hat{\Lambda}} \mu H \overline{\phi}\, d\boldsymbol{x} = V \overline{\phi}_{|\Lambda_{\mathrm{ext}}} \qquad \forall\, \phi \in \mathcal{H}. \end{cases} \tag{3.31}$$

Theorem 3.2.4. *Let $V \in \mathbb{C}$ be given. Then, there is a unique $H \in \mathcal{H}$ that satisfies (3.31).*

Proof. This is obtained by a direct application of the Lax–Milgram theorem (Theorem 1.2.1). Obviously, the left-hand side of (3.31) is a continuous sesquilinear form on $\mathcal{H} \times \mathcal{H}$ and the right-hand side is a continuous antilinear form on $\mathcal{H}$. In addition, the coercivity is obtained by using (2.11) and writing

$$\mathrm{Re}\Big(\int_{\Lambda} \sigma^{-1} |\nabla H|^2\, d\boldsymbol{x} + \mathrm{i}\omega \int_{\hat{\Lambda}} \mu |H|^2\, d\boldsymbol{x} \Big) = \int_{\Lambda} \sigma^{-1} |\nabla H|^2\, d\boldsymbol{x} \geq \sigma_M^{-1} |H|^2_{\mathcal{H}}.$$ □

An energy interpretation can be derived from the formulation (3.31). We have indeed by choosing $\phi = H$ the identity

$$\int_{\Lambda} \sigma^{-1} |\nabla H|^2 \, d\boldsymbol{x} + \mathrm{i}\omega \int_{\hat{\Lambda}} \mu |H|^2 \, d\boldsymbol{x} = V \overline{H}_{|\Lambda_{\mathrm{ext}}}.$$

Or, equivalently from (3.1) and (3.7),

$$\int_{\Lambda} \tilde{\boldsymbol{J}} \cdot \overline{\tilde{\boldsymbol{E}}} \, d\boldsymbol{x} + \mathrm{i}\omega \int_{\hat{\Lambda}} \mu |H|^2 \, d\boldsymbol{x} = V \overline{H}_{|\Lambda_{\mathrm{ext}}}.$$

Note that, thanks to (3.21), the number $H_{|\Lambda_{\mathrm{ext}}}$ is actually the linear current density (Current by unit length) that flows in Λ_1 in a section of unit length (in the Ox_3–direction). Therefore the above equation is in fact, an energy identity (cf. [74]) in the form

Electric Energy + *Magnetic Energy* = *Voltage* × *Current Intensity*.

The above identity can also be seen as an equivalent definition of the voltage V.

Remark 3.2.3. A link between the voltage and the total current formulations can be established in the following way: Let H denote the solution of (3.31) and define $H_0 := H_{|\Lambda_{\mathrm{ext}}}$ and $\hat{H} := H/H_0$. Then obviously $\hat{H}_{|\Lambda_{\mathrm{ext}}} = 1$. The normalized magnetic field $\hat{H}$ can be computed by solving decoupled problems (3.22), (3.23) with H replaced by $\hat{H}$ and H_0 replaced by 1. The value of H_0 is then obtained by using (3.31) with ϕ such that $\phi_{|\Lambda_{\mathrm{ext}} \cup \overline{\Lambda}_2} = 1$, i.e.

$$\int_{\Lambda_1} \sigma^{-1} \nabla \hat{H} \cdot \nabla \overline{\phi} \, d\boldsymbol{x} + \mathrm{i}\omega \int_{\Lambda_1} \mu \hat{H} \overline{\phi} \, d\boldsymbol{x} + \mathrm{i}\omega\mu_0 \, |\Lambda_{\mathrm{ext}}| + \mathrm{i}\omega \int_{\Lambda_2} \mu \hat{H} \, d\boldsymbol{x} = \frac{V}{H_0}.$$

To show that the value of H_0 does not depend on this choice of the test function ϕ, for all ϕ with $\phi = 1$ on $\Lambda_{\mathrm{ext}} \cup \overline{\Lambda}_2$, we write by using the Green formula and (3.4),

$$\begin{aligned}\int_{\Lambda_1} \sigma^{-1} \nabla \hat{H} \cdot \nabla \overline{\phi} \, d\boldsymbol{x} &= -\int_{\Lambda_1} \mathrm{div}(\sigma^{-1} \nabla \hat{H}) \overline{\phi} \, d\boldsymbol{x} + \int_{\gamma_1^-} \sigma^{-1} \frac{\partial \hat{H}}{\partial n} \, ds \\ &= -\mathrm{i}\omega \int_{\Lambda_1} \mu \hat{H} \overline{\phi} \, d\boldsymbol{x} + \int_{\gamma_1^-} \sigma^{-1} \frac{\partial \hat{H}}{\partial n} \, ds.\end{aligned}$$

We then obtain the formula

$$\frac{V}{H_0} = \mathrm{i}\omega \int_{\Lambda_2} \mu \hat{H} \, d\boldsymbol{x} + \mathrm{i}\omega\mu_0 \, |\Lambda_{\mathrm{ext}}| + \int_{\gamma_1^-} \sigma^{-1} \frac{\partial \hat{H}}{\partial n} \, ds, \tag{3.32}$$

from which we deduce H_0 in function of $\hat{H}$ and V.

Remark 3.2.4. The generalization to a configuration with N conductors $\Lambda_1, \ldots, \Lambda_N$ is straightforward.

3.2.3 *Power Data*

In electrotechnical devices the active power is defined by the relation (cf. [126]),

$$P = \frac{\omega}{2\pi} \int_0^{\frac{2\pi}{\omega}} \mathrm{Re}(e^{\mathrm{i}\omega t} V)\, \mathrm{Re}(e^{\mathrm{i}\omega t} I)\, dt,$$

where I is the total current in the inductor per unit length in the Ox_3–direction, i.e. $I = H_0$ (thanks to (3.21)). In other words, the power is averaged on one time period. We obtain

$$P = \frac{1}{2} \mathrm{Re}(V \overline{H}_0).$$

Multiplying (3.32) by $|H_0|^2$ and taking the real part, we get

$$P = \frac{1}{2} |H_0|^2 \,\mathrm{Re} \left(\mathrm{i}\omega \int_{\Lambda_2} \mu \hat{H}\, d\boldsymbol{x} + \mathrm{i}\omega\mu_0 |\Lambda_{\mathrm{ext}}| + \int_{\gamma_1^-} \sigma^{-1} \frac{\partial \hat{H}}{\partial n}\, ds \right).$$

This shows that the power depends on the modulus of H_0 only. The above relation enables prescribing a value of H_0 for a given power P. Note that, as it may be expected, the active power depends on the modulus of H.

3.3 A Transversal Model

Let us now address another type of two-dimensional models. We shall show that the choice of a new disposition of the conductors and of the current leads to a model where the magnetic field is not constant in the vacuum but derives from a scalar potential. This feature makes the mathematical modelling more delicate.

We consider a typical configuration with three disjointed domains (conductors) $\Omega_k = \Lambda_k \times \mathbb{R}$, $k = 1, 2, 3$, and denote by Λ the union $\Lambda := \Lambda_1 \cup \Lambda_2 \cup \Lambda_3$, by γ its boundary and by Λ_{ext} the vacuum $\Lambda_{\mathrm{ext}} := \mathbb{R}^2 \setminus \overline{\Lambda}$. Here the inductor is the union of the two domains $\Omega_2 = \Lambda_2 \times \mathbb{R}$ and $\Omega_3 = \Lambda_3 \times \mathbb{R}$.

Due to the geometry in Fig. 3.3, we seek an electric current density of the form

$$\boldsymbol{J}(x_1, x_2, x_3) = J(x_1, x_2)\, \boldsymbol{e}_3, \tag{3.33}$$

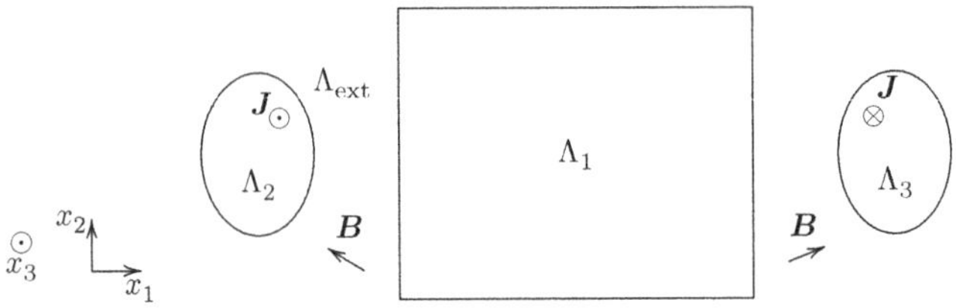

Fig. 3.3 A typical configuration of the conductors for the transversal model

i.e. $\tilde{\boldsymbol{J}} = 0$. We next assume that the 3-D inductor forms a "loop at the infinity", in such a way that the total current flowing in the inductor satisfies the identities

$$\int_{\Lambda_2} J(\boldsymbol{x})\,d\boldsymbol{x} = -\int_{\Lambda_3} J(\boldsymbol{x})\,d\boldsymbol{x} = I. \tag{3.34}$$

This means that the total current flowing inward Λ_2 (in the Ox_3–direction) is the same as the one flowing outward Λ_3. This assumption is compatible with the current conservation principle since these two conductors are assumed to be linked at the infinity in the Ox_3–direction. Following the same approach, a conductor which is not inductor must be assumed to have zero total current, i.e.

$$\int_{\Lambda_1} J(\boldsymbol{x})\,d\boldsymbol{x} = 0. \tag{3.35}$$

Using (3.2), (3.6), (3.5) and Theorem 1.4.2, we obtain the existence of a scalar potential $A = A(x_1, x_2)$ such that (3.14) holds, and consequently

$$-\operatorname{div}(\mu^{-1}\nabla A) = J \qquad \text{in } \mathbb{R}^2. \tag{3.36}$$

Therefore, since μ is constant in Λ_{ext},

$$\Delta A = 0 \qquad \text{in } \Lambda_{\text{ext}}. \tag{3.37}$$

Now clearly from the setting (3.33) and (3.1) and since $\tilde{\boldsymbol{J}} = 0$ in $\mathbb{R}^2$, we deduce that $\operatorname{\mathbf{curl}} H = 0$ and then H is constant. The assumption (3.9) yields then $H = 0$ in $\mathbb{R}^2$.

In summary, we have

$$\boldsymbol{H} = \tilde{\boldsymbol{H}} = \mu^{-1}\operatorname{\mathbf{curl}} A \quad \text{with } A = A(x_1, x_2).$$

Our aim is to obtain a well posed problem for eddy currents with the setting (3.33). For this, we shall describe hereafter the derivation of a model using the potential A. Let us define

$$I_k = \int_{\Lambda_k} J(\boldsymbol{x})\,d\boldsymbol{x} \qquad k = 1,2,3,$$

and set

$$I_1 = 0,\ I_2 = I,\ I_3 = -I,$$

and define for a function $\psi : \mathbb{R}^2 \to \mathbb{C}$, its mean value on Λ_k,

$$M_k(\psi) := \frac{1}{|\Lambda_k|}\int_{\Lambda_k} \psi\,d\boldsymbol{x} \qquad 1 \le k \le 3,$$

where $|\Lambda_k|$ is the measure (area) of the domain Λ_k. We have the following result.

Theorem 3.3.1. *Assume that a total current of amount I flows in the inductor (i.e. hypotheses* (3.34)–(3.35)*). Then, the scalar potential A is solution of the following problem:*

$$-\operatorname{div}(\mu^{-1}\nabla A) + \mathrm{i}\omega\sigma\Big(A - \frac{M_k(\sigma A)}{M_k(\sigma)}\Big)$$

$$= \frac{\sigma I_k}{|\Lambda_k| M_k(\sigma)} \qquad \textit{in } \Lambda_k,\ k = 1,2,3, \tag{3.38}$$

$$\Delta A = 0 \qquad \textit{in } \Lambda_{ext}, \tag{3.39}$$

$$[A] = \Big[\mu^{-1}\frac{\partial A}{\partial n}\Big] = 0 \qquad \textit{on } \gamma, \tag{3.40}$$

$$A(\boldsymbol{x}) = \beta + \mathcal{O}(|\boldsymbol{x}|^{-1}) \qquad |\boldsymbol{x}| \to \infty. \tag{3.41}$$

Proof. Let us derive equations in each connected component of the domain.

(i) Equation (3.39) is already obtained in (3.37).
(ii) In $\Lambda_k, k = 1,2,3$, we use (3.3), (3.7), and (3.14) to get

$$\mathbf{curl}\,(\mathrm{i}\omega A + \sigma^{-1} J) = 0.$$

In two dimensions, this implies the existence of constants $C_1, C_2, C_3 \in \mathbb{C}$ such that

$$\mathrm{i}\omega\sigma A + J = \sigma C_k \qquad \text{in } \Lambda_k, \quad k = 1,2,3. \tag{3.42}$$

In addition, we deduce from (3.36) that

$$-\operatorname{div}(\mu^{-1}\nabla A) + \mathrm{i}\omega\sigma A = \sigma C_k \qquad \text{in } \Lambda_k, \quad k = 1,2,3. \tag{3.43}$$

The constants C_k can be related to the prescribed current I as follows: Integrating (3.42) on Λ_k and using (3.34), (3.35), we obtain

$$I_k = |\Lambda_k| \, (C_k \, M_k(\sigma) - \mathrm{i}\omega \, M_k(\sigma A)), \quad k = 1,2,3.$$

From this we deduce

$$C_k = \frac{I_k + \mathrm{i}\omega \, |\Lambda_k| \, M_k(\sigma A)}{|\Lambda_k| \, M_k(\sigma)}, \quad k = 1,2,3. \tag{3.44}$$

Replacing these expressions in (3.43), we finally retrieve (3.38).

(iii) From (3.11) and (3.14), we deduce

$$0 = [\mathbf{curl}\, A \cdot \boldsymbol{n}]_\gamma = [\nabla A \cdot \boldsymbol{t}]_\gamma,$$

Therefore, the jump of A is constant across γ and can be chosen equal to zero. Furthermore, we have from (3.10), (3.14) and (3.6):

$$\left[\mu^{-1} \frac{\partial A}{\partial n}\right]_\gamma = [\mu^{-1} \, \mathbf{curl}\, A \cdot \boldsymbol{t}]_\gamma = 0. \qquad \square$$

Let us look for solutions of (3.38)–(3.41) in the space $\mathcal{W}^1(\mathbb{R}^2)$. Since these solutions are known up to an additive constant (A =Const. is indeed solution of (3.38)–(3.41)), we consider the space

$$\mathcal{V} := \{\phi \in \mathcal{W}^1(\mathbb{R}^2);\ M_1(\phi) = 0\},$$

and define on it the sesquilinear and antilinear forms:

$$\mathscr{B}(A,\phi) := \int_{\mathbb{R}^2} \mu^{-1} \nabla A \cdot \nabla \overline{\phi} \, d\boldsymbol{x} + \mathrm{i}\omega \int_\Lambda \sigma A \overline{\phi} \, d\boldsymbol{x}$$
$$- \mathrm{i}\omega \sum_{k=1}^{3} \frac{M_k(\sigma A) M_k(\sigma \overline{\phi})}{M_k(\sigma)} \, |\Lambda_k|,$$
$$\mathscr{L}(\phi) := \sum_{k=1}^{3} \frac{M_k(\sigma \overline{\phi})}{M_k(\sigma)} \, I_k.$$

We have for (3.38)–(3.41), the variational problem,

$$\begin{cases} \text{Find } A \in \mathcal{V} \quad \text{such that} \\ \mathscr{B}(A,\phi) = \mathscr{L}(\phi) \qquad \forall\, \phi \in \mathcal{V}. \end{cases} \tag{3.45}$$

We have the following result.

Theorem 3.3.2. *Problem* (3.45) *has a unique solution* $A \in \mathcal{V}$.

Proof. We use for this the Lax–Milgram theorem (Theorem 1.2.1). The continuity of the sesquilinear form $\mathscr{B}$ is obtained by using hypotheses on σ (2.11) which imply for $\phi, \psi \in \mathcal{V}$:

$$\begin{aligned}\int_{\Lambda_k} \Big|\sigma\Big(\phi - \frac{M_k(\sigma\phi)}{M_k(\sigma)}\Big)\overline{\psi}\Big|\,d\boldsymbol{x} &\le \sigma_M \Big(\|\phi\|_{\mathcal{L}^2(\Lambda_k)}\,\|\psi\|_{\mathcal{L}^2(\Lambda_k)} \\ &\qquad + \sigma_m^{-1}\,\sigma_M\,\|\phi\|_{\mathcal{L}^1(\Lambda_k)}\,\|\psi\|_{\mathcal{L}^1(\Lambda_k)}\Big) \\ &\le C_1\,\|\phi\|_{\mathcal{H}^1(\Lambda_k)}\,\|\psi\|_{\mathcal{H}^1(\Lambda_k)}. \end{aligned} \tag{3.46}$$

Therefore, from (2.7) and (3.46), we deduce

$$\begin{aligned}|\mathscr{B}(\phi,\psi)| &\le C_2\,\|\nabla\phi\|_{\mathcal{L}^2(\mathbb{R}^2)^2}\,\|\nabla\psi\|_{\mathcal{L}^2(\mathbb{R}^2)^2} + C_3\,\|\phi\|_{\mathcal{H}^1(\Lambda)}\|\psi\|_{\mathcal{H}^1(\Lambda)} \\ &\le C_4\,\|\phi\|_{\mathcal{W}^1(\mathbb{R}^2)}\,\|\psi\|_{\mathcal{W}^1(\mathbb{R}^2)}.\end{aligned}$$

To prove the coercivity of $\mathscr{B}$, we have

$$\mathscr{B}(\phi,\phi) = \int_{\mathbb{R}^2} \mu^{-1}|\nabla\phi|^2\,d\boldsymbol{x} + \mathrm{i}\omega\sum_{k=1}^{3}\Big(\int_{\Lambda_k}\sigma|\phi|^2\,d\boldsymbol{x} - |\Lambda_k|\frac{|M_k(\sigma\phi)|^2}{M_k(\sigma)}\Big).$$

Let us mention a variant of the norm equivalence result that is proved in [149]. It reads

$$\|\phi\|_{\mathcal{W}^1(\mathbb{R}^2)} \le C\,|\phi|_{\mathcal{W}^1(\mathbb{R}^2)} \qquad \forall\,\phi \in \mathcal{V}. \tag{3.47}$$

Combining this result with Hypothesis (2.7), we obtain

$$\mathrm{Re}\,(\mathscr{B}(\phi,\phi)) = \int_{\mathbb{R}^2}\mu^{-1}|\nabla\phi|^2\,d\boldsymbol{x} \ge C\,\|\phi\|^2_{\mathcal{W}^1(\mathbb{R}^2)}.$$

Therefore

$$|\mathscr{B}(\phi,\phi)| \ge \mathrm{Re}\,(\mathscr{B}(\phi,\phi)) \ge C\,\|\phi\|^2_{\mathcal{W}^1(\mathbb{R}^2)}.$$

The continuity of the antilinear form $\mathscr{L}$ is obtained thanks to Property (2.11), i.e.

$$|\mathscr{L}(\phi)| \le \frac{\sigma_M}{\sigma_m}\sum_{k=1}^{3}\|\phi\|_{\mathcal{L}^1(\Lambda_k)} \le C\,\|\phi\|_{\mathcal{W}^1(\mathbb{R}^2)}.$$

Existence and uniqueness of a solution of (3.45) is then established. □

Clearly (3.45) is not appropriated for numerical treatment. This one is indeed formulated in an unbounded domain. A simple idea for solving this problem would consist in approximating the plane $\mathbb{R}^2$ by a "sufficiently large domain" and design some *transparent* boundary condition to replace the condition at the infinity. This approach maybe in some cases loosely accurate. Rather, we shall resort to a boundary integral formulation for the exterior domain. In other words, the external Laplace equation is represented by an integral formulation on the boundary γ. This results in a coupling between partial differential equations in the conductors Λ_k and an integral equation on γ. This coupling can be achieved in many ways and we shall expose hereafter some coupling formulations and discuss their advantages and drawbacks.

3.3.1 A Formulation Using the Steklov–Poincaré Operator

Let us consider the exterior Steklov–Poincaré operator P defined in Sect. 1.4.4. We have, from (3.38) and (3.40) by using the Green formula on $\mathbb{R}^2$:

$$\int_\Lambda \mu^{-1}\nabla A\cdot\nabla\overline{\phi}\,d\boldsymbol{x} + \mathrm{i}\omega\Big(\int_\Lambda \sigma A\overline{\phi}\,d\boldsymbol{x} - \sum_{k=1}^{3}\frac{M_k(\sigma A)M_k(\sigma\overline{\phi})}{M_k(\sigma)}\,|\Lambda_k|\Big)$$
$$-\,\mu_0^{-1}\int_\gamma \frac{\partial A^+}{\partial n}\,\overline{\phi}\,ds = \sum_{k=1}^{3}\frac{M_k(\sigma\overline{\phi})}{M_k(\sigma)}\,I_k. \tag{3.48}$$

Using the exterior Steklov–Poincaré operator we obtain the variational formulation

$$\begin{cases}\text{Find } A\in\mathcal{W} \text{ such that}\\ \mathscr{B}_\Lambda(A,\phi)+\mathscr{P}(A,\phi)=\mathscr{L}(\phi) \qquad \forall\,\phi\in\mathcal{W},\end{cases} \tag{3.49}$$

where

$$\mathcal{W} := \{\phi\in\mathcal{H}^1(\Lambda);\ M_1(\phi)=0\},$$

and

$$\mathscr{B}_\Lambda(\phi,\psi) := \int_\Lambda \mu^{-1}\nabla\phi\cdot\nabla\overline{\psi}\,d\boldsymbol{x} + \mathrm{i}\omega\Big(\int_\Lambda \sigma\phi\overline{\psi}\,d\boldsymbol{x} - \sum_{k=1}^{3}\frac{M_k(\sigma\phi)M_k(\sigma\overline{\psi})}{M_k(\sigma)}\,|\Lambda_k|\Big),$$
$$\mathscr{P}(\phi,\psi) := \mu_0^{-1}\int_\gamma P\phi\,\overline{\psi}\,ds.$$

Let us recall, for the sake of completeness, that the operator P is given by

$$P = (-\frac{1}{2}I + R')\,K^{-1},$$

where I is the identity operator, K is given by (1.60), R' is the adjoint operator of R given by (1.65), and G is the Green function defined by (1.50).

Theorem 3.3.3. *Problem* (3.49) *has a unique solution. Moreover, by extending A outside Λ by a harmonic function with the same trace as A on γ, we obtain the solution of* (3.45)*. Conversely, if A is a solution of* (3.45)*, then $A_{|\Lambda}$ is solution of* (3.49)*.*

Proof. The proof of this theorem is an immediate consequence of the above considerations. However, we can directly prove the existence and uniqueness of a solution of (3.49) by using the Lax–Milgram theorem (Theorem 1.2.1).

Using the continuity of the operator P and the trace inequality (1.3) and (2.7), we obtain,

$$\begin{aligned}\mathscr{P}(\phi,\psi) &\leq \mu_0^{-1} \|P\phi\|_{\mathcal{H}^{-\frac{1}{2}}(\gamma)} \|\psi\|_{\mathcal{H}^{\frac{1}{2}}(\gamma)} \\ &\leq C_1 \|\phi\|_{\mathcal{H}^{\frac{1}{2}}(\gamma)} \|\psi\|_{\mathcal{H}^{\frac{1}{2}}(\gamma)} \\ &\leq C_2 \|\phi\|_{\mathcal{H}^1(\Lambda)} \|\psi\|_{\mathcal{H}^1(\Lambda)}.\end{aligned} \tag{3.50}$$

Combining inequalities (3.46) and (3.50), we obtain the continuity of $\mathscr{B}_\Lambda + \mathscr{P}$. The coercivity of $\mathscr{B}_\Lambda + \mathscr{P}$ is established by using Theorem 1.4.10. We have thanks to the 2-D version of the Poincaré–Friedrichs inequality (1.5) and (2.7):

$$\begin{aligned}\mathrm{Re}(\mathscr{B}_\Lambda(\phi,\phi)) + \mathrm{Re}(\mathscr{P}(\phi,\phi)) &= \int_\Lambda \mu^{-1}|\nabla\phi|^2\,d\boldsymbol{x} + C_1\,\|\phi\|^2_{\mathcal{H}^{\frac{1}{2}}(\gamma)} \\ &\geq \mu_M^{-1}\,\|\phi\|^2_{\mathcal{H}^1(\Lambda)} + C_2\,\|\phi\|^2_{\mathcal{H}^1(\Lambda)}.\end{aligned}$$

Existence and uniqueness of a solution of (3.49) is then achieved. □

3.3.2 A Formulation Using Simple Layer Potentials

As we have already seen in Chap. 1, Sect. 1.4.4, the representation of the exterior Steklov–Poincaré operator by double layer potentials may present some difficulties from the implementation point of view. To avoid this, we use the technique used in Sect. 1.4.4 that involves the Poisson integral. Let us present this in the context of our problem.

Let us assume that (3.38)–(3.41) is solved and let ψ denote a solution of the interior problem

$$\begin{cases} \Delta\psi = 0 & \text{in } \Lambda, \\ \psi = A & \text{on } \gamma. \end{cases}$$

It follows, by using (1.56), that for $\boldsymbol{x} \in \gamma$, we have

$$A(\boldsymbol{x}) = -\int_\gamma \Big(\frac{\partial A^+}{\partial n}(\boldsymbol{y}) - \frac{\partial \psi}{\partial n}(\boldsymbol{y})\Big)\, G(\boldsymbol{x},\boldsymbol{y})\, ds(\boldsymbol{y}) + \xi,$$

where $\xi \in \mathbb{C}$ and where we recall that A^+ (*resp.* A^-) is the external (*resp.* internal) value of A, $G(\boldsymbol{x},\boldsymbol{y}) = -\frac{1}{2\pi}\ln|\boldsymbol{x}-\boldsymbol{y}|$.

By (3.40) we have

$$\frac{\partial A^+}{\partial n} = \frac{\mu_0}{\mu}\frac{\partial A^-}{\partial n}$$

and (3.38)–(3.41) can then be written in the form:

$$-\operatorname{div}(\mu^{-1}\nabla A) + \mathrm{i}\omega\sigma\Big(A - \frac{M_k(\sigma A)}{M_k(\sigma)}\Big) = \frac{\sigma I_k}{|\Lambda_k|\, M_k(\sigma)} \qquad \text{in } \Lambda_k, \quad (3.51)$$

$$\Delta\psi = 0 \qquad \text{in } \Lambda_k, \quad (3.52)$$

$$A(\boldsymbol{x}) = \psi(\boldsymbol{x}) = -\int_\gamma \Big(\mu_0 p(\boldsymbol{y}) - \frac{\partial \psi}{\partial n}(\boldsymbol{y})\Big)\, G(\boldsymbol{x},\boldsymbol{y})\, ds(\boldsymbol{y}) + \xi \qquad \boldsymbol{x} \in \gamma, \quad (3.53)$$

$$p = \mu^{-1}\frac{\partial A^-}{\partial n} = \mu_0^{-1}\frac{\partial A^+}{\partial n} \qquad \text{on } \gamma, \quad (3.54)$$

for $k = 1, 2, 3$.

Remark 3.3.1. It is easy to see that if (A, ψ, p) is a solution of (3.51), (3.54) with given $\xi \in \mathbb{C}$, then $(A - \xi, \psi - \xi, p)$ is also a solution when we fix in (3.53) the constant ξ to 0. This is the reason for which we choose in the following $\xi = 0$ and relax the condition $M_1(A) = 0$ in (3.45).

A weak formulation of (3.51)–(3.54) with $\xi = 0$ consists in finding $A \in \mathcal{H}^1(\Lambda)$, $\psi \in \mathcal{H}^1(\Lambda)$ and $p \in \mathcal{H}^{-\frac{1}{2}}(\gamma)$ that satisfy:

$$\begin{aligned}
&\int_\Lambda \mu^{-1}\nabla A\cdot\nabla\overline{\phi}\, d\boldsymbol{x} \\
&\quad + \mathrm{i}\omega\Big(\int_\Lambda \sigma A\overline{\phi}\, d\boldsymbol{x} - \sum_{k=1}^{3}\frac{M_k(\sigma A)\, M_k(\sigma\overline{\phi})}{M_k(\sigma)}|\Lambda_k|\Big) \\
&\quad - \int_\gamma p\overline{\phi}\, ds = \sum_{k=1}^{3}\frac{M_k(\sigma\overline{\phi})}{M_k(\sigma)} I_k \qquad \forall\, \phi \in \mathcal{H}^1(\Lambda), \quad (3.55)
\end{aligned}$$

$$\int_\Lambda \nabla\psi\cdot\nabla\overline{\phi}\, d\boldsymbol{x} = 0 \qquad \forall\, \phi \in \mathcal{H}^1_0(\Lambda), \quad (3.56)$$

$$\int_\gamma \psi \overline{q}\, ds = \int_\gamma A\overline{q}\, ds =$$

$$\int_\gamma\int_\gamma \Big(\frac{\partial \psi}{\partial n}(y) - \mu_0 p(y)\Big)\, G(x,y)\overline{q}(x)\, ds(x)\, ds(y) \quad \forall\, q \in \mathcal{H}^{-\frac{1}{2}}(\gamma). \quad (3.57)$$

Theorem 3.3.4. *Problem* (3.55)–(3.57) *has a unique solution* (A,ψ,p). *Moreover, A is the solution, up to a constant, of* (3.49) *and* $p = \mu^{-1}\dfrac{\partial A}{\partial n}$. *Conversely, if A is the solution of* (3.49), *and if* ψ *and* p *are defined by*

$$\begin{cases} \Delta\psi = 0 & \text{in } \Lambda, \\ \psi = A & \text{on } \gamma, \\ p = \mu^{-1}\dfrac{\partial A}{\partial n} & \text{on } \gamma, \end{cases}$$

then there exists $\xi \in \mathbb{C}$ *such that the triple* $(A+\xi, \psi+\xi, p)$ *is solution of* (3.55)–(3.57).

Proof. The proof follows directly from above calculations and Theorems 3.3.2 and 3.3.3. □

An variant of the formulation (3.51)–(3.54), can be stated by using Remark 3.3.1. This one consists in looking for $(A,\psi,p) \in \mathcal{W}\times\mathcal{W}\times\tilde{\mathcal{H}}^{-\frac{1}{2}}(\gamma)$ such that

$$\int_\Lambda \mu^{-1}\nabla A\cdot\nabla\overline{\phi}\, dx$$
$$+ i\omega\Big(\int_\Lambda \sigma A\overline{\phi}\, dx - \sum_{k=1}^{3} \frac{M_k(\sigma A)\, M_k(\sigma\overline{\phi})}{M_k(\sigma)}\, |\Lambda_k|\Big)$$
$$- \int_\gamma p\overline{\phi}\, ds = \sum_{k=1}^{3} \frac{M_k(\sigma\overline{\phi})}{M_k(\sigma)}\, I_k \qquad \forall\, \phi\in\mathcal{W}, \quad (3.58)$$

$$\int_\Lambda \nabla\psi\cdot\nabla\overline{\phi}\, dx = 0 \qquad \forall\, \phi \in \mathcal{H}^1_0(\Lambda), \quad (3.59)$$

$$\int_\gamma \psi\overline{q}\, ds = \int_\gamma A\overline{q}\, ds$$
$$= \int_\gamma\int_\gamma \Big(\frac{\partial\psi}{\partial n}(y) - \mu_0 p(y)\Big)\, G(x,y)\overline{q}(x)\, ds(x)\, ds(y) \forall\, q \in \tilde{\mathcal{H}}^{-\frac{1}{2}}(\gamma). \quad (3.60)$$

The formulation (3.58)–(3.60) is however less advantageous from a practical point of view, although a Lagrange multiplier can be used to the handle the γ–null integral condition.

Theorem 3.3.5. *Problem* (3.58)–(3.60) *has a unique solution* (A, ψ, p). *Moreover, A is the solution of* (3.49) *on* Λ *and we have*

$$\begin{cases} \Delta\psi = 0 & \text{in } \Lambda, \\ p = \mu^{-1}\dfrac{\partial A}{\partial n} & \text{on } \gamma, \end{cases}$$

Conversely, if A is the solution of (3.49), *if* ψ *and* p *satisfy*

$$\begin{cases} \Delta\psi = 0 & \text{in } \Lambda, \\ \psi = A + \xi & \text{on } \gamma, \\ M_1(\psi) = 0, & \\ p = \mu^{-1}\dfrac{\partial A}{\partial n} & \text{on } \gamma, \end{cases}$$

with $\xi \in \mathbb{C}$, *then the triple* (A, ψ, p) *is a solution of* (3.58)–(3.60).

Proof. Let A be the solution of (3.49). By setting $p = \mu^{-1}\partial A/\partial n$ and ψ such that $\Delta\psi = 0$ in Λ, $\psi = A + \xi$ on γ with ξ chosen such that $M_1(\psi) = 0$, we obtain (3.59)–(3.60). More precisely, we obtain $\psi \in \mathcal{W}$ and

$$\int_\Lambda \nabla\psi \cdot \nabla\overline{\phi}\, d\boldsymbol{x} = 0 \qquad \forall\, \phi \in \mathcal{H}_0^1(\Lambda).$$

In addition, we have $(\psi - A)_{|\gamma} = \xi \in \mathbb{C}$.

Let us define A^+ in $\overline{\Lambda}_{\text{ext}}$ by $\Delta A^+ = 0$ in Λ_{ext}, $A^+ = A$ on γ and by considering that $A = \psi - \xi$ on γ with $\Delta(\psi - \xi) = 0$ in Λ, we obtain by (1.56),

$$A(\boldsymbol{x}) = \int_\gamma \Big(\frac{\partial\psi}{\partial n}(\boldsymbol{y}) - \frac{\partial A^+}{\partial n}(\boldsymbol{y})\Big) G(\boldsymbol{x}, \boldsymbol{y})\, ds(\boldsymbol{y}) \qquad \forall\, \boldsymbol{x} \in \gamma,$$

or by setting $p = \mu_0^{-1}\dfrac{\partial A^+}{\partial n}$ on γ,

$$A(\boldsymbol{x}) = \int_\gamma \Big(\frac{\partial\psi}{\partial n}(\boldsymbol{y}) - \mu_0 p(\boldsymbol{y})\Big) G(\boldsymbol{x}, \boldsymbol{y})\, ds(\boldsymbol{y}) \qquad \forall\, \boldsymbol{x} \in \gamma.$$

Equation (3.49) implies that $\mu_0^{-1}A^+ = \mu^{-1}A$ and we obtain (3.58).

From the above arguments, the reverse proposition is easy to prove. □

3.3.3 A Formulation Using a Simple–Double Layer Potential

An alternative variational formulation can be considered by writing an integral representation of $A_{|\Lambda_{\text{ext}}} = A^+$. For this, we consider (3.39) and extend A^+ by zero to Λ. Using (3.41) and (1.56), we have the representation

$$\frac{1}{2}A^+(\boldsymbol{x}) = \int_\gamma A^+(\boldsymbol{y})\,\frac{\partial G}{\partial n_{\boldsymbol{y}}}(\boldsymbol{x},\boldsymbol{y})\,ds(\boldsymbol{y}) - \int_\gamma \frac{\partial A^+}{\partial n}(\boldsymbol{y})\,G(\boldsymbol{x},\boldsymbol{y})\,ds(\boldsymbol{y}) + \xi \quad (3.61)$$

for $\boldsymbol{x} \in \gamma$, where ξ is a constant. We can then construct a variational formulation that couples external and internal problems via the normal derivative of A (following [26]).

By using (3.38)–(3.39), we obtain

$$\mathscr{B}(A,\phi) = \mathscr{B}_\Lambda(A,\phi) - \int_\gamma p\overline{\phi}\,ds,$$

where $p = \mu_0^{-1}\dfrac{\partial A^+}{\partial n}$ on γ. Multiplying (3.61) by $\overline{q} \in \tilde{\mathcal{H}}^{-\frac{1}{2}}(\gamma)$ where

$$\tilde{\mathcal{H}}^{-\frac{1}{2}}(\gamma) = \{q \in \mathcal{H}^{-\frac{1}{2}}(\gamma);\ \int_\gamma q\,ds = 0\},$$

and integrating on γ (recall that the boundary integrals actually denote duality pairings), results in

$$\begin{aligned}\frac{1}{2}\int_\gamma A\overline{q}\,ds = &\int_\gamma\int_\gamma A(\boldsymbol{y})\frac{\partial G}{\partial n_{\boldsymbol{y}}}(\boldsymbol{x},\boldsymbol{y})\overline{q}(\boldsymbol{x})\,ds(\boldsymbol{y})\,ds(\boldsymbol{x})\\ &- \mu_0\int_\gamma\int_\gamma p(\boldsymbol{y})G(\boldsymbol{x},\boldsymbol{y})\overline{q}(\boldsymbol{x})\,ds(\boldsymbol{y})\,ds(\boldsymbol{x}).\end{aligned}$$

Let us define the sesquilinear forms:

$$\begin{aligned}
\mathscr{C}(\phi,q) &:= \int_\gamma \phi\,\overline{q}\,ds,\\
\mathscr{R}(\phi,q) &:= 2\int_\gamma R\phi\,\overline{q}\,ds := 2\int_\gamma\int_\gamma \frac{\partial G}{\partial n_{\boldsymbol{y}}}(\boldsymbol{x},\boldsymbol{y})\,\phi(\boldsymbol{y})\,\overline{q}(\boldsymbol{x})\,ds(\boldsymbol{y})\,ds(\boldsymbol{x}),\\
\mathscr{D}(p,q) &:= 2\mu_0\int_\gamma Kp\,\overline{q}\,ds := 2\mu_0\int_\gamma\int_\gamma G(\boldsymbol{x},\boldsymbol{y})\,p(\boldsymbol{y})\,\overline{q}(\boldsymbol{x})\,ds(\boldsymbol{y})\,ds(\boldsymbol{x}).
\end{aligned}$$

We obtain the formulation

$$\begin{cases} \text{Find } (A,p) \in \mathcal{W} \times \tilde{\mathcal{H}}^{-\frac{1}{2}}(\gamma) \text{ such that:} & \\ \mathscr{B}_\Lambda(A,\phi) - \overline{\mathscr{C}(\phi,p)} = \mathscr{L}(\phi) & \forall\, \phi \in \mathcal{W}, \\ \mathscr{D}(p,q) + \mathscr{C}(A,q) - \mathscr{R}(A,q) = 0 & \forall\, q \in \tilde{\mathcal{H}}^{-\frac{1}{2}}(\gamma), \end{cases} \tag{3.62}$$

where we recall that

$$\mathcal{W} = \{\phi \in \mathcal{H}^1(\Lambda);\ M_1(\phi) = 0\}.$$

It is easy to check that if (A,p) is a solution to (3.62), then we have

$$p = \mu_0^{-1} \frac{\partial A^+}{\partial n}\Big|_\gamma,$$

where A^+ is defined by:

$$\begin{cases} \Delta A^+ = 0 & \text{in } \Lambda_{\text{ext}}, \\ A^+ = A & \text{on } \gamma, \\ A(\boldsymbol{x}) = \alpha + \mathcal{O}(|\boldsymbol{x}|^{-1}) & |\boldsymbol{x}| \to \infty, \end{cases}$$

with $\alpha \in \mathbb{C}$. The mixed formulation (3.62) was first derived, for the Laplace equation, by Johnson et al. [108], where a finite element approximation is obtained and for which convergence is proved. Note also that an alternative mixed formulation can be found in [87].

Theorem 3.3.6. *Problem* (3.62) *has a unique solution* (A,p). *Moreover,* A *is the solution of* (3.49). *Conversely, if* $A \in \mathcal{W}$ *is a solution of* (3.49) *and if* $p := PA$ *where* P *is the Steklov–Poincaré operator, then the pair* (A,p) *is the solution of* (3.62).

Proof. Existence and uniqueness are consequences of the equivalence between problems (3.49)–(3.62). More precisely, let $A \in \mathcal{W}$ denote a solution of (3.49) and let $p = PA$, then $p \in \tilde{\mathcal{H}}^{-\frac{1}{2}}(\gamma)$, and the first variational equation of (3.62) is valid. To prove the second equation, we recall the same developments that led to this formulation. □

3.3.4 A Formulation Using the Poisson Formula

The use of simple and double layer integrals in the formulations given in Sects. 3.3.1 and 3.3.3 may exhibit some difficulties related to the integration of singular functions. We can then resort to an alternative formulation that uses the so-called *Poisson formula.* To present this formulation, we denote by B_r the open disk centered at 0 with radius $r > 0$. We denote by S_r the boundary of B_r. We assume furthermore that r is chosen in such a way that $\overline{\Lambda} \subset S_r$. Then, using the Poisson integral formula (see [62], Vol. 1, p. 249), we have

$$A(\boldsymbol{x}) = \frac{r^2 - |\boldsymbol{x}|^2}{2\pi r} \int_{S_r} \frac{A(\boldsymbol{y})}{|\boldsymbol{x} - \boldsymbol{y}|^2} \, ds(\boldsymbol{y}) \qquad \text{for } |\boldsymbol{x}| > r.$$

Let B_R stand for the open disk with radius $R > r$, centered at 0 and with boundary S_R. Problem (3.38)–(3.41) can be casted in the following variational form:

Find $A \in \mathcal{H}^1(B_R)$ such that:

$$\int_{B_R} \mu^{-1} \nabla A \cdot \nabla \overline{\psi} \, d\boldsymbol{x} + \mathrm{i}\omega \Big(\int_{\Lambda_k} \sigma A \overline{\psi} \, d\boldsymbol{x} - \sum_{k=1}^{3} \frac{M_k(\sigma A)\, M_k(\sigma \overline{\psi})}{M_k(\sigma)} |\Lambda_k| \Big)$$

$$= \sum_{k=1}^{3} \frac{M_k(\sigma \overline{\psi})}{M_k(\sigma)} I_k \qquad \forall\, \psi \in \mathcal{H}_0^1(B_R), \tag{3.63}$$

$$\int_{S_R} A(\boldsymbol{x}) \overline{q}(\boldsymbol{x}) \, ds(\boldsymbol{x}) - \frac{r^2 - R^2}{2\pi r} \int_{S_R} \Big(\overline{q}(\boldsymbol{x}) \int_{S_r} \frac{A(\boldsymbol{y})}{|\boldsymbol{x} - \boldsymbol{y}|^2} \, ds(\boldsymbol{y}) \Big) \, ds(\boldsymbol{x})$$

$$= 0 \qquad \forall\, q \in \mathcal{H}^{-\frac{1}{2}}(S_R). \tag{3.64}$$

We recall that we have $\mu = \mu_0$ outside $\overline{\Lambda}$.

Obviously (3.63)–(3.64) is equivalent to (3.38)–(3.41). Note, in addition, that the formulation (3.63)–(3.64) requires choosing the values of the radii r and R. This issue has to be carefully examined when investigating numerical approximation.

3.3.5 *Other Formulations*

More elaborate variational formulations of (3.38)–(3.41) can be found in [58, 100, 113]. These formulations have the advantage of yielding symmetric bilinear forms in the case of the Laplace operator. Here, the operator in the inner domain gives a non hermitian form. For this reason, we believe the usage of such formulations is of limited interest.

Chapter 4
Three-Dimensional Models

We consider in the present chapter, the fully three-dimensional case defined by (2.62)–(2.68). We shall consider here, for simplicity, a unique connected domain of genus 1 (i.e. a torus-like domain, see Fig. 1.1). This one is denoted by Ω. We also consider, as in Chap. 1, S as a cut in Ω, i.e. $\Omega \setminus S$ is simply connected, and denote by Ω_{ext} the set $\mathbb{R}^3 \setminus \overline{\Omega}$ and by Σ a cut in Ω_{ext}, i.e. a surface in Ω_{ext} such that $\Omega_{\text{ext}} \setminus \Sigma$ is simply connected (See Fig. 1.1). All the notations and assumptions on this domain are already described in Chap. 1. Let us recall the set of time-harmonic eddy current equations (2.62)–(2.68):

$$\mathbf{curl}\, \boldsymbol{H} = \boldsymbol{J} \qquad \text{in } \mathbb{R}^3, \tag{4.1}$$

$$\mathrm{i}\omega \boldsymbol{B} + \mathbf{curl}\, \boldsymbol{E} = 0 \qquad \text{in } \mathbb{R}^3 \setminus S, \tag{4.2}$$

$$\boldsymbol{B} = \mu \boldsymbol{H} \qquad \text{in } \mathbb{R}^3, \tag{4.3}$$

$$\boldsymbol{J} = \sigma \boldsymbol{E} \qquad \text{in } \mathbb{R}^3, \tag{4.4}$$

$$\operatorname{div} \boldsymbol{B} = 0 \qquad \text{in } \mathbb{R}^3, \tag{4.5}$$

$$|\boldsymbol{H}(\boldsymbol{x})| + |\boldsymbol{E}(\boldsymbol{x})| = \mathcal{O}(|\boldsymbol{x}|^{-1}) \qquad |\boldsymbol{x}| \to \infty, \tag{4.6}$$

and recall also that the coefficients σ and μ are assumed to fulfill the conditions (2.11) and (2.7) and satisfy (see Sect. 2.2.2)

$$\sigma = 0, \quad \mu = \mu_0 = \text{Const.} \qquad \text{in } \Omega_{\text{ext}}. \tag{4.7}$$

We consider in addition the vector potential $\boldsymbol{A}$ defined by (2.12), (2.13) and $\boldsymbol{A} \in \mathcal{W}^1(\mathbb{R}^3)$, i.e.

$$\boldsymbol{B} = \mathbf{curl}\, \boldsymbol{A} \qquad \text{in } \mathbb{R}^3, \tag{4.8}$$

$$\operatorname{div} \boldsymbol{A} = 0 \qquad \text{in } \mathbb{R}^3, \tag{4.9}$$

$$|\boldsymbol{A}(\boldsymbol{x})| = \mathcal{O}(|\boldsymbol{x}|^{-1}) \qquad |\boldsymbol{x}| \to \infty. \tag{4.10}$$

R. Touzani and J. Rappaz, *Mathematical Models for Eddy Currents and Magnetostatics: With Selected Applications*, Scientific Computation, DOI 10.1007/978-94-007-0202-8_4,

Remark 4.1. The Faraday equation (4.2) is assumed to be valid in $\mathbb{R}^3 \setminus S$ only. If this one was valid in $\mathbb{R}^3$, we should obtain owing to (4.8),

$$\mathbf{curl}\,(\mathrm{i}\omega \boldsymbol{A} + \boldsymbol{E}) = 0 \quad \text{in } \mathbb{R}^3.$$

Recalling that the current voltage is defined by

$$V := \int_\gamma (\mathrm{i}\omega \boldsymbol{A} + \boldsymbol{E}) \cdot \boldsymbol{t}\, ds, \tag{4.11}$$

where γ is any closed curve on Γ that intersects the cut S at one point and $\boldsymbol{t}$ is a unit tangential vector to γ_0. In this case, we should deduce from the Stokes theorem that the voltage is null. This is absurd. In fact, we are in presence of a paradox because a torus device does not allow for applying a current or voltage. To avoid this obstacle, we "cut" the torus to connect it to a source.

In the following, we first present a model based on the current density vector field. The numerical approximation of this model turns out to be rather prohibitive in terms of computational time since it leads to a nonlocal problem in the whole domain. Its statement is however useful to introduce a more elaborate model that is formulated in terms of the magnetic field $\boldsymbol{H}$. This one is described in detail and its well–posedness is proved.

4.1 A Current Density Formulation

This model was first derived by Bossavit [32, 34]. Using (4.2), (4.4) and (4.8), we obtain

$$\mathbf{curl}\,(\mathrm{i}\omega \boldsymbol{A} + \sigma^{-1}\boldsymbol{J}) = 0 \qquad \text{in } \Omega \setminus S.$$

From Theorem 1.3.5, we deduce the existence of a function $\varphi \in \mathcal{H}^1(\Omega)$ and a complex number V such that

$$\mathrm{i}\omega \boldsymbol{A} + \sigma^{-1}\boldsymbol{J} = \nabla\varphi + V\,\nabla q \qquad \text{in } \Omega \setminus S, \tag{4.12}$$

where $q \in \mathcal{H}^1(\Omega \setminus S)$ is a solution to (1.17). Let us define an operator that maps the current density $\boldsymbol{J}$ to the vector potential $\boldsymbol{A}$ by means of (2.19)–(2.21). For this end we define the spaces:

$$\begin{aligned}
\mathcal{X} &:= \{\, \boldsymbol{v} \in \mathcal{H}(\mathrm{div}, \Omega);\ \mathrm{div}\, \boldsymbol{v} = 0 \text{ in } \Omega,\ \boldsymbol{v} \cdot \boldsymbol{n} = 0 \text{ on } \Gamma \,\},\\
\mathcal{Y} &:= \{\, \boldsymbol{v} \in \mathcal{H}(\mathrm{div}, \Omega) \cap \mathcal{H}(\mathbf{curl}, \Omega);\ \mathrm{div}\, \boldsymbol{v} = 0 \text{ in } \Omega \,\},\\
\mathcal{V} &:= \{\, \boldsymbol{v} \in \mathcal{W}^1(\mathbb{R}^3);\ \mathrm{div}\, \boldsymbol{v} = 0 \text{ in } \mathbb{R}^3 \,\}.
\end{aligned}$$

Clearly, $\mathcal{X}$, $\mathcal{Y}$ and $\mathcal{V}$ are Hilbert spaces when endowed respectively with the norms:

$$\|v\|_{\mathcal{X}} := \|v\|_{\mathcal{L}^2(\Omega)},$$
$$\|v\|_{\mathcal{Y}} := \left(\|v\|^2_{\mathcal{L}^2(\Omega)} + \|\,\mathbf{curl}\,v\|^2_{\mathcal{L}^2(\Omega)}\right)^{\frac{1}{2}},$$
$$\|v\|_{\mathcal{V}} := \|\,\mathbf{curl}\,v\|_{\mathcal{L}^2(\mathbb{R}^3)},$$

the latter being justified by Theorem 1.3.2.

We define the linear mapping

$$T : g \in \mathcal{L}^2(\Omega) \mapsto u = Tg \in \mathcal{Y},$$

where u is the restriction to Ω of the unique solution of (2.22) with J replaced by g, i.e.,

$$\begin{cases} \text{Find } u \in \mathcal{V} \text{ such that} \\ \displaystyle\int_{\mathbb{R}^3} \mu^{-1}\,\mathbf{curl}\,u \cdot \mathbf{curl}\,\overline{v}\,dx = \int_{\Omega} g \cdot \overline{v}\,dx \qquad \forall\, v \in \mathcal{V}. \end{cases} \tag{4.13}$$

Following the conclusion of Theorem 2.3.1 where A is replaced by u and J by g, we have $Tg \in \mathcal{Y}$ if $g \in \mathcal{X}$.

From (4.12), (4.1) and the interface condition derived from (4.1) (see Theorem 1.3.1), we deduce the problem for J:

$$\mathrm{i}\omega\, T J + \sigma^{-1} J = \nabla\varphi + V\,\nabla q \qquad \text{in } \Omega \setminus S, \tag{4.14}$$
$$\operatorname{div} J = 0 \qquad \text{in } \Omega, \tag{4.15}$$
$$J \cdot n = 0 \qquad \text{on } \Gamma. \tag{4.16}$$

The complex constant V that appears in (4.12) is precisely the current voltage. To see this, let us consider the boundary γ of the surface Σ. Clearly, γ is a closed curve that intersects the cut S at one point (See Fig. 1.1). We have by integrating (4.12) along this curve, using (4.11) and setting $E = \sigma^{-1} J$,

$$\int_\gamma (\mathrm{i}\omega A + E)\cdot t\,ds = \int_\gamma \nabla\varphi \cdot t\,ds + V \int_\gamma \nabla q \cdot t\,ds = V$$

since φ is continuous and q has a jump equal to unity on S.

In order to derive a variational formulation of (4.14)–(4.16), we multiply the right-hand side of (4.14) by $\overline{v} \in \mathcal{X}$ and integrate over $\Omega \setminus S$. We obtain by the Green formula and (1.17),

$$\int_\Omega \nabla\varphi\cdot\overline{\boldsymbol{v}}\,d\boldsymbol{x} + V\int_{\Omega\setminus S}\nabla q\cdot\overline{\boldsymbol{v}}\,d\boldsymbol{x} = -\int_\Omega \varphi\,\operatorname{div}\overline{\boldsymbol{v}}\,d\boldsymbol{x} - V\int_{\Omega\setminus S} q\,\operatorname{div}\overline{\boldsymbol{v}}\,d\boldsymbol{x}$$
$$+\int_\Gamma \varphi\,\overline{\boldsymbol{v}}\cdot\boldsymbol{n}\,ds + V\int_\Gamma q\,\overline{\boldsymbol{v}}\cdot\boldsymbol{n}\,ds$$
$$+V\int_S [q]\,\overline{\boldsymbol{v}}\cdot\boldsymbol{n}\,ds$$
$$= V\int_S \overline{\boldsymbol{v}}\cdot\boldsymbol{n}\,ds.$$

From this we deduce the problem:

$$\begin{cases} \text{Find } \boldsymbol{J}\in\boldsymbol{\mathcal{X}} \text{ such that} \\ \displaystyle\int_\Omega (\mathrm{i}\omega\,\boldsymbol{T}\boldsymbol{J} + \sigma^{-1}\boldsymbol{J})\cdot\overline{\boldsymbol{v}}\,d\boldsymbol{x} = V\int_S \overline{\boldsymbol{v}}\cdot\boldsymbol{n}\,ds \qquad \forall\,\boldsymbol{v}\in\boldsymbol{\mathcal{X}}. \end{cases} \tag{4.17}$$

Remark 4.1.1. An energy identity can be deduced from (4.17) by choosing $\boldsymbol{v} = \boldsymbol{J}$,

$$\int_\Omega (\mathrm{i}\omega\,\boldsymbol{T}\boldsymbol{J}\cdot\overline{\boldsymbol{J}} + \sigma^{-1}|\boldsymbol{J}|^2)\,d\boldsymbol{x} = V\int_S \overline{\boldsymbol{J}}\cdot\boldsymbol{n}\,ds. \tag{4.18}$$

Let us treat the first term in the left-hand side of (4.18). Using the Green formula, Eqs. (4.1), (4.3) and (4.8), we deduce

$$\int_\Omega \boldsymbol{T}\boldsymbol{J}\cdot\overline{\boldsymbol{J}}\,d\boldsymbol{x} = \int_\Omega \boldsymbol{A}\cdot\overline{\boldsymbol{J}}\,d\boldsymbol{x}$$
$$= \int_\Omega \boldsymbol{A}\cdot\mathbf{curl}\,\overline{\boldsymbol{H}}\,d\boldsymbol{x}$$
$$= \int_\Omega \mathbf{curl}\,\boldsymbol{A}\cdot\overline{\boldsymbol{H}}\,d\boldsymbol{x} - \int_\Gamma \boldsymbol{A}\times\boldsymbol{n}\cdot\overline{\boldsymbol{H}}\,ds$$
$$= \int_\Omega \boldsymbol{B}\cdot\overline{\boldsymbol{H}}\,d\boldsymbol{x} + \int_{\Omega_{\mathrm{ext}}} \overline{\boldsymbol{H}}\cdot\mathbf{curl}\,\boldsymbol{A}\,d\boldsymbol{x}$$
$$= \int_{\mathbb{R}^3} \boldsymbol{B}\cdot\overline{\boldsymbol{H}}\,d\boldsymbol{x}.$$

Using this result and (4.4), Identity (4.18) can then be written

$$\mathrm{i}\omega\int_{\mathbb{R}^3} \boldsymbol{B}\cdot\overline{\boldsymbol{H}}\,d\boldsymbol{x} + \int_\Omega \overline{\boldsymbol{J}}\cdot\boldsymbol{E}\,d\boldsymbol{x} = V\int_S \overline{\boldsymbol{J}}\cdot\boldsymbol{n}\,ds. \tag{4.19}$$

According to [139] and [74], the left-hand side of (4.19) is the total energy in the considered system (i.e. the sum of magnetic and electric energies), and the right-hand side is the product of the voltage V by the total current flowing in the conductor Ω.

We now turn to the proof of existence and uniqueness of a solution to (4.17). We have the following result.

Theorem 4.1.1. *Problem* (4.17) *has a unique solution* $\boldsymbol{J} \in \boldsymbol{\mathcal{X}}$.

Proof. The proof uses the Lax–Milgram theorem (Theorem 1.2.1).
Denoting by $\mathscr{B}$ the sesquilinear form

$$\mathscr{B}(\boldsymbol{J}, \boldsymbol{v}) := \mathrm{i}\omega \int_{\Omega} \boldsymbol{T}\boldsymbol{J} \cdot \overline{\boldsymbol{v}}\, d\boldsymbol{x} + \int_{\Omega} \sigma^{-1} \boldsymbol{J} \cdot \overline{\boldsymbol{v}}\, d\boldsymbol{x},$$

and by $\mathcal{L}$ the antilinear form

$$\mathcal{L}(\boldsymbol{v}) := V \int_{S} \overline{\boldsymbol{v}} \cdot \boldsymbol{n}\, ds,$$

Problem (4.17) reads

$$\mathscr{B}(\boldsymbol{J}, \boldsymbol{v}) = \mathscr{L}(\boldsymbol{v}) \qquad \forall\, \boldsymbol{v} \in \boldsymbol{\mathcal{X}}.$$

Now we deduce from (2.23) that the mapping $\boldsymbol{T}$ is continuous, that is,

$$\|\boldsymbol{T}\boldsymbol{g}\|_{\boldsymbol{\mathcal{Y}}} \le C\, \|\boldsymbol{g}\|_{\boldsymbol{\mathcal{X}}}.$$

This implies by using (2.11) that

$$\begin{aligned}
|\mathscr{B}(\boldsymbol{v}, \boldsymbol{w})| &\le \omega\, \|\boldsymbol{T}\boldsymbol{v}\|_{\boldsymbol{\mathcal{L}}^2(\Omega)}\, \|\boldsymbol{w}\|_{\boldsymbol{\mathcal{L}}^2(\Omega)} + \sigma_m^{-1}\, \|\boldsymbol{v}\|_{\boldsymbol{\mathcal{L}}^2(\Omega)}\, \|\boldsymbol{w}\|_{\boldsymbol{\mathcal{L}}^2(\Omega)} \\
&\le \omega\, \|\boldsymbol{T}\boldsymbol{v}\|_{\boldsymbol{\mathcal{Y}}}\, \|\boldsymbol{w}\|_{\boldsymbol{\mathcal{L}}^2(\Omega)} + \sigma_m^{-1}\, \|\boldsymbol{v}\|_{\boldsymbol{\mathcal{X}}}\, \|\boldsymbol{w}\|_{\boldsymbol{\mathcal{X}}} \\
&\le C_1\, \|\boldsymbol{v}\|_{\boldsymbol{\mathcal{X}}}\, \|\boldsymbol{w}\|_{\boldsymbol{\mathcal{X}}}.
\end{aligned}$$

Therefore the form $\mathscr{B}$ is continuous. Moreover, the continuity of $\mathscr{L}$ is obtained from the following bound for a $\boldsymbol{v} \in \boldsymbol{\mathcal{X}}$,

$$\begin{aligned}
|\mathscr{L}(\boldsymbol{v})| &\le |V|\, \|\boldsymbol{v} \cdot \boldsymbol{n}\|_{\mathcal{H}^{-\frac{1}{2}}(\Gamma)} \\
&\le C\, \big(\|\boldsymbol{v}\|^2_{\boldsymbol{\mathcal{L}}^2(\Omega)} + \|\operatorname{div} \boldsymbol{v}\|^2_{\mathcal{L}^2(\Omega)}\big)^{\frac{1}{2}} \\
&= C\, \|\boldsymbol{v}\|_{\boldsymbol{\mathcal{L}}^2(\Omega)}.
\end{aligned}$$

Let us prove that the form $\mathscr{B}$ is coercive on $\boldsymbol{\mathcal{X}} \times \boldsymbol{\mathcal{X}}$. Let $\boldsymbol{v}$ denote an element of $\boldsymbol{\mathcal{X}}$. We have

$$\mathscr{B}(\boldsymbol{v}, \boldsymbol{v}) = \mathrm{i}\omega \int_{\Omega} \boldsymbol{T}\boldsymbol{v} \cdot \overline{\boldsymbol{v}}\, d\boldsymbol{x} + \int_{\Omega} \sigma^{-1} |\boldsymbol{v}|^2\, d\boldsymbol{x}.$$

Seting $\boldsymbol{u} = \boldsymbol{T}\,\boldsymbol{v}$, we have from (4.13),

$$\int_{\mathbb{R}^3} \mu^{-1} |\,\mathbf{curl}\,\boldsymbol{u}|^2\,d\boldsymbol{x} = \int_{\Omega} \boldsymbol{v}\cdot\overline{\boldsymbol{u}}\,d\boldsymbol{x} = \int_{\Omega} \boldsymbol{v}\cdot\boldsymbol{T}\overline{\boldsymbol{v}}\,d\boldsymbol{x} = \int_{\Omega} \overline{\boldsymbol{v}}\cdot\boldsymbol{T}\,\boldsymbol{v}\,d\boldsymbol{x}.$$

It follows that

$$\mathscr{B}(\boldsymbol{v},\boldsymbol{v}) = \mathrm{i}\omega \int_{\mathbb{R}^3} \mu^{-1} |\,\mathbf{curl}\,\boldsymbol{u}|^2\,d\boldsymbol{x} + \int_{\Omega} \sigma^{-1} |\boldsymbol{v}|^2\,d\boldsymbol{x}.$$

Hence, by using (2.11),

$$\begin{aligned} |\mathscr{B}(\boldsymbol{v},\boldsymbol{v})|^2 &= \omega^2 \Big(\int_{\mathbb{R}^3} \mu^{-1} |\,\mathbf{curl}\,\boldsymbol{u}|^2\,d\boldsymbol{x} \Big)^2 + \Big(\int_{\Omega} \sigma^{-1} |\boldsymbol{v}|^2\,d\boldsymbol{x} \Big)^2 \\ &\geq \sigma_M^{-2}\, \|\boldsymbol{v}\|_{\boldsymbol{\mathcal{X}}}^4. \end{aligned}$$

The existence and uniqueness of a solution is then proved. □

Problem (4.17) has the advantage to be formulated in a bounded domain. The major difficulty relies on the evaluation of the operator $\boldsymbol{T}$.

A particular interesting case is the one of nonmagnetic materials, i.e. when $\mu = \mu_0$. In this case, $\boldsymbol{T}$ can be expressed as an integral operator by using (2.24). We have indeed in this case

$$\boldsymbol{T}\,\boldsymbol{J}(\boldsymbol{x}) = \mu_0 \int_{\Omega} G(\boldsymbol{x},\boldsymbol{y})\,\boldsymbol{J}(\boldsymbol{y})\,d\boldsymbol{y} \qquad \boldsymbol{x}\in\Omega,$$

where G is the Green function defined by (1.21). The current density vector field can then be obtained by solving the integral equation

$$\mathrm{i}\omega\mu_0 \int_{\Omega} G(\boldsymbol{x},\boldsymbol{y})\,\boldsymbol{J}(\boldsymbol{y})\,d\boldsymbol{y} + \sigma^{-1}\boldsymbol{J}(\boldsymbol{x}) = \nabla\varphi(\boldsymbol{x}) + V\nabla q(\boldsymbol{x}) \quad \boldsymbol{x}\in\Omega\setminus S. \tag{4.20}$$

This problem can be written in the variational formulation:

$$\left\{ \begin{aligned} &\text{Find } \boldsymbol{J}\in\boldsymbol{\mathcal{X}} \text{ such that} \\ &\int_{\Omega} \Big(\mathrm{i}\omega\mu_0 \int_{\Omega} G(\boldsymbol{x},\boldsymbol{y})\,\boldsymbol{J}(\boldsymbol{y})\,d\boldsymbol{y} + \sigma^{-1}\boldsymbol{J}(\boldsymbol{x}) \Big)\cdot\overline{\boldsymbol{v}}(\boldsymbol{x})\,d\boldsymbol{x} \\ &\qquad\qquad = V \int_S \overline{\boldsymbol{v}}\cdot\boldsymbol{n}\,ds \qquad \forall\,\boldsymbol{v}\in\boldsymbol{\mathcal{X}}. \end{aligned} \right. \tag{4.21}$$

For completeness let us interpret (4.21) as a partial differential equation in the whole space. We clearly have

$$\mathrm{i}\omega\boldsymbol{T}\,\boldsymbol{J} + \sigma^{-1}\boldsymbol{J} = V\boldsymbol{\delta}_S \qquad \text{in } \mathbb{R}^3,$$

where δ_S is the distribution supported by the surface S and defined by

$$\int_S \delta_S \cdot \overline{v}\, ds := \int_S \overline{v} \cdot n\, ds \qquad \forall\, v \in \mathscr{D}(\mathbb{R}^3), \tag{4.22}$$

where the left-hand side integral stands for the duality pairing between $\mathscr{D}'(\mathbb{R}^3)$ and $\mathscr{D}(\mathbb{R}^3)$.

The main difficulty in solving (4.21) comes here from the fact that we are in presence of a nonlocal problem formulated in Ω. Using a numerical method will then result in a dense matrix which severely restricts the performances of the method.

4.2 A Magnetic Field Formulation

Although the model presented in the previous section is inefficient from a computational point of view, its setting is helpful to derive a more feasible formulation. This one is due to Bossavit and Vérité (see [33, 34, 36, 171, 172]). We note that if $v \in \mathcal{X}$, then v can be extended to $\mathbb{R}^3$ by setting $v = 0$ in Ω_{ext} and then $\operatorname{div} v = 0$ in $\mathbb{R}^3$ since $v \cdot n = 0$ on Γ (see Theorem 1.3.1). We thus define the space

$$\tilde{\mathcal{X}} = \{v \in \mathcal{H}(\text{div}, \mathbb{R}^3);\ \operatorname{div} v = 0 \text{ in } \mathbb{R}^3,\ v = 0 \text{ in } \Omega_{\text{ext}}\}.$$

Let v denote a function in this space. We deduce from Theorem 1.3.4 the existence of a function $w \in \mathcal{W}^1(\mathbb{R}^3)$ such that

$$v = \operatorname{\mathbf{curl}} w,\ \operatorname{div} w = 0 \quad \text{in } \mathbb{R}^3.$$

By replacing v by $\operatorname{\mathbf{curl}} w$ and evaluating the first terms of (4.17) in function of the magnetic field instead of J, we have by setting $A = TJ$ and using (4.8), (4.3) and the Green formula in Ω and Ω_{ext} successively,

$$\begin{aligned}\int_\Omega TJ \cdot \operatorname{\mathbf{curl}} \overline{w}\, dx &= \int_\Omega \operatorname{\mathbf{curl}} A \cdot \overline{w}\, dx + \int_\Gamma A \times n \cdot \overline{w}\, ds \\ &= \int_\Omega \operatorname{\mathbf{curl}} A \cdot \overline{w}\, dx + \int_{\Omega_{\text{ext}}} \operatorname{\mathbf{curl}} A \cdot \overline{w}\, dx \\ &\quad - \int_{\Omega_{\text{ext}}} A \cdot \operatorname{\mathbf{curl}} \overline{w}\, dx \\ &= \int_{\mathbb{R}^3} \mu H \cdot \overline{w}\, dx.\end{aligned}$$

Finally, we have

$$\mathrm{i}\omega \int_{\mathbb{R}^3} \mu \boldsymbol{H} \cdot \overline{\boldsymbol{w}}\, d\boldsymbol{x} + \int_{\Omega} \sigma^{-1}\, \mathbf{curl}\, \boldsymbol{H} \cdot \mathbf{curl}\, \overline{\boldsymbol{w}}\, d\boldsymbol{x} = \tilde{\mathscr{L}}(\boldsymbol{w}),$$

where, according to (1.11),

$$\tilde{\mathscr{L}}(\boldsymbol{w}) := \mathscr{L}(\mathbf{curl}\, \boldsymbol{w}) = V \int_S \mathrm{curl}_S\, \overline{\boldsymbol{w}}\, ds.$$

Note that the appropriate choice for the magnetic field is the space

$$\mathcal{H} := \{\boldsymbol{v} \in \mathcal{H}(\mathbf{curl}, \mathbb{R}^3);\ \mathbf{curl}\, \boldsymbol{v} = 0 \text{ in } \Omega_{\mathrm{ext}}\}, \tag{4.23}$$

endowed with the norm

$$\|\boldsymbol{v}\|_{\mathcal{H}} := \left(\|\boldsymbol{v}\|^2_{\mathcal{L}^2(\mathbb{R}^3)} + \|\,\mathbf{curl}\, \boldsymbol{v}\|^2_{\mathcal{L}^2(\Omega)}\right)^{\frac{1}{2}}.$$

This yields the variational formulation:

$$\begin{cases} \text{Find } \boldsymbol{H} \in \mathcal{H} \text{ such that} \\ \mathrm{i}\omega \displaystyle\int_{\mathbb{R}^3} \mu \boldsymbol{H} \cdot \overline{\boldsymbol{w}}\, d\boldsymbol{x} + \int_{\Omega} \sigma^{-1}\, \mathbf{curl}\, \boldsymbol{H} \cdot \mathbf{curl}\, \overline{\boldsymbol{w}}\, d\boldsymbol{x} \\ \qquad\qquad = \tilde{\mathscr{L}}(\boldsymbol{w}) \quad \forall\, \boldsymbol{w} \in \mathcal{H}. \end{cases} \tag{4.24}$$

Remark 4.2.1. As it was previously mentioned, the integral that defines the form $\tilde{\mathscr{L}}$ actually stands for a duality pairing. The integrand is indeed well defined since for a function $\boldsymbol{w} \in \mathcal{H}(\mathbf{curl}, \mathbb{R}^3)$, we have $\mathrm{div}\, \mathbf{curl}\, \boldsymbol{w} = 0$ in the sense of distributions and then by Theorem 1.3.1, the function $\mathbf{curl}\, \boldsymbol{w} \cdot \boldsymbol{n}$ is well defined on S.

Theorem 4.2.1. *Problem* (4.24) *admits a unique solution* $\boldsymbol{H}$. *Moreover, if we define the fields* $\boldsymbol{J}, \boldsymbol{B}, \boldsymbol{E}$ *by* $\boldsymbol{J} = \mathbf{curl}\, \boldsymbol{H}$ *in* $\mathbb{R}^3$, $\boldsymbol{B} = \mu \boldsymbol{H}$ *in* $\mathbb{R}^3$ *and* $\boldsymbol{E} = \sigma^{-1} \boldsymbol{J}$ *in* Ω, *then the functions* $\boldsymbol{H}, \boldsymbol{J}, \boldsymbol{B}, \boldsymbol{E}$ *solve the system of equations* (4.1)–(4.6).

Proof. Let us define the sesquilinear form on $\mathcal{H}$,

$$\tilde{\mathscr{B}}(\boldsymbol{v}, \boldsymbol{w}) := \mathrm{i}\omega \int_{\mathbb{R}^3} \mu\, \boldsymbol{v} \cdot \overline{\boldsymbol{w}}\, d\boldsymbol{x} + \int_{\Omega} \sigma^{-1}\, \mathbf{curl}\, \boldsymbol{v} \cdot \mathbf{curl}\, \overline{\boldsymbol{w}}\, d\boldsymbol{x}.$$

We deduce the variational formulation

$$\text{Find } \boldsymbol{H} \in \mathcal{H} \quad \text{such that} \quad \tilde{\mathscr{B}}(\boldsymbol{H}, \boldsymbol{w}) = \tilde{\mathscr{L}}(\boldsymbol{w}) \qquad \forall\, \boldsymbol{w} \in \mathcal{H}.$$

Clearly, from (2.7) and (2.11), we deduce that the form $\mathscr{B}$ is continuous and coercive, i.e.

$$
\begin{aligned}
|\tilde{\mathscr{B}}(\boldsymbol{v},\boldsymbol{v})|^2 &= \omega^2\Big(\int_{\mathbb{R}^3}\mu|\boldsymbol{v}|^2\,d\boldsymbol{x}\Big)^2 + \Big(\int_{\Omega}\sigma^{-1}|\,\mathbf{curl}\,\boldsymbol{v}|^2\,d\boldsymbol{x}\Big)^2\\
&\geq \min\big(\omega^2\mu_m^2,\sigma_M^{-2}\big)\,\|\boldsymbol{v}\|_{\mathcal{H}}^4,
\end{aligned}
$$

where the constants μ_m and σ_M are defined in (2.7) and (2.11) respectively. The antilinear form $\tilde{\mathscr{L}}$ is continuous since

$$\operatorname{div}\mathbf{curl}\,\boldsymbol{w} = 0,$$

implies that $\mathbf{curl}\,\overline{\boldsymbol{w}} \in \mathcal{H}(\operatorname{div},\mathbb{R}^3)$ and then $\mathbf{curl}\,\overline{\boldsymbol{w}}\cdot\boldsymbol{n}_{|S}$ is well defined (see Theorem 1.3.1).

Existence and uniqueness are obtained thanks to the Lax–Milgram theorem (Theorem 1.2.1).

Let us now interpret this solution as the one of a set of partial differential equations. Taking in (4.24) a test function $\boldsymbol{v}\in\mathscr{D}(\mathbb{R}^3)$ with support contained in $\Omega\setminus S$, we obtain in the sense of distributions:

$$\mathrm{i}\omega\mu\boldsymbol{H} + \mathbf{curl}\,(\sigma^{-1}\,\mathbf{curl}\,\boldsymbol{H}) = 0 \quad \text{in } \Omega\setminus S.$$

Defining $\boldsymbol{E} = \sigma^{-1}\,\mathbf{curl}\,\boldsymbol{H}$, $\boldsymbol{B} = \mu\boldsymbol{H}$ in Ω and $\boldsymbol{J} = \sigma\boldsymbol{E}$ in $\mathbb{R}^3$, we retrieve (4.1) and (4.2).

Furthermore, since

$$
\begin{aligned}
\boldsymbol{B}_{|\Omega_{\text{ext}}} &= \mu_0\boldsymbol{H}_{|\Omega_{\text{ext}}} \in \mathcal{H}(\mathbf{curl},\Omega_{\text{ext}}),\\
\operatorname{div}\boldsymbol{H} &= \mu_0^{-1}\operatorname{div}\boldsymbol{B} = 0 \qquad \text{in } \Omega_{\text{ext}},
\end{aligned}
$$

then Theorem 1.3.2 implies $\boldsymbol{H}_{|\Omega_{\text{ext}}} \in \mathcal{W}^1(\Omega_{\text{ext}})$ and

$$|\boldsymbol{H}(\boldsymbol{x})| = \mathcal{O}(|\boldsymbol{x}|^{-1}) \qquad \text{when } |\boldsymbol{x}|\to\infty.$$

It remains to interpret the constant V. For this, we take $\boldsymbol{w} = \boldsymbol{H}$ in (4.24) and recall (4.19). □

It may be instructive to interpret (4.24) as a set of partial differential equation. For this, we start by writing for any smooth vector field $\boldsymbol{w}$, using the Stokes theorem,

$$\int_S \operatorname{curl}_S \boldsymbol{w}\,ds = \int_{\partial S} \boldsymbol{w}\cdot\boldsymbol{t}\,dl$$

where ∂S is the closed curve that represents the boundary of the surface S and $\boldsymbol{t}$ is the unit tangent vector to ∂S oriented according to the Ampère rule. A classical interpretation of (4.24) leads to:

$$\mathrm{i}\omega\mu\boldsymbol{H} + \mathbf{curl}\,\boldsymbol{E} = V\,\boldsymbol{\delta}_{\partial S}, \tag{4.25}$$

$$\mathbf{curl}\, \boldsymbol{H} = \boldsymbol{J}, \tag{4.26}$$

$$\boldsymbol{J} = \sigma \boldsymbol{E} \tag{4.27}$$

in $\mathbb{R}^3$, where $\boldsymbol{\delta}_{\partial S}$ is the vector delta distribution supported by the curve ∂S defined by

$$\int_{\mathbb{R}^3} \boldsymbol{\delta}_{\partial S} \cdot \boldsymbol{w}\, d\boldsymbol{x} = \int_{\partial S} \boldsymbol{w} \cdot \boldsymbol{t}\, d\ell \qquad \forall\, \boldsymbol{w} \in \mathscr{D}(\mathbb{R}^3). \tag{4.28}$$

We recall here that the integral on the left-hand side actually stands for the duality pairing between a distribution and a test function.

Problem (4.24) is not well suited for numerical treatment. The integral on $\mathbb{R}^3$ must indeed be transformed. To handle this drawback we consider integral representation of the external magnetic field $\boldsymbol{H}_{|\Omega_{\text{ext}}}$. We adopt for this end the technique used by Bossavit and Vérité in [36, 171, 172].

Let $\boldsymbol{w}$ denote a function in $\boldsymbol{\mathcal{H}}$. Following Theorem 1.3.6, there exist a function $\psi \in \mathcal{W}^1(\Omega_{\text{ext}})$ and a complex number β such that

$$\boldsymbol{w} = \nabla \psi + \beta \nabla p \qquad \text{in } \Omega_{\text{ext}} \setminus \Sigma,$$

where p is the unique solution of (1.20). We may then write for the solution $\boldsymbol{H} \in \boldsymbol{\mathcal{H}}$ the form

$$\boldsymbol{H} = \nabla \phi + \alpha \nabla p \qquad \text{in } \Omega_{\text{ext}} \setminus \Sigma, \tag{4.29}$$

where $\phi \in \mathcal{W}^1(\Omega_{\text{ext}})$ and $\alpha \in \mathbb{C}$. The first integral in (4.24) can be expanded as follows:

$$\begin{aligned}\int_{\mathbb{R}^3} \mu \boldsymbol{H} \cdot \overline{\boldsymbol{w}}\, d\boldsymbol{x} &= \int_{\Omega} \mu \boldsymbol{H} \cdot \overline{\boldsymbol{w}}\, d\boldsymbol{x} \\ &\quad + \mu_0 \overline{\beta} \int_{\Omega_{\text{ext}} \setminus \Sigma} \nabla \phi \cdot \nabla p\, d\boldsymbol{x} + \mu_0 \alpha \int_{\Omega_{\text{ext}} \setminus \Sigma} \nabla p \cdot \nabla \overline{\psi}\, d\boldsymbol{x} \\ &\quad + \mu_0 \int_{\Omega_{\text{ext}}} \nabla \phi \cdot \nabla \overline{\psi}\, d\boldsymbol{x} + \mu_0 \alpha \overline{\beta} \int_{\Omega_{\text{ext}} \setminus \Sigma} |\nabla p|^2\, d\boldsymbol{x}.\end{aligned}$$

Let us evaluate each of the above integrals. Since the set Ω_{ext} has as boundary $\Gamma \cup \Sigma$, we have by the Green formula and from the definition of (1.20):

$$\int_{\Omega_{\text{ext}} \setminus \Sigma} \nabla \phi \cdot \nabla p\, d\boldsymbol{x} = -\int_{\Omega_{\text{ext}} \setminus \Sigma} \phi\, \Delta p\, d\boldsymbol{x} - \int_{\Gamma} \phi \frac{\partial p}{\partial n}\, ds + \int_{\Sigma} \phi \left[\frac{\partial p}{\partial n}\right] ds = 0.$$

In the same way,

$$\int_{\Omega_{\text{ext}} \setminus \Sigma} \nabla p \cdot \nabla \overline{\psi}\, d\boldsymbol{x} = 0.$$

Finally, using again (1.20) and the Green formula,

$$\int_{\Omega_{\text{ext}}\setminus\Sigma} |\nabla p|^2\, d\boldsymbol{x} = -\int_{\Omega_{\text{ext}}\setminus\Sigma} p\,\Delta p\, d\boldsymbol{x} - \int_\Gamma \frac{\partial p}{\partial n}\, p\, ds + \int_\Sigma \frac{\partial p}{\partial n}\,[p]\, ds$$
$$= \int_\Sigma \frac{\partial p}{\partial n}\, ds.$$

The quantity

$$L := \mu_0 \int_{\Omega_{\text{ext}}\setminus\Sigma} |\nabla p|^2\, d\boldsymbol{x} = \mu_0 \int_\Sigma \frac{\partial p}{\partial n}\, ds \tag{4.30}$$

is called *self inductance* of the inductor Ω.

Clearly we have replaced the unknown magnetic field $\boldsymbol{H}$ in (4.24) by the unknowns $(\boldsymbol{H}_{|\Omega}, \phi, \alpha)$. In order to derive a new variational formulation, the space $\mathcal{H}$ is to be replaced by the following one:

$$\mathcal{K} := \big\{(\boldsymbol{w}, \psi, \beta) \in \mathcal{H}(\mathbf{curl}, \Omega) \times \mathcal{W}^1(\Omega_{\text{ext}}) \times \mathbb{C};$$
$$\boldsymbol{w} \times \boldsymbol{n} + \mathbf{curl}_\Gamma\, \psi + \beta\, \mathbf{curl}_\Gamma\, p = 0 \text{ on } \Gamma\big\}. \tag{4.31}$$

Note that the interface condition that operates in the definition of $\mathcal{K}$ is well defined by similar arguments to those of Remark 4.2.1. In fact, since $\psi \in \mathcal{W}^1(\Omega_{\text{ext}})$, then $\mathbf{curl}\,\nabla\psi = 0$ in the sense of distributions in Ω_{ext} and then $\nabla\psi \times \boldsymbol{n}$ is well defined owing to Theorem 1.3.1.

We obtain the equivalent variational formulation:

$$\begin{cases} \text{Find } (\boldsymbol{H}, \phi, \alpha) \in \mathcal{K} \text{ such that} \\ \mathrm{i}\omega \displaystyle\int_\Omega \mu \boldsymbol{H} \cdot \overline{\boldsymbol{w}}\, d\boldsymbol{x} + \mathrm{i}\omega\mu_0 \int_{\Omega_{\text{ext}}} \nabla\phi \cdot \nabla\overline{\psi}\, d\boldsymbol{x} + \mathrm{i}\omega L \alpha \overline{\beta} \\ \qquad + \displaystyle\int_\Omega \sigma^{-1}\, \mathbf{curl}\, \boldsymbol{H} \cdot \mathbf{curl}\, \overline{\boldsymbol{w}}\, d\boldsymbol{x} = \tilde{\mathscr{L}}(\boldsymbol{w}) \quad \forall\, (\boldsymbol{w}, \psi, \beta) \in \mathcal{K}. \end{cases} \tag{4.32}$$

Naturally, the above formulation is still not ready for numerical approximation. The integral over Ω_{ext} is indeed to be replaced by an integral representation. A classical interpretation of this formulation by choosing $\boldsymbol{w} = 0$, $\beta = 0$ and $\psi \in \mathscr{D}(\Omega_{\text{ext}})$ shows that ψ is actually harmonic in Ω_{ext} and then it can be represented on Γ by an integral equation. This will be done after proving that the formulation (4.32) is well posed.

Theorem 4.2.2. *Problem* (4.32) *has a unique solution. Moreover, if* $\boldsymbol{H}$ *is the solution of* (4.24), *then the triple* $(\boldsymbol{H}_{|\Omega}, \phi, \alpha)$, *where* $\phi \in \mathcal{W}^1(\Omega_{ext})$ *and* $\alpha \in \mathbb{C}$ *are such that* $\boldsymbol{H} = \nabla\phi + \alpha\nabla p$ *in* $\Omega_{ext} \setminus \Sigma$, *is solution of* (4.32). *Conversely, if*

$(\boldsymbol{H}, \phi, \alpha)$ *is solution of* (4.32), *then the function* $\tilde{\boldsymbol{H}}$ *defined by* $\tilde{\boldsymbol{H}} = \boldsymbol{H}$ *in* Ω *and* $\tilde{\boldsymbol{H}} = \nabla\phi + \alpha\nabla p$ *in* $\Omega_{ext} \setminus \Sigma$ *is solution of* (4.24).

Proof. Let $\boldsymbol{H}$ be the unique solution of (4.24). We have already proven that there exists $\phi \in \mathcal{W}^1(\mathbb{R}^3)$ and $\alpha \in \mathbb{C}$ such that the triple $(\boldsymbol{H}_{|\Omega_{\text{ext}}}, \phi, \alpha)$ is solution of (4.32).

Let now $(\boldsymbol{H}_{|\Omega_{\text{ext}}}, \phi, \alpha)$ denote a solution of (4.32) and define $\tilde{\boldsymbol{H}}$ by $\tilde{\boldsymbol{H}} = \boldsymbol{H}$ in Ω, $\tilde{\boldsymbol{H}} = \nabla\phi + \alpha\nabla p$ in $\Omega_{\text{ext}} \setminus \Sigma$. We remark that $\tilde{\boldsymbol{H}}$ can be extended as a function of $\mathcal{H}(\mathbf{curl}, \Omega_{\text{ext}})$ and we have $\mathbf{curl}\, \tilde{\boldsymbol{H}} = 0$ in Ω_{ext}. Moreover since $\boldsymbol{H} \times \boldsymbol{n} = (\nabla\phi + \alpha\nabla p) \times \boldsymbol{n}$ on Γ, we have $\tilde{\boldsymbol{H}} \in \mathcal{H}(\mathbf{curl}, \mathbb{R}^3)$ and with $\mathbf{curl}\, \tilde{\boldsymbol{H}} = 0$ in Ω_{ext} we obtain $\tilde{\boldsymbol{H}} \in \mathcal{H}$.

Now if $\boldsymbol{w} \in \mathcal{H}$, Theorem 1.3.6 implies that there exist $\psi \in \mathcal{W}^1(\mathbb{R}^3)$ and $\beta \in \mathbb{C}$ such that $\boldsymbol{w} = \nabla\psi + \beta\nabla p$ in $\Omega_{\text{ext}} \setminus \Sigma$. By computing

$$\begin{aligned} \mathrm{i}\omega\mu_0 \int_{\Omega_{\text{ext}}} \tilde{\boldsymbol{H}} \cdot \overline{\boldsymbol{w}}\, d\boldsymbol{x} &= \mathrm{i}\omega\mu_0 \int_{\Omega_{\text{ext}}} (\nabla\phi + \alpha\nabla p) \cdot (\nabla\overline{\psi} + \overline{\beta}\nabla p)\, d\boldsymbol{x} \\ &= \mathrm{i}\omega\mu_0 \int_{\Omega_{\text{ext}}} \nabla\phi \cdot \nabla\overline{\psi}\, d\boldsymbol{x} + \mathrm{i}\omega L\alpha\overline{\beta}, \end{aligned}$$

and with (4.32), we obtain that $\tilde{\boldsymbol{H}}$ is a solution of (4.24). □

Remark 4.2.2. The constant α in the variational formulation (4.32) can be interpreted as follows: The total current in the inductor is defined by

$$I = \int_S \mathbf{curl}\, \boldsymbol{H} \cdot \boldsymbol{n}\, ds = \int_S \mathrm{curl}_S\, \boldsymbol{H}\, ds.$$

Let ∂S stand for the boundary of the surface S and let $\boldsymbol{t}$ denote the unit tangent vector to $\boldsymbol{t}$ oriented according to the Ampère rule. We have from the interface condition on Γ since the vector $\boldsymbol{H} \times \boldsymbol{n}$ is tangential to Γ,

$$\boldsymbol{H} \cdot \boldsymbol{t} = \nabla\phi \cdot \boldsymbol{t} + \alpha\nabla p \cdot \boldsymbol{t} \qquad \text{on } \partial S.$$

We thus obtain by the Stokes theorem since p has a jump of unity on $\partial S \cap \overline{\Sigma}$,

$$\begin{aligned} I &= \int_{\partial S} \boldsymbol{H} \cdot \boldsymbol{t}\, ds \\ &= \int_{\partial S} \nabla\phi \cdot \boldsymbol{t}\, ds + \alpha \int_{\partial S} \nabla p \cdot \boldsymbol{t}\, ds \\ &= \alpha. \end{aligned}$$

The number α is then the total current in the inductor Ω.

We may note that although (4.32) is stated in Ω and Ω_{ext}, the unknown function ϕ is harmonic in Ω_{ext} and thus can be represented by an integral formulation over Γ. For this end, we resort to using the Steklov–Poincaré operator as defined in Sect. 1.3.5. From the expansion (4.29) we deduce since div $\boldsymbol{H} = 0$ in Ω_{ext}, that ϕ is harmonic in Ω_{ext} and then by the Green formula,

$$\int_{\Omega_{\text{ext}}} \nabla\phi \cdot \nabla\overline{\psi}\, dx = -\int_\Gamma \frac{\partial\phi}{\partial n}\,\overline{\psi}\, ds = \int_\Gamma (P\phi)\,\overline{\psi}\, ds,$$

where the operator P is defined by (1.38).

Let us rewrite (4.32) using this. We first replace the space $\mathcal{K}$ by the following one:

$$\mathcal{K}_\Gamma := \big\{(\boldsymbol{w}, \psi, \beta) \in \boldsymbol{\mathcal{H}}(\mathbf{curl}, \Omega) \times \mathcal{H}^{\frac{1}{2}}(\Gamma) \times \mathbb{C};$$
$$\boldsymbol{w} \times \boldsymbol{n} + \mathbf{curl}_\Gamma\, \psi + \beta\, \mathbf{curl}_\Gamma\, p = 0 \text{ on } \Gamma\big\}. \quad (4.33)$$

Note that the surface functions $\mathbf{curl}_\Gamma\, \psi$ and $\mathbf{curl}_\Gamma\, p$ are well defined for $\psi \in \mathcal{H}^{\frac{1}{2}}(\Gamma)$ and $p \in \mathcal{H}^{\frac{1}{2}}(\Gamma \setminus \partial\Sigma)$ (see [62], Vol. 4, p. 136 for instance).

We have the equivalent variational formulation to (4.32),

$$\begin{cases} \text{Find } (\boldsymbol{H}, \phi, \alpha) \in \mathcal{K}_\Gamma \text{ such that} \\ \mathrm{i}\omega \displaystyle\int_\Omega \mu \boldsymbol{H} \cdot \overline{\boldsymbol{w}}\, d\boldsymbol{x} + \mathrm{i}\omega\mu_0 \int_\Gamma (P\phi)\overline{\psi}\, ds + \mathrm{i}\omega L\alpha\overline{\beta} \\ \qquad + \displaystyle\int_\Omega \sigma^{-1}\, \mathbf{curl}\, \boldsymbol{H} \cdot \mathbf{curl}\, \overline{\boldsymbol{w}}\, d\boldsymbol{x} = \tilde{\mathscr{L}}(\boldsymbol{w}) \quad \forall\, (\boldsymbol{w}, \psi, \beta) \in \mathcal{K}_\Gamma. \end{cases} \quad (4.34)$$

It remains now to give a more easy-to-compute formula to calculate the inductance coefficient. Let us for this define the "Surface Current"

$$\boldsymbol{J}_\Gamma := \mathbf{curl}_\Gamma\, p = \boldsymbol{n} \times \nabla p \quad \text{on } \Gamma.$$

We next prove the following preliminary lemma:

Lemma 4.2.1. *Let* $\boldsymbol{a}$ *stand for the vector field*

$$\boldsymbol{a}(\boldsymbol{x}) = \int_\Gamma \boldsymbol{J}_\Gamma(\boldsymbol{y}) G(\boldsymbol{x}, \boldsymbol{y})\, ds(\boldsymbol{y}) \quad \boldsymbol{x} \in \Omega_{ext}, \quad (4.35)$$

$$\text{with } G(\boldsymbol{x}, \boldsymbol{y}) = \frac{1}{4\pi} \frac{1}{|\boldsymbol{x} - \boldsymbol{y}|}.$$

We have the following properties:

$$\operatorname{div} \boldsymbol{a} = 0, \quad (4.36)$$

$$\boldsymbol{curl}\, \boldsymbol{a} = \nabla p, \tag{4.37}$$

in Ω_{ext}.

Proof. We have for $\boldsymbol{x} \in \Omega_{\text{ext}}$ by the Green formula:

$$\operatorname{div} \boldsymbol{a}(\boldsymbol{x}) = \int_\Gamma \nabla_x \cdot (\boldsymbol{J}_\Gamma(\boldsymbol{y})\, G(\boldsymbol{x}, \boldsymbol{y}))\, ds(\boldsymbol{y}) = \int_\Gamma \nabla_x G(\boldsymbol{x}, \boldsymbol{y}) \cdot \boldsymbol{J}_\Gamma(\boldsymbol{y})\, ds(\boldsymbol{y}).$$

Using the identity

$$\nabla_y G(\boldsymbol{x}, \boldsymbol{y}) = -\nabla_x G(\boldsymbol{x}, \boldsymbol{y}),$$

and the property $[\nabla p \times \boldsymbol{n}] = 0$ on Γ, we obtain, again by the Green formula, for $\boldsymbol{x} \in \Omega_{\text{ext}}$,

$$\begin{aligned}
\operatorname{div} \boldsymbol{a}(\boldsymbol{x}) &= -\int_\Gamma \nabla_y G(\boldsymbol{x}, \boldsymbol{y}) \cdot (\nabla p(\boldsymbol{y}) \times \boldsymbol{n}(\boldsymbol{y}))\, ds(\boldsymbol{y}) \\
&= \int_{\Omega_{\text{ext}}} \operatorname{div}_y \big(\nabla_y G(\boldsymbol{x}, \boldsymbol{y}) \times \nabla p(\boldsymbol{y})\big)\, d\boldsymbol{y} \\
&= -\int_{\Omega_{\text{ext}}} \big(\nabla_y G(\boldsymbol{x}, \boldsymbol{y}) \cdot \mathbf{curl}\, \nabla p(\boldsymbol{y}) - \nabla p(\boldsymbol{y}) \cdot \mathbf{curl}_y\, \nabla_y G(\boldsymbol{x}, \boldsymbol{y})\big)\, d\boldsymbol{y} \\
&= 0.
\end{aligned}$$

This proves (4.36).

Defining $\boldsymbol{w} = \mathbf{curl}\, \boldsymbol{a}$, we get

$$\begin{aligned}
\boldsymbol{w}(\boldsymbol{x}) &= \int_\Gamma \mathbf{curl}_x \big(G(\boldsymbol{x}, \boldsymbol{y})\, \boldsymbol{J}_\Gamma(\boldsymbol{y})\big)\, ds(\boldsymbol{y}) \\
&= \int_\Gamma \nabla_x G(\boldsymbol{x}, \boldsymbol{y}) \times \boldsymbol{J}_\Gamma(\boldsymbol{y})\, ds(\boldsymbol{y}) \\
&= \int_\Gamma (\nabla p(\boldsymbol{y}) \cdot \nabla_x G(\boldsymbol{x}, \boldsymbol{y}))\, \boldsymbol{n}(\boldsymbol{y})\, ds(\boldsymbol{y}) \\
&\quad - \int_\Gamma \big(\boldsymbol{n}(\boldsymbol{y}) \cdot \nabla_x G(\boldsymbol{x}, \boldsymbol{y})\big)\, \nabla p(\boldsymbol{y})\, ds(\boldsymbol{y}).
\end{aligned}$$

Since

$$\nabla_y G(\boldsymbol{x}, \boldsymbol{y}) = -\nabla_x G(\boldsymbol{x}, \boldsymbol{y}),$$

then

$$w(x) = -\int_\Gamma (\nabla p(y) \cdot \nabla_y G(x, y))\, n(y)\, ds(y) + \int_\Gamma \frac{\partial G}{\partial n_y}(x, y)\, \nabla p(y)\, ds(y)$$
$$= -I_1(x) + I_2(x). \tag{4.38}$$

To evaluate I_1, we first make use of the Gradient theorem to get

$$\begin{aligned} I_1(x) &= -\int_{\Omega_{\text{ext}}} \nabla_y(\nabla p(y) \cdot \nabla_y G(x, y))\, dy \\ &= -\int_{\Omega_{\text{ext}}} (\nabla p(y) \cdot \nabla_y)\, \nabla_y G(x, y)\, dy \\ &\quad - \int_{\Omega_{\text{ext}}} (\nabla_y G(x, y) \cdot \nabla_y)\, \nabla p(y)\, dy. \end{aligned} \tag{4.39}$$

We have for the first integral in I_1 by the Green formula and (1.20),

$$\begin{aligned} &\int_{\Omega_{\text{ext}}} (\nabla p(y) \cdot \nabla_y)\, \nabla_y G(x, y)\, dy \\ &= -\int_{\Omega_{\text{ext}} \setminus \Sigma} \nabla_y G(x, y)\, \Delta p(y)\, dy - \int_\Gamma \frac{\partial p}{\partial n}(y)\, \nabla_y G(x, y)\, ds(y) \\ &\quad - \int_\Sigma \Big[\frac{\partial p}{\partial n}(y)\Big] \nabla_y G(x, y)\, ds(y) = 0. \end{aligned} \tag{4.40}$$

For I_2, we have by the Green formula

$$\begin{aligned} I_2(x) &= -\int_{\Omega_{\text{ext}}} \nabla p(y)\, \Delta_y G(x, y)\, dy - \int_{\Omega_{\text{ext}}} (\nabla_y G(x, y) \cdot \nabla_y)\, \nabla p(y)\, dy \\ &= \nabla p(x) - \int_{\Omega_{\text{ext}}} (\nabla_y G(x, y) \cdot \nabla_y)\, \nabla p(y)\, dy. \end{aligned} \tag{4.41}$$

Collecting (4.39), (4.40) and (4.41) in (4.38), we obtain $w = \nabla p$ in Ω_{ext} and then (4.37). □

Now we have the following result (See also Bossavit–Vérité [36]).

Theorem 4.2.3. *The self inductance L defined by* (4.30), *is also given by*

$$L = \frac{\mu_0}{4\pi} \int_\Gamma \int_\Gamma \frac{J_\Gamma(x) \cdot J_\Gamma(y)}{|x - y|}\, ds(x)\, ds(y). \tag{4.42}$$

Proof. We first note that since the function p is in $\mathcal{H}^1_{\text{loc}}(\Omega_{\text{ext}})$, then $\mathbf{curl}\,\nabla p = 0$ and then, thanks to (1.3.1), the trace $\nabla p \times \boldsymbol{n}$ on Γ is well defined.

Let us use Lemma 4.2.1 and let $\boldsymbol{a}$ be the vector field defined by (4.35), we can write using (4.30), (4.37) and the Green formula,

$$\begin{aligned}
L &= \mu_0 \int_{\Omega_{\text{ext}}\setminus\Sigma} |\nabla p|^2 \, d\boldsymbol{y} \\
&= \mu_0 \int_{\Omega_{\text{ext}}\setminus\Sigma} |\,\mathbf{curl}\,\boldsymbol{a}|^2 \, d\boldsymbol{y} \\
&= \mu_0 \int_{\Omega_{\text{ext}}} \operatorname{div}(\boldsymbol{a} \times \mathbf{curl}\,\boldsymbol{a}) \, d\boldsymbol{y} + \mu_0 \int_{\Omega_{\text{ext}}} \boldsymbol{a} \cdot \mathbf{curl}\,\mathbf{curl}\,\boldsymbol{a} \, d\boldsymbol{y} \\
&= \mu_0 \int_{\Omega_{\text{ext}}} \operatorname{div}(\boldsymbol{a} \times \mathbf{curl}\,\boldsymbol{a}) \, d\boldsymbol{y} + \mu_0 \int_{\Omega_{\text{ext}}} \boldsymbol{a} \cdot \mathbf{curl}\,\nabla p \, d\boldsymbol{y} \\
&= -\mu_0 \int_{\Gamma} (\boldsymbol{a} \times \mathbf{curl}\,\boldsymbol{a}) \cdot \boldsymbol{n} \, ds \\
&= -\mu_0 \int_{\Gamma} (\boldsymbol{a} \times \nabla p) \cdot \boldsymbol{n} \, ds \\
&= \mu_0 \int_{\Gamma} \boldsymbol{a} \cdot \boldsymbol{J}_\Gamma \, ds \\
&= \frac{\mu_0}{4\pi} \int_{\Gamma}\int_{\Gamma} \frac{\boldsymbol{J}_\Gamma(\boldsymbol{x}) \cdot \boldsymbol{J}_\Gamma(\boldsymbol{y})}{|\boldsymbol{x} - \boldsymbol{y}|} \, ds(\boldsymbol{y}) \, ds(\boldsymbol{x}). \qquad \square
\end{aligned}$$

4.3 An Electric Field Model

In a similar way to the magnetic field model, we can derive a model using electric field as unknown. Electric field models are difficult to use in the cases where an unknown conductor zone is present. The obtained model involves indeed the electric conductivity σ rather than its inverse (electric resistivity) and the equation naturally makes sense in the zone where $\sigma = 0$, i.e. in nonconducting regions.

We have chosen here to present the simple model where a source current density is prescribed. We show later in a remark that prescribing the voltage in such models is more delicate since the right-hand side of the resulting variational formulation is not well defined without modification of the used functional space.

The starting point is the system of equations (2.57)–(2.61). Written in terms of the electric field $\boldsymbol{E}$, we obtain the problem:

$$\mathrm{i}\omega\sigma \boldsymbol{E} + \mathbf{curl}\,(\mu^{-1}\,\mathbf{curl}\,\boldsymbol{E}) = -\mathrm{i}\omega \boldsymbol{J}_0 \qquad \text{in } \mathbb{R}^3, \tag{4.43}$$

$$|\boldsymbol{E}(\boldsymbol{x})| + |\,\mathbf{curl}\,\boldsymbol{E}(\boldsymbol{x})| = \mathcal{O}(|\boldsymbol{x}|^{-1}) \qquad \text{for } |\boldsymbol{x}| \to \infty. \tag{4.44}$$

Here the source current $\boldsymbol{J}_0$ is a function with support in the conductors Ω, given in $\boldsymbol{\mathcal{L}}^2(\mathbb{R}^3)$ with $\operatorname{div} \boldsymbol{J}_0 = 0$.

It is clear that Eqs. (4.43)–(4.44) are not enough to ensure uniqueness of a solution. The electric field is indeed determined up to a gradient in the exterior domain (where $\sigma = 0$). The following gauge conditions are generally supplied for this:

$$\operatorname{div} \boldsymbol{E} = 0 \qquad \text{in } \Omega_{\text{ext}}, \tag{4.45}$$

$$\int_{\Gamma_k} \boldsymbol{E}_{|\Omega_{\text{ext}}} \cdot \boldsymbol{n}\, ds = 0 \qquad k = 1, 2, \ldots \tag{4.46}$$

where the Γ_k are the boundaries of the connected components Ω_k of Ω. We note that Condition (4.45) is natural since we have from the Maxwell equations (see (2.6), (2.9)) the equation

$$\operatorname{div}(\varepsilon \boldsymbol{E}) = \varrho_q \qquad \text{in } \mathbb{R}^3.$$

Assuming that the empty space contains no charges ($\varrho_q = 0$) and that the electric permittivity is constant ($\varepsilon = \varepsilon_0$), we retrieve (4.45).

In the following we shall use the formulation studied by Ammari et al. [13] and later by Hiptmair [100].

4.3.1 A Formulation in the Whole Space

Following [100], we define the space

$$\boldsymbol{\mathcal{V}} := \Big\{ \boldsymbol{v};\ \frac{\boldsymbol{v}}{1 + |\boldsymbol{x}|} \in \boldsymbol{\mathcal{L}}^2(\mathbb{R}^3);\ \mathbf{curl}\, \boldsymbol{v} \in \boldsymbol{\mathcal{L}}^2(\mathbb{R}^3), \\ \operatorname{div} \boldsymbol{v} = 0 \text{ in } \Omega_{\text{ext}},\ \int_{\Gamma_k} \boldsymbol{v} \cdot \boldsymbol{n}\, ds = 0,\ k = 1, 2, \ldots \Big\}.$$

We have for $\boldsymbol{E}$ the following variational formulation:

$$\begin{cases} \text{Find } \boldsymbol{E} \in \boldsymbol{\mathcal{V}} \text{ such that} \\ \mathrm{i}\omega \displaystyle\int_{\Omega} \sigma \boldsymbol{E} \cdot \overline{\boldsymbol{v}}\, d\boldsymbol{x} + \int_{\mathbb{R}^3} \mu^{-1}\, \mathbf{curl}\, \boldsymbol{E} \cdot \mathbf{curl}\, \overline{\boldsymbol{v}}\, d\boldsymbol{x} \\ \qquad = -\mathrm{i}\omega \displaystyle\int_{\Omega} \boldsymbol{J}_0 \cdot \overline{\boldsymbol{v}}\, d\boldsymbol{x} \qquad \forall\, \boldsymbol{v} \in \boldsymbol{\mathcal{V}}. \end{cases} \tag{4.47}$$

Remark 4.3.1. We have chosen for this problem the use of the weighted Sobolev space $\mathcal{W}(\mathbf{curl}, \mathbb{R}^3)$ where

$$\mathcal{W}(\mathbf{curl}, X) := \Big\{ \boldsymbol{v};\ \frac{\boldsymbol{v}}{1+|\boldsymbol{x}|} \in \mathcal{L}^2(X),\ \mathbf{curl}\, \boldsymbol{v} \in \mathcal{L}^2(X) \Big\},$$

in which lie functions that behave as $\mathcal{O}(|\boldsymbol{x}|^{-1})$ when $|\boldsymbol{x}| \to \infty$. It can be shown however (see [13]) that the resulting electric field behaves like $|\boldsymbol{x}|^{-2}$ when $|\boldsymbol{x}| \to \infty$.

The following result is proved in [100].

Theorem 4.3.1. *Problem* (4.47) *has a unique solution* $\boldsymbol{E} \in \mathcal{V}$.

Like for the $\boldsymbol{H}$–model, the variational formulation is not ready for producing a practical numerical scheme. The integral over $\mathbb{R}^3$ has to be decomposed indeed into integrals over Ω and Ω_{ext} and the latter one is difficult to handle. In [100] considers a symmetric coupling of the interior and the exterior fields. We have chosen here to present a formulation that couples the electric field in the conductor with the magnetic field in the free space. This is presented in the next section.

4.3.2 A Coupled Interior–Exterior Formulation

We consider an original formulation of the electric field model that couples a scalar potential related to the magnetic field in the vacuum to the $\boldsymbol{E}$–formulation in the conductors. This formulation is borrowed from [158] which is obtained for a bounded domain, and adapted here to the whole space.

We start from the variational formulation (4.47) and consider the case without a source current but a prescribed voltage or total current. Using the Green formula and the property

$$\mathbf{curl}\,\mathbf{curl}\, \boldsymbol{E} = 0 \qquad \text{in } \Omega_{\text{ext}},$$

we obtain for all $\boldsymbol{v} \in \mathcal{W}(\mathbf{curl}, \Omega_{\text{ext}})$

$$\int_{\Omega_{\text{ext}}} \mathbf{curl}\, \boldsymbol{E} \cdot \mathbf{curl}\, \overline{\boldsymbol{v}}\, dx = - \int_{\Gamma} \mathbf{curl}\, \boldsymbol{E} \times \boldsymbol{n} \cdot \overline{\boldsymbol{v}}\, ds.$$

We recall that $\mu = \mu_0$ is constant in Ω_{ext} (See Chap. 2). Then (4.47) becomes, with $\boldsymbol{J}_0 = 0$:

$$\int_{\Omega} \big(\mathrm{i}\omega\sigma \boldsymbol{E} \cdot \overline{\boldsymbol{v}} + \mu^{-1}\, \mathbf{curl}\, \boldsymbol{E} \cdot \mathbf{curl}\, \overline{\boldsymbol{v}}\big)\, dx \\ - \int_{\Gamma} \mu^{-1}\, \mathbf{curl}\, \boldsymbol{E} \times \boldsymbol{n} \cdot \overline{\boldsymbol{v}}\, ds = 0 \qquad \forall\, \boldsymbol{v} \in \mathcal{H}(\mathbf{curl}, \Omega).$$

Using the Faraday equation (4.2) and the interface condition $[\boldsymbol{H} \times \boldsymbol{n}] = 0$ on Γ, we obtain for all $\boldsymbol{v} \in \mathcal{H}(\mathbf{curl}, \Omega)$,

$$\int_{\Omega} (\mathrm{i}\omega\sigma \boldsymbol{E} \cdot \overline{\boldsymbol{v}} + \mu^{-1} \,\mathbf{curl}\, \boldsymbol{E} \cdot \mathbf{curl}\, \overline{\boldsymbol{v}}) d\boldsymbol{x} - \mathrm{i}\omega \int_{\Gamma} \overline{\boldsymbol{v}} \times \boldsymbol{n} \cdot \boldsymbol{H}_{|\Omega_{\text{ext}}} \, ds = 0.$$

We use the decomposition (4.29) to obtain for all $\boldsymbol{v} \in \mathcal{H}(\mathbf{curl}, \Omega)$,

$$\int_{\Omega} \left(\mathrm{i}\omega\sigma \boldsymbol{E} \cdot \overline{\boldsymbol{v}} + \mu^{-1} \,\mathbf{curl}\, \boldsymbol{E} \cdot \mathbf{curl}\, \overline{\boldsymbol{v}}\right) d\boldsymbol{x} - \mathrm{i}\omega \int_{\Gamma} \overline{\boldsymbol{v}} \times \boldsymbol{n} \cdot \nabla\phi \, ds - \mathrm{i}\omega\alpha \int_{\Gamma} \overline{\boldsymbol{v}} \times \boldsymbol{n} \cdot \nabla p \, ds = 0. \quad (4.48)$$

Let us consider the equation in Ω_{ext}. We have in the sense of distributions, when a voltage V is applied, the Eq. (4.25). Taking the scalar product of (4.25) with $\nabla\psi$, where $\psi \in \mathcal{W}^1(\Omega_{\text{ext}})$, we get

$$\mathrm{i}\omega \int_{\Omega_{\text{ext}}} \mu \boldsymbol{H} \cdot \nabla\overline{\psi} \, d\boldsymbol{x} + \int_{\Omega_{\text{ext}}} \mathbf{curl}\, \boldsymbol{E} \cdot \nabla\overline{\psi} \, d\boldsymbol{x} = V \int_{\partial S} \nabla\overline{\psi} \cdot \boldsymbol{t} \, d\ell = 0. \quad (4.49)$$

Using again the decomposition (4.29) and the Green formula, we get for all $\psi \in \mathcal{W}^1(\Omega_{\text{ext}})$

$$\mathrm{i}\omega \int_{\Omega_{\text{ext}}} \mu \nabla\phi \cdot \nabla\overline{\psi} \, d\boldsymbol{x} + \int_{\Gamma} \boldsymbol{E} \times \boldsymbol{n} \cdot \nabla\overline{\psi} \, ds = 0. \quad (4.50)$$

Let us finally take the scalar product of (4.25) with ∇p where p is defined by (1.20), we have

$$\mathrm{i}\omega \int_{\Omega_{\text{ext}} \setminus \Sigma} \mu \boldsymbol{H} \cdot \nabla p \, d\boldsymbol{x} + \int_{\Omega_{\text{ext}} \setminus \Sigma} \mathbf{curl}\, \boldsymbol{E} \cdot \nabla p \, d\boldsymbol{x} = V \int_{\partial S} \nabla p \cdot \boldsymbol{t} \, d\ell = V.$$

The decomposition (4.29) yields

$$\mathrm{i}\omega \int_{\Omega_{\text{ext}} \setminus \Sigma} \mu \nabla\phi \cdot \nabla p \, d\boldsymbol{x} + \mathrm{i}\omega\alpha \int_{\Omega_{\text{ext}} \setminus \Sigma} \mu \, |\nabla p|^2 \, d\boldsymbol{x} + \int_{\Gamma} \boldsymbol{E} \times \boldsymbol{n} \cdot \nabla p \, ds = V.$$

Using the Green formula, the definition (1.20) of p and the definition (4.30) of the self-inductance we get

$$\mathrm{i}\omega\alpha L + \int_{\Gamma} \boldsymbol{E} \times \boldsymbol{n} \cdot \nabla p \, ds = V. \quad (4.51)$$

We now gather (4.48)–(4.51) to obtain for all $(\boldsymbol{v}, \psi) \in \boldsymbol{\mathcal{H}}(\mathbf{curl}, \Omega) \times W^1(\Omega_{\mathrm{ext}})$:

$$\mathrm{i}\omega \int_\Omega \sigma \boldsymbol{E} \cdot \overline{\boldsymbol{v}}\, d\boldsymbol{x} + \int_\Omega \mu^{-1}\, \mathbf{curl}\, \boldsymbol{E} \cdot \mathbf{curl}\, \overline{\boldsymbol{v}}\, d\boldsymbol{x} - \mathrm{i}\omega \int_\Gamma \overline{\boldsymbol{v}} \times \boldsymbol{n} \cdot \nabla\phi\, ds = \mathrm{i}\omega\alpha \int_\Gamma \overline{\boldsymbol{v}} \times \boldsymbol{n} \cdot \nabla p\, ds, \tag{4.52}$$

$$\mathrm{i}\omega \int_{\Omega_{\mathrm{ext}}} \mu \nabla\phi \cdot \nabla\overline{\psi}\, d\boldsymbol{x} + \int_\Gamma \boldsymbol{E} \times \boldsymbol{n} \cdot \nabla\overline{\psi}\, ds = 0, \tag{4.53}$$

$$\mathrm{i}\omega\alpha L + \int_\Gamma \boldsymbol{E} \times \boldsymbol{n} \cdot \nabla p = V. \tag{4.54}$$

We note that the formulation (4.52)–(4.54) enables prescribing either current voltage V or total current intensity α.

The total current model is thus given by looking for $(\boldsymbol{E}, \phi) \in \boldsymbol{\mathcal{H}}(\mathbf{curl}, \Omega) \times W^1(\Omega_{\mathrm{ext}})$ such that for all $(\boldsymbol{v}, \psi) \in \boldsymbol{\mathcal{H}}(\mathbf{curl}, \Omega) \times W^1(\Omega_{\mathrm{ext}})$,

$$\mathrm{i}\omega \int_\Omega \sigma \boldsymbol{E} \cdot \overline{\boldsymbol{v}}\, d\boldsymbol{x} + \int_\Omega \mu^{-1}\, \mathbf{curl}\, \boldsymbol{E} \cdot \mathbf{curl}\, \overline{\boldsymbol{v}}\, d\boldsymbol{x} - \mathrm{i}\omega \int_\Gamma \overline{\boldsymbol{v}} \times \boldsymbol{n} \cdot \nabla\phi\, ds = \mathrm{i}\omega\alpha \int_\Gamma \overline{\boldsymbol{v}} \times \boldsymbol{n} \cdot \nabla p\, ds, \tag{4.55}$$

$$\omega^2 \int_{\Omega_{\mathrm{ext}}} \mu \nabla\phi \cdot \nabla\overline{\psi}\, d\boldsymbol{x} - \mathrm{i}\omega \int_\Gamma \boldsymbol{E} \times \boldsymbol{n} \cdot \nabla\overline{\psi}\, ds = 0, \tag{4.56}$$

where $\alpha \in \mathbb{C}$ is the applied total current. Likewise, we can prescribe the voltage $V \in \mathbb{C}$ and look for $(\boldsymbol{E}, \phi, \alpha) \in \boldsymbol{\mathcal{H}}(\mathbf{curl}, \Omega) \times W^1(\Omega_{\mathrm{ext}}) \times \mathbb{C}$ such that for all $(\boldsymbol{v}, \psi, \beta) \in \boldsymbol{\mathcal{H}}(\mathbf{curl}, \Omega) \times W^1(\Omega_{\mathrm{ext}}) \times \mathbb{C}$,

$$\mathrm{i}\omega \int_\Omega \sigma \boldsymbol{E} \cdot \overline{\boldsymbol{v}}\, d\boldsymbol{x} + \int_\Omega \mu^{-1}\, \mathbf{curl}\, \boldsymbol{E} \cdot \mathbf{curl}\, \overline{\boldsymbol{v}}\, d\boldsymbol{x} - \mathrm{i}\omega \int_\Gamma \overline{\boldsymbol{v}} \times \boldsymbol{n} \cdot \nabla\phi\, ds - \mathrm{i}\omega\alpha \int_\Gamma \overline{\boldsymbol{v}} \times \boldsymbol{n} \cdot \nabla p\, ds = 0, \tag{4.57}$$

$$\omega^2 \int_{\Omega_{\mathrm{ext}}} \mu \nabla\phi \cdot \nabla\overline{\psi}\, d\boldsymbol{x} - \mathrm{i}\omega \int_\Gamma \boldsymbol{E} \times \boldsymbol{n} \cdot \nabla\overline{\psi}\, ds = 0, \tag{4.58}$$

$$\omega^2 L\alpha\overline{\beta} - \mathrm{i}\omega\overline{\beta} \int_\Gamma \boldsymbol{E} \times \boldsymbol{n} \cdot \nabla p\, ds = -\mathrm{i}\omega\overline{\beta}\, V. \tag{4.59}$$

Let us prove that the variational formulations (4.55)–(4.56) and (4.57)–(4.59) define well posed problems.

Theorem 4.3.2. *Problem* (4.55)–(4.56) *has a unique solution*

$$(\boldsymbol{E}, \phi) \in \boldsymbol{\mathcal{H}}(\boldsymbol{curl}, \Omega) \times W^1(\Omega_{ext}).$$

Proof. In view of applying the Lax-Milgram theorem, we consider an equivalent of the variational formulation (4.55)–(4.56) obtained by multiplying (4.55) by the complex number $(1 - \mathrm{i})$, (4.56) by $(1 + \mathrm{i})$ and adding. We define the sesquilinear form

$$\begin{aligned}\mathscr{B}((\boldsymbol{w}, \theta); (\boldsymbol{v}, \psi)) := (1+\mathrm{i})\,\omega \int_\Omega \sigma \boldsymbol{w} \cdot \overline{\boldsymbol{v}}\, d\boldsymbol{x} + (1-\mathrm{i}) \int_\Omega \mu^{-1}\, \mathbf{curl}\, \boldsymbol{w} \cdot \mathbf{curl}\, \overline{\boldsymbol{v}}\, d\boldsymbol{x} \\ + (1+\mathrm{i})\,\omega^2 \int_{\Omega_{\mathrm{ext}}} \mu\, \nabla\theta \cdot \nabla\overline{\psi}\, d\boldsymbol{x} \\ + \omega \left((1-\mathrm{i}) \int_\Gamma \boldsymbol{w} \times \boldsymbol{n} \cdot \nabla\overline{\psi}\, ds - (1+\mathrm{i}) \int_\Gamma \overline{\boldsymbol{v}} \times \boldsymbol{n} \cdot \nabla\theta\, ds \right),\end{aligned}$$

and the antilinear form

$$\mathscr{L}((\boldsymbol{v}, \psi)) := (1+\mathrm{i})\,\omega\alpha \int_\Gamma \overline{\boldsymbol{v}} \times \boldsymbol{n} \cdot \nabla p\, ds.$$

Problem (4.55)–(4.56) reads then

$$\mathscr{B}((\boldsymbol{E}, \phi); (\boldsymbol{v}, \psi)) = \mathscr{L}((\boldsymbol{v}, \psi)) \qquad \forall\ (\boldsymbol{v}, \psi) \in \boldsymbol{\mathcal{H}}(\mathbf{curl}, \Omega) \times \mathcal{W}^1(\Omega_{\mathrm{ext}}).$$

We have for $(\boldsymbol{v}, \psi) \in \boldsymbol{\mathcal{H}}(\mathbf{curl}, \Omega) \times \mathcal{W}^1(\Omega_{\mathrm{ext}})$

$$\begin{aligned}\operatorname{Re} \mathscr{B}((\boldsymbol{v}, \psi); (\boldsymbol{v}, \psi)) = \omega \int_\Omega \sigma |\boldsymbol{v}|^2\, d\boldsymbol{x} + \int_\Omega \mu^{-1} |\,\mathbf{curl}\, \boldsymbol{v}|^2\, d\boldsymbol{x} \\ + \omega^2 \int_{\Omega_{\mathrm{ext}}} \mu |\nabla\psi|^2\, d\boldsymbol{x}.\end{aligned}$$

Therefore by using (2.7), (2.11) and (1.6),

$$\begin{aligned}|\mathscr{B}((\boldsymbol{v}, \psi); (\boldsymbol{v}, \psi))| &\geq \operatorname{Re} \mathscr{B}((\boldsymbol{v}, \psi); (\boldsymbol{v}, \psi)) \\ &\geq \omega\sigma_m \|\boldsymbol{v}\|^2_{\boldsymbol{\mathcal{L}}^2(\Omega)} + \mu_M^{-1} \|\,\mathbf{curl}\, \boldsymbol{v}\,\|^2_{\boldsymbol{\mathcal{L}}^2(\Omega)} \\ &\quad + \omega^2 \mu_m \|\nabla\psi\|^2_{\boldsymbol{\mathcal{L}}^2(\Omega_{\mathrm{ext}})} \\ &\geq C \left(\|\boldsymbol{v}\|^2_{\boldsymbol{\mathcal{H}}(\mathbf{curl}, \Omega)} + \|\psi\|^2_{\mathcal{W}^1(\Omega_{\mathrm{ext}})} \right).\end{aligned}$$

The form $\mathscr{B}$ is thus coercive.

For the form $\mathscr{L}$ we have by the same method

$$\mathscr{L}((\boldsymbol{v}, \psi)) = \mathrm{i}\omega\alpha \int_\Gamma \overline{\boldsymbol{v}} \times \boldsymbol{n} \cdot \nabla p\, ds = -\mathrm{i}\omega\alpha \int_{\Omega_{\mathrm{ext}} \setminus \Sigma} \mathbf{curl}\, \overline{\tilde{\boldsymbol{v}}} \cdot \nabla p\, d\boldsymbol{x},$$

where $\tilde{v}$ is an extension of v to Ω_{ext} such that $\tilde{v} \in \mathcal{H}(\mathbf{curl}, \mathbb{R}^3)$. Such an extension exists and we have (see [100]):

$$\|\,\mathbf{curl}\,\tilde{v}\|_{\mathcal{L}^2(\mathbb{R}^3)} \leq C_1\|v \times n\|_{\mathcal{H}^{-\frac{1}{2}}(\Gamma)} \leq C_2\|v\|_{\mathcal{H}(\mathbf{curl},\Omega)}. \tag{4.60}$$

We have thanks to (4.60), for all $(v,\psi) \in \mathcal{H}(\mathbf{curl},\Omega) \times \mathcal{W}^1(\Omega_{\text{ext}})$,

$$\begin{aligned}|\mathscr{L}((v,\psi))| &\leq \omega\,|\alpha|\left|\int_{\Omega_{\text{ext}}\setminus\Sigma} \mathbf{curl}\,\tilde{v}\cdot\nabla p\,dx\right| \\ &\leq C_3\|\nabla p\|_{\mathcal{L}^2(\Omega_{\text{ext}}\setminus\Sigma)}\|v\|_{\mathcal{H}(\mathbf{curl},\Omega_{\text{ext}})}.\end{aligned}$$

Therefore $\mathscr{L}$ is continuous and the Lax–Milgram theorem (Theorem 1.2.1) ensures then existence and uniqueness of (E,ϕ). □

Remark 4.3.2. Once the variational problem is solved, the voltage can be deduced from (4.54).

Theorem 4.3.3. *Problem* (4.57)–(4.59) *has a unique solution*

$$(E,\phi,\alpha) \in \mathcal{H}(\mathbf{curl},\Omega) \times \mathcal{W}^1(\Omega_{ext}) \times \mathbb{C}.$$

Proof. Multiplying (4.57) by the complex number (1−i), (4.58) by (1+i) and (4.59) by (1 + i) and adding the three equations, we are led to the variational formulation:

$$\mathscr{B}((E,\phi,\alpha);(v,\psi,\beta)) = \mathscr{L}((v,\psi,\beta))$$

for all $(v,\psi,\beta) \in \mathcal{H}(\mathbf{curl},\Omega) \times \mathcal{W}^1(\Omega_{\text{ext}}) \times \mathbb{C}$ where

$$\begin{aligned}\mathscr{B}((w,\theta,\gamma);(v,\psi,\beta)) &= (1+\mathrm{i})\,\omega\int_\Omega \sigma w\cdot\overline{v}\,dx \\ &+ (1-\mathrm{i})\int_\Omega \mu^{-1}\,\mathbf{curl}\,w\cdot\mathbf{curl}\,\overline{v}\,dx - (1+\mathrm{i})\,\omega\int_\Gamma \overline{v}\times n\cdot\nabla\theta\,ds \\ &- (1+\mathrm{i})\,\omega\alpha\int_\Gamma \overline{v}\times n\cdot\nabla p\,ds + (1+\mathrm{i})\,\omega^2\int_{\Omega_{\text{ext}}} \mu\,\nabla\theta\cdot\nabla\overline{\psi}\,dx \\ &+ (1-\mathrm{i})\,\omega\int_\Gamma w\times n\cdot\nabla\overline{\psi}\,ds + (1+\mathrm{i})\,\omega^2 L\gamma\overline{\beta} \\ &+ (1-\mathrm{i})\,\omega\overline{\beta}\int_\Gamma w\times n\cdot\nabla p\,ds, \\ \mathscr{L}((v,\psi,\beta)) &= (1-\mathrm{i})\,\omega\overline{\beta}V.\end{aligned}$$

Using the same tools as for the proof of Theorem 4.3.2, we can see that the antilinear form $\mathscr{L}$ and the sesquilinear form $\mathscr{B}$ are continuous on the space $\mathcal{H}(\mathbf{curl},\Omega)\,\times$

$\mathcal{W}^1(\Omega_{\text{ext}}) \times \mathbb{C}$. Let us prove the coercivity of $\mathscr{B}$. We have

$$\operatorname{Re} \mathscr{B}\big((\boldsymbol{v}, \psi, \beta); (\boldsymbol{v}, \psi, \beta)\big) = \omega \int_{\Omega} \sigma |\boldsymbol{v}|^2 \, d\boldsymbol{x} + \int_{\Omega} \mu^{-1} |\, \mathbf{curl}\, \boldsymbol{v}|^2 \, d\boldsymbol{x}$$
$$+ \omega^2 \int_{\Omega_{\text{ext}}} \mu \, |\nabla \psi|^2 \, d\boldsymbol{x} + \omega^2 L \, |\beta|^2.$$

Thus, by (2.7), (2.11) and (1.6),

$$\begin{aligned}\big|\mathscr{B}\big((\boldsymbol{v}, \psi, \beta); (\boldsymbol{v}, \psi, \beta)\big)\big| &\geq \operatorname{Re} \mathscr{B}\big((\boldsymbol{v}, \psi, \beta); (\boldsymbol{v}, \psi, \beta)\big) \\ &\geq \omega \sigma_m \|\boldsymbol{v}\|^2_{\boldsymbol{\mathcal{L}}^2(\Omega)} + \mu_M^{-1} \|\, \mathbf{curl}\, \boldsymbol{v}\|^2_{\boldsymbol{\mathcal{L}}^2(\Omega)} \\ &\quad + \omega^2 \mu_m \|\nabla \psi\|^2_{\boldsymbol{\mathcal{L}}^2(\Omega_{\text{ext}})} + \omega^2 L^2 |\beta|^2 \\ &\geq C \, \big(\|\boldsymbol{v}\|^2_{\boldsymbol{\mathcal{H}}(\mathbf{curl}, \Omega)} + \|\psi\|^2_{\mathcal{W}^1(\Omega_{\text{ext}})} + |\beta|^2\big).\end{aligned}$$

This proves that $\mathscr{B}$ is coercive. The antilinear form $\mathscr{L}$ is continuous since we have

$$\big|\mathscr{L}\big((\boldsymbol{v}, \psi, \beta)\big)\big| \leq \omega |V| \, |\beta|.$$

The Lax–Milgram theorem (Theorem 1.2.1) enables to conclude. □

We may observe that, just like for (4.47), problems (4.55)–(4.56) and (4.57)–(4.59) are still posed in the whole space $\mathbb{R}^3$, the difference being that we have a scalar elliptic problem in the vacuum. This difficulty can be removed, like for the magnetic field model (4.34) by using the Steklov–Poincaré operator which enables representing the scalar potential ϕ by an integral equation on Γ. We omit the details since the approach is similar.

Remark 4.3.3. The variational formulations (4.55)–(4.56) and (4.57)–(4.59) can be viewed as variants of the ones given in [158] (problems (17) and (18)) for the case of unbounded domains.

Chapter 5
Axisymmetric Models

The present chapter deals with the derivation of some mathematical models for eddy current setups with symmetry of rotation. The symmetry is assumed for the geometry and data as well. For a better clarity, we proceed as we have done so far: we present a formal derivation of the models and then, once a mathematical problem is defined, we establish a rigorous mathematical result of existence and uniqueness of a solution.

We present in the following two mathematical models:

- The first one is obtained as an adapted version of the model developed in Chap. 4 to the axisymmetric case. This problem has as unknown the magnetic field. We shall see that the axisymmetric problem involves only two components of this field.
- The second one uses the vector potential $\boldsymbol{A}$ as unknown. This choice is motivated by the fact that, in the case of axial symmetry, $\boldsymbol{A}$ has only one nonvanishing component and can then be identified to a scalar field.

5.1 Axisymmetric Setting

Let us consider the cylindrical coordinate system

$$\boldsymbol{\Phi} : (r, \theta, z) \in \mathbb{R}^+ \times [0, 2\pi) \times \mathbb{R} \mapsto \boldsymbol{\Phi}(r, \theta, z) = \boldsymbol{x} = r \cos\theta \boldsymbol{e}_1 + r \sin\theta \boldsymbol{e}_2 + z\boldsymbol{e}_3,$$

where we recall that $(\boldsymbol{e}_i)$ is the canonical orthogonal basis of $\mathbb{R}^3$. Let now $(\boldsymbol{e}_r, \boldsymbol{r}_\theta, \boldsymbol{e}_z)$ stand for the normalized natural tangent system associated to the cylindrical system, that is

$$\begin{aligned} \boldsymbol{e}_r &= \cos\theta\, \boldsymbol{e}_1 + \sin\theta\, \boldsymbol{e}_2, \\ \boldsymbol{e}_\theta &= -\sin\theta\, \boldsymbol{e}_1 + \cos\theta\, \boldsymbol{e}_2, \\ \boldsymbol{e}_z &= \boldsymbol{e}_3. \end{aligned}$$

R. Touzani and J. Rappaz, *Mathematical Models for Eddy Currents and Magnetostatics: With Selected Applications*, Scientific Computation, DOI 10.1007/978-94-007-0202-8_5,

To any scalar field $u : \mathbb{R}^3 \to \mathbb{C}$, we associate the function

$$\breve{u}(r,\theta,z) := u(\boldsymbol{\Phi}(r,\theta,z)) \qquad (r,\theta,z) \in \mathbb{R}^+ \times [0,2\pi) \times \mathbb{R}.$$

To any vector field $\boldsymbol{v} : \mathbb{R}^3 \to \mathbb{C}^3$, we associate the vector function

$$\breve{\boldsymbol{v}}(r,\theta,z) := \boldsymbol{v}(\boldsymbol{\Phi}(r,\theta,z)) \qquad (r,\theta,z) \in \mathbb{R}^+ \times [0,2\pi) \times \mathbb{R},$$

that is

$$\begin{aligned}\breve{\boldsymbol{v}}(r,\theta,z) &= \breve{v}_r(r,\theta,z)\,\boldsymbol{e}_r + \breve{v}_\theta(r,\theta,z)\,\boldsymbol{e}_\theta + \breve{v}_z(r,\theta,z)\,\boldsymbol{e}_z \\ &= v_1(\boldsymbol{x})\,\boldsymbol{e}_1 + v_2(\boldsymbol{x})\,\boldsymbol{e}_2 + v_3(\boldsymbol{x})\,\boldsymbol{e}_3,\end{aligned}$$

where $\boldsymbol{x} = \boldsymbol{\Phi}(r,\theta,z)$. Let us recall the well known formulae:

$$\begin{aligned}\nabla u &= \frac{\partial \breve{u}}{\partial r}\,\boldsymbol{e}_r + \frac{1}{r}\frac{\partial \breve{u}}{\partial \theta}\,\boldsymbol{e}_\theta + \frac{\partial \breve{u}}{\partial z}\,\boldsymbol{e}_z, \\ \operatorname{div} \boldsymbol{v} &= \frac{1}{r}\Big(\frac{\partial}{\partial r}(r\breve{v}_r) + \frac{\partial \breve{v}_\theta}{\partial \theta} + r\frac{\partial \breve{v}_z}{\partial z}\Big), \\ \mathbf{curl}\,\boldsymbol{v} &= \frac{1}{r}\Big(\frac{\partial \breve{v}_z}{\partial \theta} - r\frac{\partial \breve{v}_\theta}{\partial z}\Big)\boldsymbol{e}_r + \Big(\frac{\partial \breve{v}_r}{\partial z} - \frac{\partial \breve{v}_z}{\partial r}\Big)\boldsymbol{e}_\theta + \frac{1}{r}\Big(\frac{\partial}{\partial r}(r\breve{v}_\theta) - \frac{\partial \breve{v}_r}{\partial \theta}\Big)\boldsymbol{e}_z, \\ \Delta u &= \frac{1}{r}\Big(\frac{\partial}{\partial r}\Big(r\frac{\partial \breve{u}}{\partial r}\Big) + \frac{1}{r}\frac{\partial^2 \breve{u}}{\partial \theta^2} + r\frac{\partial^2 \breve{u}}{\partial z^2}\Big).\end{aligned}$$

We shall now give some definitions on the notion of axisymmetry applied to functions, domains and function spaces.

Definition 5.1.1. The scalar field u (*resp.* vector field $\boldsymbol{v}$) is said to be *Axisymmetric* if $\breve{u}$ (*resp.* $\breve{v}_r, \breve{v}_\theta, \breve{v}_z$) does not depend on θ. In this case we write simply

$$\breve{u} = \breve{u}(r,z) \quad \text{and} \quad \breve{\boldsymbol{v}} = \breve{v}_r(r,z)\,\boldsymbol{e}_r + \breve{v}_\theta(r,z)\,\boldsymbol{e}_\theta + \breve{v}_z(r,z)\,\boldsymbol{e}_z.$$

Note that if u and $\boldsymbol{v}$ are axisymmetric then we have

$$\begin{aligned}\nabla u &= \frac{\partial \breve{u}}{\partial r}\,\boldsymbol{e}_r + \frac{\partial \breve{u}}{\partial z}\,\boldsymbol{e}_z, \\ \operatorname{div} \boldsymbol{v} &= \frac{1}{r}\frac{\partial}{\partial r}(r\breve{v}_r) + \frac{\partial \breve{v}_z}{\partial z}, \\ \mathbf{curl}\,\boldsymbol{v} &= -\frac{\partial \breve{v}_\theta}{\partial z}\,\boldsymbol{e}_r + \Big(\frac{\partial \breve{v}_r}{\partial z} - \frac{\partial \breve{v}_z}{\partial r}\Big)\boldsymbol{e}_\theta + \frac{1}{r}\frac{\partial}{\partial r}(r\breve{v}_\theta)\,\boldsymbol{e}_z, \\ \Delta u &= \frac{1}{r}\frac{\partial}{\partial r}\Big(r\frac{\partial \breve{u}}{\partial r}\Big) + \frac{\partial^2 \breve{u}}{\partial z^2},\end{aligned}$$

$$\begin{aligned}\nabla \boldsymbol{v} &= \frac{\partial \breve{v}_r}{\partial r}\, \boldsymbol{e}_r \otimes \boldsymbol{e}_r - \frac{\breve{v}_\theta}{r}\, \boldsymbol{e}_r \otimes \boldsymbol{e}_\theta + \frac{\partial \breve{v}_r}{\partial z}\, \boldsymbol{e}_r \otimes \boldsymbol{e}_z \\ &+ \frac{\partial \breve{v}_\theta}{\partial r}\, \boldsymbol{e}_\theta \otimes \boldsymbol{e}_r + \frac{\breve{v}_r}{r}\, \boldsymbol{e}_\theta \otimes \boldsymbol{e}_\theta + \frac{\partial \breve{v}_\theta}{\partial z}\, \boldsymbol{e}_\theta \otimes \boldsymbol{e}_z \\ &+ \frac{\partial \breve{v}_z}{\partial r}\, \boldsymbol{e}_z \otimes \boldsymbol{e}_r + \frac{\partial \breve{v}_z}{\partial z}\, \boldsymbol{e}_z \otimes \boldsymbol{e}_z.\end{aligned}$$

Definition 5.1.2. We shall say that a set $\Omega \subset \mathbb{R}^3$ is *axisymmetric* if, for any $\boldsymbol{x} = \boldsymbol{\Phi}(r, \theta, z) \in \Omega$ we have

$$\boldsymbol{\Phi}(r, \zeta, z) \in \Omega \qquad \forall\, \zeta \in [0, 2\pi).$$

When this is the case, we define the set

$$\begin{aligned}\breve{\Omega} &= \{(r, z) \in \mathbb{R}^+ \times \mathbb{R};\ \boldsymbol{\Phi}(r, \zeta, z) \in \Omega,\ \forall\, \zeta \in [0, 2\pi)\} \\ &= \{(r, z) \in \mathbb{R}^+ \times \mathbb{R};\ \boldsymbol{\Phi}(r, 0, z) \in \Omega\}.\end{aligned}$$

Furthermore, if Ω is axisymmetric and Γ is its boundary then we define

$$\begin{aligned}\breve{\Gamma} &= \{(r, z) \in \mathbb{R}^+ \times \mathbb{R};\ \boldsymbol{\Phi}(r, \zeta, z) \in \Gamma,\ \forall\, \zeta \in [0, 2\pi)\} \\ &= \{(r, z) \in \mathbb{R}^+ \times \mathbb{R};\ \boldsymbol{\Phi}(r, 0, z) \in \Gamma\}.\end{aligned}$$

Definition 5.1.3. Let $\Omega \subset \mathbb{R}^3$ denote an axisymmetric domain and let $\mathcal{W}$ stand for a Banach space, equipped with the norm $\|\cdot\|_{\mathcal{W}}$, of axisymmetric scalar functions $u : \Omega \to \mathbb{C}$ (*resp.* $\boldsymbol{\mathcal{W}}$ is a Banach space of vector functions $\boldsymbol{v} : \Omega \to \mathbb{C}^3$ equipped with the norm $\|\cdot\|_{\boldsymbol{\mathcal{W}}}$). We define the space

$$\breve{\mathcal{W}} := \{\breve{u} : \breve{\Omega} \to \mathbb{C};\ \exists\, u \in \mathcal{W} \text{ with } \breve{u}(r, z) = u(\boldsymbol{\Phi}(r, \theta, z))\ \forall\, \theta \in [0, 2\pi)\}$$

with its norm $\|\breve{u}\|_{\breve{\mathcal{W}}} := \|u\|_{\mathcal{W}}$, (respectively

$$\breve{\boldsymbol{\mathcal{W}}} := \{\breve{\boldsymbol{v}} : \breve{\Omega} \to \mathbb{C}^3;\ \exists\, \boldsymbol{v} \in \boldsymbol{\mathcal{W}} \text{ with } \breve{\boldsymbol{v}}(r, z) = \boldsymbol{v}(\boldsymbol{\Phi}(r, \theta, z))\ \forall\, \theta \in [0, 2\pi)\},$$

with its norm $\|\breve{\boldsymbol{u}}\|_{\breve{\boldsymbol{\mathcal{W}}}} := \|\boldsymbol{u}\|_{\boldsymbol{\mathcal{W}}}$.)

Thus, if $u \in \mathcal{L}^2(\Omega)$ is an axisymmetric function, we have

$$\int_{\Omega} |u(\boldsymbol{x})|^2\, d\boldsymbol{x} = 2\pi \int_{\breve{\Omega}} |\breve{u}(r, z)|^2\, r\, dr\, dz.$$

This introduces the space

$$\breve{\mathcal{L}}^2(\breve{\Omega}) = \Big\{\breve{u} : \breve{\Omega} \to \mathbb{C};\ \int_{\breve{\Omega}} |\breve{u}(r,z)|^2\, r\, dr\, dz < \infty\Big\},$$

endowed with the norm

$$\|\breve{u}\|_{\breve{\mathcal{L}}^2(\breve{\Omega})} := \Big(2\pi \int_{\breve{\Omega}} |\breve{u}(r,z)|^2\, r\, dr\, dz\Big)^{\frac{1}{2}}.$$

As an example, if Ω is an axisymmetric bounded domain of $\mathbb{R}^3$ and $\Omega_{\text{ext}} := \mathbb{R}^3 \setminus \overline{\Omega}$, then Ω_{ext} is axisymmetric and

$$\breve{\mathcal{H}}^1(\breve{\Omega}) = \Big\{\breve{u} : \breve{\Omega} \to \mathbb{C};\ \breve{u},\ \frac{\partial \breve{u}}{\partial r}, \frac{\partial \breve{u}}{\partial z} \in \breve{\mathcal{L}}^2(\breve{\Omega})\Big\},$$

$$\breve{\mathcal{W}}^1(\breve{\Omega}_{\text{ext}}) = \left\{\breve{u} : \breve{\Omega}_{\text{ext}} \to \mathbb{C};\ \frac{|\breve{u}|}{1 + (r^2 + z^2)^{\frac{1}{2}}},\ \frac{\partial \breve{u}}{\partial r}, \frac{\partial \breve{u}}{\partial z} \in \breve{\mathcal{L}}^2(\breve{\Omega}_{\text{ext}})\right\}.$$

The spaces $\breve{\mathcal{H}}^1(\breve{\Omega})$ and $\breve{\mathcal{W}}^1(\breve{\Omega}_{\text{ext}})$ are respectively equipped with the norms

$$\|\breve{u}\|_{\breve{\mathcal{H}}^1(\breve{\Omega})} := \|u\|_{\mathcal{H}^1(\Omega)}, \quad \|\breve{u}\|_{\breve{\mathcal{W}}^1(\breve{\Omega}_{\text{ext}})} := \|u\|_{\mathcal{W}^1(\Omega_{\text{ext}})}.$$

5.2 A Magnetic Field Model

Let us consider a model configuration of conductors as presented in Chap. 4 with axisymmetric domains. Figure 5.1 illustrates a typical configuration in the space of cartesian coordinates and its corresponding configuration in the (r,z)–space. We recall that $\Omega = \Omega_1 \cup \Omega_2$ where Ω_1 is the inductor and Ω_2 is the workpiece.

We seek axisymmetric solutions of (2.62)–(2.68) that experience the property

$$\begin{aligned} &\breve{\boldsymbol{J}}(r,z) = J_\theta(r,z)\, \boldsymbol{e}_\theta \quad \text{for } (r,z) \in \mathbb{R}^+ \times \mathbb{R}, \\ &\text{with } J_\theta(r,z) = 0 \text{ for } (r,z) \in \breve{\Omega}_{\text{ext}}. \end{aligned} \tag{5.1}$$

Remark that this choice readily implies in particular $\operatorname{div} \boldsymbol{J} = 0$ in $\mathbb{R}^3$.

We start by proving the following result.

Theorem 5.2.1. *Let $\boldsymbol{J}, \boldsymbol{E}, \boldsymbol{B}, \boldsymbol{H}$ denote smooth axisymmetric vector fields that satisfy* (2.62)–(2.68) *such that $\boldsymbol{J}$ satisfies* (5.1). *Then we have for $\breve{\boldsymbol{E}}$ and $\breve{\boldsymbol{H}}$ the expressions,*

$$\breve{\boldsymbol{E}}(r,z) = E_\theta(r,z)\, \boldsymbol{e}_\theta \qquad \textit{for } (r,z) \in \breve{\Omega}_1 \cup \breve{\Omega}_2, \tag{5.2}$$

$$\breve{\boldsymbol{H}}(r,z) = H_r(r,z)\, \boldsymbol{e}_r + H_z(r,z)\, \boldsymbol{e}_z \qquad \textit{for } (r,z) \in \mathbb{R}^+ \times \mathbb{R}. \tag{5.3}$$

Fig. 5.1 An axisymmetric configuration of conductors

Proof. From (5.1), (2.66) implies readily (5.2). In addition, from (2.62) and since H_r, H_θ and H_z are independent of θ, we obtain

$$\frac{\partial H_\theta}{\partial r} = \frac{\partial H_\theta}{\partial z} = 0,$$

which implies, since $H_\theta = \mathcal{O}((r^2 + z^2)^{-\frac{1}{2}})$ when $(r^2 + z^2) \to \infty$, that $H_\theta = 0$ in $\mathbb{R}^+ \times \mathbb{R}$. □

Let us present a mathematical model that is a simple adaptation of the three-dimensional model obtained in Chap. 4 to the axisymmetric case. Let us for this, consider (4.24) and use the particular form (5.3) of the magnetic field.

Combining the expression of **curl** $\boldsymbol{H}$ for an axisymmetric magnetic field $\boldsymbol{H}$ satisfying (5.3), the variational equation (4.24) can be written

$$\begin{aligned}
2\mathrm{i}\pi\omega \int_{\mathbb{R}^+ \times \mathbb{R}} \mu\,(H_r \overline{v}_r + H_z \overline{v}_z)\, r\, dr\, dz \\
+ 2\pi \int_{\breve{\Omega}} \sigma^{-1} \Big(\frac{\partial H_r}{\partial z} - \frac{\partial H_z}{\partial r}\Big)\Big(\frac{\partial \overline{v}_r}{\partial z} - \frac{\partial \overline{v}_z}{\partial r}\Big)\, r\, dr\, dz \\
= V \int_{\breve{\Omega}_1} \Big(\frac{\partial \overline{v}_r}{\partial z} - \frac{\partial \overline{v}_z}{\partial r}\Big)\, dr\, dz,
\end{aligned}$$

for all smooth vector fields (v_r, v_z) that satisfy

$$\frac{\partial v_r}{\partial z} - \frac{\partial v_z}{\partial r} = 0 \qquad \text{in } \breve{\Omega}_{\text{ext}}.$$

Here we recall that V stands for current voltage. Note that the cut in the inductor Ω_1 is chosen as the surface is defined by

$$S = \{\boldsymbol{\Phi}(r, 0, z);\ (r, z) \in \breve{\Omega}_1\},$$

and the unit normal to this surface is the vector $\boldsymbol{e}_\theta$.

To complete the mathematical setting, we define the function spaces

$$\breve{\mathcal{U}} := \Big\{\boldsymbol{v} = v_r\boldsymbol{e}_r + v_z\boldsymbol{e}_z \in \breve{\mathcal{H}}(\text{curl}; \mathbb{R}^+\!\times \mathbb{R});\ \frac{\partial v_r}{\partial z} - \frac{\partial v_z}{\partial r} = 0 \text{ in } \breve{\Omega}_{\text{ext}}\Big\},$$

where

$$\begin{aligned}\breve{\mathcal{H}}(\text{curl}; X) &:= \{\breve{\boldsymbol{v}} = v_r\,\boldsymbol{e}_r + v_z\,\boldsymbol{e}_z \in \breve{\mathcal{H}}(\textbf{curl}; X)\}\\ &= \Big\{\boldsymbol{v} = v_r\,\boldsymbol{e}_r + v_z\,\boldsymbol{e}_z;\ \breve{\boldsymbol{v}} \in \breve{\mathcal{L}}^2(X),\ \frac{\partial v_r}{\partial z} - \frac{\partial v_z}{\partial r} \in \breve{L}^2(X)\Big\}.\end{aligned}$$

This space is clearly a Hilbert space when endowed with the norm

$$\begin{aligned}\|\breve{\boldsymbol{v}}\|_{\breve{\mathcal{U}}} &:= \Big(\int_{\mathbb{R}^+\!\times\mathbb{R}} (|v_r|^2 + |v_z|^2)\, r\,dr\,dz + \int_{\breve{\Omega}} \Big|\frac{\partial v_r}{\partial z} - \frac{\partial v_z}{\partial r}\Big|^2 r\,dr\,dz\Big)^{\frac{1}{2}}\\ &= \Big(\|\breve{\boldsymbol{v}}\|^2_{\breve{\mathcal{L}}^2(X)} + \Big\|\frac{\partial v_r}{\partial z} - \frac{\partial v_z}{\partial r}\Big\|^2_{\breve{L}^2(X)}\Big)^{\frac{1}{2}}.\end{aligned}$$

The variational problem (4.24) has then as axisymmetric version:

$$\begin{cases}\text{Find } (H_r, H_z) \in \breve{\mathcal{U}} \text{ such that}\\ 2i\pi\omega \displaystyle\int_{\mathbb{R}^+\!\times\mathbb{R}} \mu\,(H_r\overline{v}_r + H_z\overline{v}_z)\, r\,dr\,dz\\ \qquad + 2\pi \displaystyle\int_{\breve{\Omega}} \sigma^{-1}\Big(\frac{\partial H_r}{\partial z} - \frac{\partial H_z}{\partial r}\Big)\Big(\frac{\partial \overline{v}_r}{\partial z} - \frac{\partial \overline{v}_z}{\partial r}\Big)\, r\,dr\,dz\\ \qquad = V \displaystyle\int_{\breve{\Omega}_1} \Big(\frac{\partial \overline{v}_r}{\partial z} - \frac{\partial \overline{v}_z}{\partial r}\Big)\,dr\,dz \qquad \forall\ (v_r, v_z) \in \breve{\mathcal{U}}.\end{cases} \tag{5.4}$$

Remark 5.2.1. It is clear that since the domain Ω_1 is the inductor, it must be a loop. Combining this argument with the symmetry of rotations we obtain the property

$$\overline{\Omega_1} \cap \{\boldsymbol{x} \in \mathbb{R}^3;\ x_1^2 + x_2^2 = 0\} = \emptyset.$$

Consequently, for $(r, z) \in \breve{\Omega}_1$, the radius r can be bounded from below by a positive constant.

Theorem 5.2.2. *Problem* (5.4) *has a unique solution* $(H_r, H_z) \in \breve{\mathcal{U}}$.

Proof. Let us define the sesquilinear and the antilinear forms:

$$\mathscr{B}((H_r, H_z); (v_r, v_z)) := 2\mathrm{i}\pi\omega \int_{\mathbb{R}^+\times\mathbb{R}} \mu(H_r\overline{v}_r + H_z\overline{v}_z)\, r\, dr\, dz$$

$$+ 2\pi \int_{\breve{\Omega}} \sigma^{-1}\Big(\frac{\partial H_r}{\partial z} - \frac{\partial H_z}{\partial r}\Big)\Big(\frac{\partial \overline{v}_r}{\partial z} - \frac{\partial \overline{v}_z}{\partial r}\Big)\, r\, dr\, dz,$$

$$\mathscr{L}((v_r, v_z)) := V \int_{\breve{\Omega}_1} \Big(\frac{\partial \overline{v}_r}{\partial z} - \frac{\partial \overline{v}_z}{\partial r}\Big)\, dr\, dz.$$

Problem (5.4) can be written in the form

$$\mathscr{B}((H_r, H_z); (v_r, v_z)) = \mathscr{L}((v_r, v_z)) \qquad \forall\, (v_r, v_z) \in \breve{\mathcal{U}}.$$

Clearly, the sesquilinear form $\mathscr{B}$ is continuous and coercive on $\breve{\mathcal{U}}$. Moreover, the form $\mathscr{L}$ is antilinear and we have thanks to Remark 5.2.1,

$$|\mathscr{L}((v_r, v_z))| \le |V|\, \|(v_r, v_z)\|_{\breve{\mathcal{U}}} \Big(\int_{\breve{\Omega}_1} \frac{1}{r}\, dr\, dz\Big)^{\frac{1}{2}} \le C\, \|(v_r, v_z)\|_{\breve{\mathcal{U}}}.$$

We can then apply the Lax-Milgram theorem (Theorem 1.2.1) to obtain existence and uniqueness. □

Remark 5.2.2. It is worth noting that an equivalent formulation to (5.4) can be given by using Remark 1.4.1. This one can be formulated as the following: Let $\breve{\mathcal{V}}$ denote the space

$$\breve{\mathcal{V}} := \Big\{ \boldsymbol{v} \in \breve{\mathcal{Z}};\ \frac{\partial v_r}{\partial r} + \frac{\partial v_z}{\partial z} = 0 \text{ in } \breve{\Omega}_{\text{ext}} \Big\},$$

where

$$\breve{\mathcal{Z}} := \Big\{ \boldsymbol{v} = (v_r, v_z);\ \boldsymbol{v} \in \breve{\mathcal{L}}^2(\mathbb{R}^+\times\mathbb{R})^2,\ \frac{\partial v_r}{\partial r} + \frac{\partial v_z}{\partial z} \in \breve{\mathcal{L}}^2(\mathbb{R}^+\times\mathbb{R}) \Big\}.$$

It is easy to see that the spaces $\breve{\mathcal{U}}$ and $\breve{\mathcal{V}}$ can be identified up to the isomorphism $(v_r, v_z) \mapsto (-v_z, v_r)$. It is also important to mention that $\breve{\mathcal{Z}}$ is not the cylindrical version of the $\mathcal{H}(\text{div}, \mathbb{R}^3)$ space. For this reason, to avoid confusion, we have chosen a different notation. Denoting by $\boldsymbol{H}^\perp$ the vector field

$$\boldsymbol{H}^\perp = H_z\, \boldsymbol{e}_r - H_r\, \boldsymbol{e}_z,$$

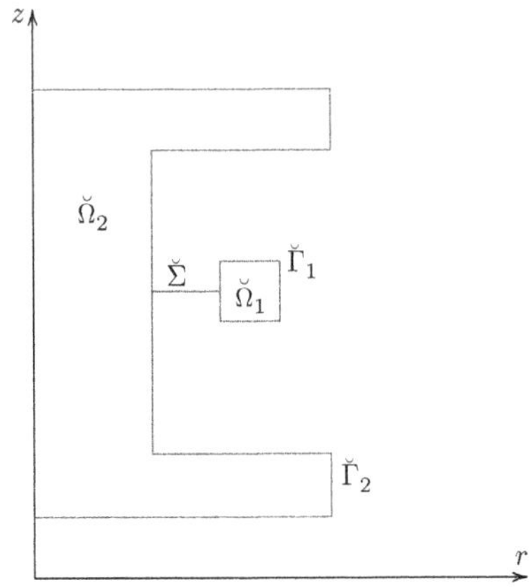

Fig. 5.2 Geometry in cylindrical coordinates

Problem (5.4) is equivalent to the following one:

$$\begin{cases} \text{Find } \boldsymbol{H}^\perp = (H_r^\perp, H_z^\perp) \in \breve{\boldsymbol{\mathcal{V}}} \text{ such that} \\ 2\mathrm{i}\pi\omega \displaystyle\int_{\mathbb{R}^+ \times \mathbb{R}} \mu\, \boldsymbol{H}^\perp \cdot \overline{\boldsymbol{v}}\, r\, dr\, dz \\ \quad + 2\pi \displaystyle\int_{\breve{\Omega}} \sigma^{-1} \Big(\frac{\partial H_r^\perp}{\partial r} + \frac{\partial H_z^\perp}{\partial z} \Big) \Big(\frac{\partial \overline{v}_r}{\partial r} + \frac{\partial \overline{v}_z}{\partial z} \Big) r\, dr\, dz \\ \qquad = V \displaystyle\int_{\breve{\Omega}_1} \Big(\frac{\partial \overline{v}_r}{\partial r} + \frac{\partial \overline{v}_z}{\partial z} \Big) dr\, dz \qquad \forall\, \boldsymbol{v} \in \breve{\boldsymbol{\mathcal{V}}}. \end{cases} \tag{5.5}$$

Problem (5.5) may be more suitable for numerical approximation since this one would be based on more classical $\boldsymbol{\mathcal{H}}(\mathrm{div}, \cdot)$ finite elements.

Naturally, as in the fully three-dimensional case, (5.4) has to be cast into a coupled interior/exterior problem by making use of a boundary integral representation. Since the Green function for the axisymmetric laplacian is rather difficult to handle, we resort to using the three dimensional integral representation and taking into account the axial symmetry in the calculation of the integrals.

Likewise, the derivation of an equivalent formulation to (5.4) that couples a partial differential equation in the conductors with an integral equation on their boundaries, requires returning back to the three dimensional formulation. More precisely, for a function $\boldsymbol{v} \in \boldsymbol{\mathcal{H}}(\mathbf{curl}, \mathbb{R}^3)$ with $\mathbf{curl}\, \boldsymbol{v} = 0$ in Ω_{ext}, Theorem 1.3.6 implies the existence of $\varphi \in \mathcal{W}^1(\Omega_{\mathrm{ext}})$ and $\beta \in \mathbb{C}$ such that

$$\boldsymbol{v} = \nabla\varphi + \beta\, \nabla p \qquad \text{in } \Omega_{\mathrm{ext}} \setminus \Sigma, \tag{5.6}$$

where p is a solution of (1.20). We naturally choose the surface Σ as a surface with symmetry of rotation such that the domain $\Omega_{\mathrm{ext}} \setminus \Sigma$ is simply connected (see Fig. 5.2).

Written in cylindrical coordinates and assuming the axisymmetry hypothesis on $\boldsymbol{v}$, a simple calculation shows that the potentials ψ and p are actually axisymmetric (do not depend on θ) and that we have

$$v_r = \frac{\partial \breve{\psi}}{\partial r} + \beta \frac{\partial \breve{p}}{\partial r}, \quad v_z = \frac{\partial \breve{\psi}}{\partial z} + \beta \frac{\partial \breve{p}}{\partial z} \qquad \text{in } \breve{\Omega}' \setminus \breve{\Sigma},$$

where $\breve{\Sigma}$ is the radial section of Σ, $\beta \in \mathbb{C}$, $\breve{\psi} \in \breve{\mathcal{W}}^1(\breve{\Omega}_{\text{ext}})$ and where $\breve{p}$ is the solution of the following problem:

$$\begin{cases} \dfrac{1}{r}\dfrac{\partial}{\partial r}\left(r\dfrac{\partial \breve{p}}{\partial r}\right) + \dfrac{\partial^2 \breve{p}}{\partial z^2} = 0 & \text{in } \breve{\Omega}_{\text{ext}} \setminus \breve{\Sigma}, \\ [\breve{p}]_{\breve{\Sigma}} = 1, & \\ \left[\dfrac{\partial \breve{p}}{\partial r}\right]_{\breve{\Sigma}} = 0, & \\ \dfrac{\partial \breve{p}}{\partial r} = 0 & \text{on } \breve{\Gamma}, \\ \breve{p}(r,z) = \mathcal{O}((r^2+z^2)^{-\frac{1}{2}}) & \text{when } (r^2+z^2) \to \infty. \end{cases}$$

For the magnetic field $\boldsymbol{H}$, we have the expansions:

$$H_r = \frac{\partial \breve{\varphi}}{\partial r} + \alpha \frac{\partial \breve{p}}{\partial r}, \quad H_z = \frac{\partial \breve{\varphi}}{\partial z} + \alpha \frac{\partial \breve{p}}{\partial z} \qquad \text{in } \breve{\Omega}_{\text{ext}} \setminus \breve{\Sigma}.$$

We can now expand the integrals in the unbounded domain by writing, using the fact that $\mu = \mu_0$ in the free space,

$$\begin{aligned} \int_{\Omega_{\text{ext}}} \mu \boldsymbol{H} \cdot \overline{\boldsymbol{v}} \, d\boldsymbol{x} &= 2\pi\mu_0 \int_{\breve{\Omega}'} (H_r \overline{v}_r + H_z \overline{v}_z)\, r\, dr\, dz \\ &= 2\pi\mu_0 \int_{\breve{\Omega}'} \left(\frac{\partial \breve{\varphi}}{\partial r}\frac{\partial \overline{\breve{\psi}}}{\partial r} + \frac{\partial \breve{\varphi}}{\partial z}\frac{\partial \overline{\breve{\psi}}}{\partial z} \right) r\, dr\, dz \\ &\quad + 2\pi\mu_0\alpha \int_{\breve{\Omega}' \setminus \breve{\Sigma}} \left(\frac{\partial \breve{p}}{\partial r}\frac{\partial \overline{\breve{\psi}}}{\partial r} + \frac{\partial \breve{p}}{\partial z}\frac{\partial \overline{\breve{\psi}}}{\partial z} \right) r\, dr\, dz \\ &\quad + 2\pi\mu_0\overline{\beta} \int_{\breve{\Omega}' \setminus \breve{\Sigma}} \left(\frac{\partial \breve{\varphi}}{\partial r}\frac{\partial \overline{\breve{p}}}{\partial r} + \frac{\partial \breve{\varphi}}{\partial z}\frac{\partial \overline{\breve{p}}}{\partial z} \right) r\, dr\, dz \\ &\quad + 2\pi\mu_0\alpha\overline{\beta} \int_{\breve{\Omega}' \setminus \breve{\Sigma}} \left(\left|\frac{\partial \breve{p}}{\partial r}\right|^2 + \left|\frac{\partial \breve{p}}{\partial z}\right|^2 \right) r\, dr\, dz. \end{aligned}$$

Using the same arguments as in Sect. 4.2, we find the second and third integrals in the right-hand side vanish and consequently

$$\int_{\Omega_{\text{ext}}} \mu \boldsymbol{H} \cdot \overline{\boldsymbol{v}}\, d\boldsymbol{x} = 2\pi\mu_0 \int_{\breve{\Omega}'} \Big(\frac{\partial \breve{\varphi}}{\partial r}\frac{\partial \overline{\breve{\psi}}}{\partial r} + \frac{\partial \breve{\varphi}}{\partial z}\frac{\partial \overline{\breve{\psi}}}{\partial z} \Big)\, r\, dr\, dz + L\,\alpha\overline{\beta}, \tag{5.7}$$

where L is the self inductance defined by

$$L := 2\pi\mu_0 \int_{\breve{\Omega}'\setminus\breve{\Sigma}} \Big(\Big|\frac{\partial \breve{p}}{\partial r}\Big|^2 + \Big|\frac{\partial \breve{p}}{\partial z}\Big|^2 \Big)\, r\, dr\, dz.$$

Defining the space $\breve{\mathcal{K}}$ as the one obtained by the coordinate transformation from (4.31), we can convert (5.5) where $\breve{\boldsymbol{H}}$ is the unknown to a problem with unknowns $(H_{r|\breve{\Omega}}, H_{z|\breve{\Omega}}, \varphi, \alpha)$. For this, we first transform (5.4) into the following one:

$$\begin{cases} \text{Find } (H_r, H_z, \breve{\varphi}, \alpha) \in \breve{\mathcal{K}} \text{ such that} \\[2mm] 2\mathrm{i}\pi\omega \displaystyle\int_{\breve{\Omega}} \mu(H_r\overline{v}_r + H_z\overline{v}_z)\, r\, dr\, dz + 2\mathrm{i}\pi\omega\mu_0 \int_{\breve{\Omega}'} \Big(\frac{\partial \breve{\varphi}}{\partial r}\frac{\partial \overline{\breve{\psi}}}{\partial r} + \frac{\partial \breve{\varphi}}{\partial z}\frac{\partial \overline{\breve{\psi}}}{\partial z} \Big)\, r\, dr\, dz \\[2mm] \qquad + 2\mathrm{i}\pi\omega L\alpha\overline{\beta} + 2\pi \displaystyle\int_{\breve{\Omega}} \sigma^{-1} \Big(\frac{\partial H_r}{\partial z} - \frac{\partial H_z}{\partial r} \Big)\Big(\frac{\partial \overline{v}_r}{\partial z} - \frac{\partial \overline{v}_z}{\partial r} \Big)\, r\, dr\, dz \\[2mm] \qquad\qquad = V \displaystyle\int_{\breve{\Omega}_1} \Big(\frac{\partial \overline{v}_r}{\partial z} - \frac{\partial \overline{v}_z}{\partial r} \Big)\, dr\, dz \qquad \forall\, (v_r, v_z, \breve{\psi}, \beta) \in \breve{\mathcal{K}}. \end{cases}$$

Since $\boldsymbol{H}$ is divergence-free in Ω_{ext}, then φ is harmonic in Ω_{ext} and we can write

$$\begin{aligned} \int_{\breve{\Omega}_{\text{ext}}} \Big(\frac{\partial \breve{\varphi}}{\partial r}\frac{\partial \overline{\breve{\psi}}}{\partial r} + \frac{\partial \breve{\varphi}}{\partial z}\frac{\partial \overline{\breve{\psi}}}{\partial z} \Big)\, r\, dr\, dz &= -\int_{\breve{\Gamma}} \Big(\frac{\partial \breve{\varphi}}{\partial r} n_r + \frac{\partial \breve{\varphi}}{\partial z} n_z \Big) \overline{\breve{\psi}}\, r\, ds(r,z) \\ &= \int_{\breve{\Gamma}} (\breve{P}\breve{\varphi})\, \overline{\breve{\psi}}\, r\, ds(r,z), \end{aligned}$$

where $\breve{P}$ is the exterior Steklov–Poincaré operator in the axisymmetric configuration and $\breve{\boldsymbol{n}}$ is the outward unit normal to Γ expressed in cylindrical coordinates given is axisymmetric configurations by

$$\breve{\boldsymbol{n}}(r,z) = n_r(r,z)\,\boldsymbol{e}_r + n_z(r,z)\,\boldsymbol{e}_z.$$

The construction of this operator is described in the next subsection. Let us notice that the interface condition specified in the space $\mathcal{K}$ in (4.31) is transformed in the space $\breve{\mathcal{K}}$ into

$$v_r = \frac{\partial \breve{\psi}}{\partial r} + \beta\frac{\partial \breve{p}}{\partial r},\quad v_z = \frac{\partial \breve{\psi}}{\partial z} + \beta\frac{\partial \breve{p}}{\partial z} \qquad \text{on } \breve{\Gamma} \qquad \forall\, (v_r, v_z, \breve{\psi}, \beta) \in \breve{\mathcal{K}}_\Gamma,$$

where the space $\breve{\mathcal{K}}_\Gamma$ is the one obtained by the coordinate transformation from $\mathcal{K}_\Gamma$ in (4.33). We eventually obtain the variational formulation:

$$\begin{cases} \text{Find } (H_r, H_z, \breve{\varphi}, \alpha) \in \breve{\mathcal{K}}_\Gamma \text{ such that} \\ 2\mathrm{i}\pi\omega \displaystyle\int_{\breve{\Omega}} \mu(H_r \overline{v}_r + H_z \overline{v}_z)\, r\, dr\, dz + 2\mathrm{i}\pi\omega\mu_0 \int_{\breve{\gamma}} (\breve{P}\breve{\varphi})\, \overline{\breve{\psi}}\, ds(r,z) \\ \quad + 2\mathrm{i}\pi\omega L\alpha\overline{\beta} + 2\pi \displaystyle\int_{\breve{\Omega}} \sigma^{-1}\Big(\frac{\partial H_r}{\partial z} - \frac{\partial H_z}{\partial r}\Big)\Big(\frac{\partial \overline{v}_r}{\partial z} - \frac{\partial \overline{v}_z}{\partial r}\Big)\, r\, dr\, dz \\ \quad = V \displaystyle\int_{\breve{\Omega}_1} \Big(\frac{\partial \overline{v}_r}{\partial z} - \frac{\partial \overline{v}_z}{\partial r}\Big)\, dr\, dz \qquad \forall\ (v_r, v_z, \breve{\psi}, \beta) \in \breve{\mathcal{K}}_\Gamma. \end{cases}$$

5.2.1 The Exterior Steklov–Poincaré Operator

According to Sect. 1.3.5, we can construct the axisymmetric version of the exterior Steklov–Poincaré operator. For this we proceed as in [146, 147]. Before giving the expression of the operator we need to express the Green function in cylindrical coordinates. Let us consider the Green function in $\mathbb{R}^3$ and take into account the axial symmetry. We define the function

$$\breve{G}((r,\theta,z),(r',\theta',z')) := G(\boldsymbol{\Phi}(r,\theta,z), \boldsymbol{\Phi}(r',\theta',z'))$$
$$= \frac{1}{4\pi} \frac{1}{\big((r\cos\theta - r'\cos\theta')^2 + (r\sin\theta - r'\sin\theta')^2 + (z-z')^2\big)^{\frac{1}{2}}}$$

for $(r,\theta,z),(r',\theta',z') \in \mathbb{R}^+ \times [0,2\pi) \times \mathbb{R}$, with $(r,\theta,z) \neq (r',\theta',z')$, where G is the Green function defined by (1.21). We also define the normal derivative $\breve{G}_n$ corresponding to G on Γ by

$$\breve{G}_n((r,\theta,z),(r',\theta',z')) := -\frac{\boldsymbol{n}(\boldsymbol{\Phi}(r',\theta',z')) \cdot (\boldsymbol{\Phi}(r,\theta,z) - \boldsymbol{\Phi}(r',\theta',z'))}{|\boldsymbol{\Phi}(r,\theta,z) - \boldsymbol{\Phi}(r',\theta',z')|^3}.$$

We can also express these functions explicitly in cylindrical coordinates:

$$\breve{G}((r,\theta,z),(r',\theta',z')) = \frac{1}{4\pi} \frac{1}{(r^2 + r'^2 - 2rr'\cos(\theta-\theta') + (z-z')^2)^{\frac{1}{2}}},$$

$$\breve{G}_n((r,\theta,z),(r',\theta',z')) = -\frac{1}{4\pi} \frac{n_r(r',z')(r\cos(\theta-\theta') - r')}{(r^2 + r'^2 - 2rr'\cos(\theta-\theta') + (z-z')^2)^{\frac{3}{2}}}$$
$$\qquad - \frac{1}{4\pi} \frac{n_z(r',z')(z-z')}{(r^2 + r'^2 - 2rr'\cos(\theta-\theta') + (z-z')^2)^{\frac{3}{2}}}.$$

Following the same formalism as in Sect. 1.3.5, we define the operators $\breve{K}$ and $\breve{R}$ respectively by

$$\int_{\breve{\Gamma}} (\breve{K}p)\,\overline{q}\,ds := \int_{\breve{\Gamma}}\int_{\breve{\Gamma}} \breve{g}((r,z),(r',z'))\,p(r',z')\,\overline{q}(r,z)\,ds(r',z')\,ds(r,z),$$

$$\breve{R}\varphi(r,z) := \int_{\breve{\Gamma}} \varphi(r',z')\,\breve{g}_n((r,z),(r',z'))\,ds(r',z') \qquad (r,z)\in\breve{\Gamma},$$

where

$$\breve{g}((r,z),(r',z')) := \int_0^{2\pi}\int_0^{2\pi} \breve{G}((r,\theta,z),(r',\theta',z'))\,d\theta'\,d\theta,$$

$$\breve{g}_n((r,z),(r',z')) := \int_0^{2\pi}\int_0^{2\pi} \breve{G}_n((r,\theta,z),(r',\theta',z'))\,d\theta'\,d\theta.$$

Following 1.3.5, the axisymmetric version of the exterior Steklov–Poincaré operator is given by

$$\breve{P} = (-\frac{1}{2}\breve{I} + \breve{R}')\,\breve{K}^{-1},$$

where $\breve{I}$ is the Identity operator and $\breve{R}'$ is the dual operator of $\breve{R}$, i.e.

$$\int_{\breve{\Gamma}} (\breve{R}'p)\,\overline{\breve{\psi}}\,ds = \int_{\breve{\Gamma}} \breve{p}\,(\overline{\breve{R}\breve{\psi}})\,ds \qquad \forall\ \breve{p}\in\breve{\mathcal{H}}^{-\frac{1}{2}}(\breve{\Gamma}),\ \forall\ \breve{\psi}\in\breve{\mathcal{H}}^{\frac{1}{2}}(\breve{\Gamma}).$$

Remark 5.2.3. The functions $\breve{g}$ and $\breve{g}_n$ cannot be evaluated explicitly. For instance, we have

$$\breve{g}((r,z),(r',z')) = \int_0^{2\pi} I((r,z),(r',z'))\,d\theta',$$

where

$$\begin{aligned} I((r,z),(r',z')) &= \int_0^{2\pi} \breve{G}((r,\theta,z),(r',\theta',z'))\,d\theta \\ &= \frac{1}{4\pi}\int_0^{2\pi} \frac{1}{(r^2+r'^2-rr'\cos\theta+(z-z')^2)^{\frac{1}{2}}}\,d\theta. \end{aligned}$$

By setting $\lambda = r'/r$, we obtain

$$
\begin{aligned}
I((r,z),(r',z'))
&= \frac{1}{4\pi r}\int_0^{2\pi}\left(1+\lambda^2-\lambda\cos\theta+\left(\frac{z-z'}{r}\right)^2\right)^{-\frac{1}{2}}\,d\theta \\
&= \frac{1}{4\pi r}\left(1+\lambda^2+\left(\frac{z-z'}{r}\right)^2\right)^{-\frac{1}{2}}\int_0^{2\pi}\left(1-\frac{\lambda\cos\theta}{1+\lambda^2+\left(\frac{z-z'}{r}\right)^2}\right)^{-\frac{1}{2}}\,d\theta.
\end{aligned}
$$

By taking

$$
\beta = \frac{\lambda}{1+\lambda^2+\left(\frac{z-z'}{r}\right)^2}
$$

we are led to the evaluation of the elliptic integral of the first kind

$$
\int_0^{2\pi}\frac{d\theta}{(1-\beta\cos\theta)^{\frac{1}{2}}}\,d\theta \quad \text{where } |\beta|<1.
$$

Such integrals can be evaluated by numerical approximation only.

5.2.2 A Formula for the Self Inductance

The calculation of the self induction coefficient can be performed by means of the formula (4.42). We simply adapt here this formula to the axisymmetric case. Let p denote the solution of (1.20). Using Theorem 1.3.7 with $p = 0$ in Ω, we find that p is solution of the integral equation

$$
\frac{1}{2}p(\boldsymbol{x}) - \int_\Gamma \frac{\partial G}{\partial \boldsymbol{n}_y}(\boldsymbol{x},\boldsymbol{y})\,p(\boldsymbol{y})\,ds(\boldsymbol{y}) = \int_\Sigma \frac{\partial G}{\partial \boldsymbol{n}_y}(\boldsymbol{x},\boldsymbol{y})\,ds(\boldsymbol{y}) \qquad \boldsymbol{x}\in\Gamma.
$$

Since p is axisymmetric, we have

$$
\frac{1}{2}\breve{p}(r,z) - \int_{\breve{\gamma}} g_n((r,z),(r',z'))\,\breve{p}(r',z')\,dr'\,dz' = \int_{\breve{\Sigma}} g_n((r,z),(r',z'))\,dr'\,dz',
$$

for $(r,z)\in\breve{\gamma}$. The surface curl $(\nabla p\times\boldsymbol{n})$ of p is expressed by the scalar function

$$
\breve{J}_{\breve{\gamma}}(r,z) := \frac{\partial\breve{p}}{\partial z}(r,z)\,n_r(r,z) - \frac{\partial\breve{p}}{\partial r}(r,z)\,n_z(r,z) \qquad (r,z)\in\breve{\gamma}.
$$

The self inductance of the conductor Ω_1 is thus obtained by

$$L = \mu_0 \int_\Gamma \int_\Gamma \boldsymbol{J}_\Gamma(\boldsymbol{x}) \cdot \overline{\boldsymbol{J}}_\Gamma(\boldsymbol{y}) G(\boldsymbol{x}, \boldsymbol{y})\, ds(\boldsymbol{y})\, ds(\boldsymbol{x})$$
$$= \mu_0 \int_{\breve{\gamma}} \int_{\breve{\gamma}} \breve{J}_{\breve{\gamma}}(r,z)\, \overline{\breve{J}}_{\breve{\gamma}}(r',z')\, g((r,z),(r',z'))\, ds(r',z')\, ds(r,z).$$

5.3 A Scalar Potential Model

This model was obtained in [48, 147] for induction heating applications (see also [21, 48]). This one is formulated in terms of the vector potential $\boldsymbol{A}$, which turns out to be scalar in the present case as we shall see.

Let $\boldsymbol{A}$ stand for the vector potential defined by (2.12), (2.13) and the property

$$|\boldsymbol{A}(\boldsymbol{x})| = \mathcal{O}(|\boldsymbol{x}|^{-1}) \qquad \text{when } |\boldsymbol{x}| \to \infty.$$

Denoting by

$$\breve{\boldsymbol{A}} = A_r\, \boldsymbol{e}_r + A_\theta\, \boldsymbol{e}_\theta + A_z\, \boldsymbol{e}_z$$

the decomposition of $\boldsymbol{A}$ in the cylindrical coordinate system, we aim at deriving a model in terms of the θ-component $\phi := A_\theta$.

The following result summarizes the obtained model:

Theorem 5.3.1. *Let $\boldsymbol{J}, \boldsymbol{E}, \boldsymbol{B}, \boldsymbol{H}$ denote smooth axisymmetric fields that satisfy* (2.62)–(2.68) *such that $\boldsymbol{J}$ experiences the property* (5.1) *and let $\boldsymbol{A}$ be the vector potential defined by* (2.12), (2.13)*. Then we have*

$$\breve{\boldsymbol{A}} = A_\theta(r,z)\, \boldsymbol{e}_\theta.$$

Furthermore, the function $\phi = A_\theta$ is solution of the following problem:

$$-\frac{\partial}{\partial r}\Big(\frac{\mu^{-1}}{r}\frac{\partial}{\partial r}(r\phi)\Big) - \frac{\partial}{\partial z}\Big(\mu^{-1}\frac{\partial \phi}{\partial z}\Big) + \mathrm{i}\omega\sigma\phi = \frac{\sigma C_k}{r} \quad \text{in } \breve{\Omega}_k, \quad k = 1,2, \tag{5.8}$$

$$-\frac{\partial}{\partial r}\Big(\frac{1}{r}\frac{\partial}{\partial r}(r\phi)\Big) - \frac{\partial^2 \phi}{\partial z^2} = 0 \qquad \text{in } \breve{\Omega}_{ext}, \tag{5.9}$$

$$[\phi]_{\breve{\Gamma}_k} = 0 \qquad k = 1,2, \tag{5.10}$$

$$\Big[\mu^{-1}\Big(\frac{1}{r}\frac{\partial}{\partial r}(r\phi)\, n_r + \frac{\partial \phi}{\partial z}\, n_z\Big)\Big]_{\breve{\Gamma}_k} = 0 \qquad k = 1,2, \tag{5.11}$$

$$\phi(r,z) = \mathcal{O}((r^2+z^2)^{-\frac{1}{2}}) \qquad (r^2+z^2) \to \infty, \quad (5.12)$$

where C_1, C_2 are complex constants.

Proof. From (2.5), (2.12) and (5.3), we have

$$\frac{\partial A_z}{\partial r} = \frac{\partial A_r}{\partial z}, \quad \mu H_r = -\frac{\partial A_\theta}{\partial z}, \quad \mu H_z = \frac{1}{r}\frac{\partial}{\partial r}\Big(rA_\theta\Big). \tag{5.13}$$

Moreover, we have using (2.13),

$$0 = \frac{\partial}{\partial r}(rA_r) + r\frac{\partial A_z}{\partial z} = r\Big(\frac{\partial A_r}{\partial r} + \frac{\partial A_z}{\partial z} + \frac{1}{r}A_r\Big). \tag{5.14}$$

Dividing by r, using (5.13) and differentiating with respect to z, we get

$$\frac{\partial^2 A_z}{\partial r^2} + \frac{\partial^2 A_z}{\partial z^2} + \frac{1}{r}\frac{\partial A_z}{\partial r} = 0$$

Then $\breve{\Delta} A_z = 0$, where $\breve{\Delta}$ is the Laplace operator in cylindrical coordinates for an axisymmetric field:

$$\breve{\Delta} f = \frac{1}{r}\frac{\partial}{\partial r}\Big(r\frac{\partial f}{\partial r}\Big) + \frac{\partial^2 f}{\partial z^2}.$$

Since $\boldsymbol{A} \in \mathcal{W}^1(\mathbb{R}^3)$, then $A_z = 0$. To prove that $A_r = 0$, we have from (5.14)

$$0 = r\frac{\partial A_r}{\partial r} + A_r.$$

This implies $A_r = C\,r^{-1}$ where $C \in \mathbb{C}$. For a smooth potential $\boldsymbol{A}$, we deduce $C = 0$, thus $A_r = 0$.

Denoting by ϕ the component A_θ, we obtain from (2.12), (2.63) and (2.65),

$$\frac{\partial}{\partial z}(\mathrm{i}\omega\phi + E_\theta) = \frac{\partial}{\partial r}\big(r(\mathrm{i}\omega\phi + E_\theta)\big) = 0 \qquad \text{in } \breve{\Omega}.$$

This implies the existence of two complex constants C_1, C_2 such that

$$\mathrm{i}\omega\phi + E_\theta = \frac{C_k}{r} \qquad \text{in } \breve{\Omega}_k, \quad k = 1,2. \tag{5.15}$$

But from (5.13) we have in $\breve{\Omega}$,

$$\frac{\partial H_r}{\partial z} = -\frac{\partial}{\partial z}\Big(\mu^{-1}\frac{\partial\phi}{\partial z}\Big), \quad \frac{\partial H_z}{\partial r} = \frac{\partial}{\partial r}\Big(\frac{\mu^{-1}}{r}\frac{\partial}{\partial r}(r\phi)\Big). \tag{5.16}$$

Furthermore, (2.62) and (2.66) give in particular in $\breve{\Omega}$,

$$\frac{\partial H_r}{\partial z} - \frac{\partial H_z}{\partial r} = \sigma\, E_\theta. \tag{5.17}$$

Combining (5.17) with (5.16), we get

$$-\frac{\partial}{\partial r}\Big(\frac{\mu^{-1}}{r}\frac{\partial}{\partial r}(r\phi)\Big) - \frac{\partial}{\partial z}\Big(\mu^{-1}\frac{\partial \phi}{\partial z}\Big) = \sigma E_\theta \qquad \text{in } \breve{\Omega}.$$

Using (5.15), we get (5.8). To obtain (5.9), we make use of the same procedure with $\mu = \mu_0$ and $\sigma = 0$ in the free space.

As far as interface conditions are concerned, we find that (5.10) is a consequence of the property $\boldsymbol{A} \in \boldsymbol{\mathcal{W}}^1(\mathbb{R}^3)$. Moreover, the θ-component of $\mathbf{curl}\,\boldsymbol{A} \times \boldsymbol{n}$ is given by

$$\Big(\frac{1}{r}\frac{\partial}{\partial r}(r\phi)\, n_r + \frac{\partial \phi}{\partial z}\, n_z\Big).$$

The interface condition derived from (4.1) results then in (5.11). □

Remark 5.3.1. The constants C_1, C_2 in Theorem 5.3.1 depend on the data to prescribe (current or voltage). To deal with this issue, we proceed in the sequel as for the two-dimensional cartesian case (see Chap. 3).

5.3.1 A Prescribed Voltage Model

We define the setup voltage as in (3.24), by the expression,

$$V := \int_{\partial \Sigma_1} (\mathrm{i}\omega \boldsymbol{A} + \boldsymbol{E})\cdot \boldsymbol{t}\, ds,$$

where $\partial\Sigma_1$ is the boundary of the cut Σ_1 defined in Fig. 1.1 and $\boldsymbol{t}$ is a unit tangent vector to the curve $\partial\Sigma_1$. Note that the choice of this tangent orientation modifies the sign of V which has no effect on the relevant quantities in eddy current processes.

To express V according to the results of Theorem 5.3.1 and (5.15), we arrange to choose a cut Σ_1 that is axisymmetric and $\boldsymbol{t} = \boldsymbol{e}_\theta$. Using Proposition 5.3.1 and relationship (5.15), we obtain

$$V = \int_0^{2\pi} (\mathrm{i}\omega\phi + E_\theta)\, r\, d\theta = 2\pi\, C_1 \qquad (r,z) \in \breve{\Omega}_1.$$

In an analogous way, the conductor $\tilde{\Omega}_2$ having no injected current, we have

$$0 = \int_0^{2\pi} (\mathrm{i}\omega\phi + E_\theta)\, r\, d\theta = 2\pi\, C_2 \qquad (r,z) \in \breve{\Omega}_2.$$

Then

$$C_1 = \frac{V}{2\pi},\; C_2 = 0. \tag{5.18}$$

The system (5.8)–(5.12) becomes then,

$$-\frac{\partial}{\partial r}\Big(\frac{\mu^{-1}}{r}\frac{\partial}{\partial r}(r\phi)\Big) - \frac{\partial}{\partial z}\Big(\mu^{-1}\frac{\partial\phi}{\partial z}\Big) + \mathrm{i}\omega\sigma\phi = \frac{\sigma V}{2\pi r} \text{ in } \breve{\Omega}_1, \tag{5.19}$$

$$-\frac{\partial}{\partial r}\Big(\frac{\mu^{-1}}{r}\frac{\partial}{\partial r}(r\phi)\Big) - \frac{\partial}{\partial z}\Big(\mu^{-1}\frac{\partial\phi}{\partial z}\Big) + \mathrm{i}\omega\sigma\phi = 0 \quad \text{in } \breve{\Omega}_2, \tag{5.20}$$

$$-\frac{\partial}{\partial r}\Big(\frac{1}{r}\frac{\partial}{\partial r}(r\phi)\Big) - \frac{\partial^2\phi}{\partial z^2} = 0 \qquad \text{in } \breve{\Omega}_{\text{ext}}, \tag{5.21}$$

$$[\phi]_{\breve{\gamma}_k} = 0 \qquad k = 1,2, \tag{5.22}$$

$$\Big[\mu^{-1}\Big(\frac{1}{r}\frac{\partial}{\partial r}(r\phi)\, n_r + \frac{\partial\phi}{\partial z} n_z\Big)\Big]_{\breve{\gamma}_k} = 0 \qquad k = 1,2, \tag{5.23}$$

$$\phi(r,z) = \mathcal{O}((r^2+z^2)^{-\frac{1}{2}}) \qquad (r^2+z^2) \to \infty. \tag{5.24}$$

In order to define a ad-hoc functional setting we prove the following result:

Theorem 5.3.2. *The space of axisymmetric vector fields* $\boldsymbol{w}$ *in* $\boldsymbol{\mathcal{W}}^{\mathbf{1}}(\mathbb{R}^3)$ *such that* $\breve{w}_r = \breve{w}_z = 0$ *is isomorphic to the following one:*

$$\breve{\mathcal{W}} := \Big\{\psi : \mathbb{R}^+\!\times \mathbb{R} \to \mathbb{C};\; \frac{1}{r}\,\psi,\, \frac{\partial\psi}{\partial r},\, \frac{\partial\psi}{\partial z} \in \breve{\mathcal{L}}^2(\mathbb{R}^+\!\times \mathbb{R})\Big\}.$$

Proof. Let $\boldsymbol{w}$ denote an axisymmetric vector field in $\boldsymbol{\mathcal{W}}^{\mathbf{1}}(\mathbb{R}^3)$ that satisfies $\breve{w}_r = \breve{w}_z = 0$. The gradient of $\boldsymbol{w}$ expressed in cylindrical coordinates is given by

$$\breve{\nabla}\breve{\boldsymbol{w}} = \begin{pmatrix} 0 & -\frac{1}{r}\breve{w}_\theta & 0 \\ \frac{\partial\breve{w}_\theta}{\partial r} & 0 & \frac{\partial\breve{w}_\theta}{\partial z} \\ 0 & 0 & 0 \end{pmatrix}.$$

We recall that the natural norm on $\mathcal{W}^1(\mathbb{R}^3)$ is

$$\|\boldsymbol{w}\|^2_{\mathcal{W}^1(\mathbb{R}^3)} = \int_{\mathbb{R}^3} \Big(\frac{|\boldsymbol{w}|^2}{(1+|\boldsymbol{x}|)^2} + \Big|\frac{\partial \boldsymbol{w}}{\partial x_1}\Big|^2 + \Big|\frac{\partial \boldsymbol{w}}{\partial x_2}\Big|^2 + \Big|\frac{\partial \boldsymbol{w}}{\partial x_3}\Big|^2 \Big)\, d\boldsymbol{x}.$$

Therefore

$$\|\boldsymbol{w}\|^2_{\mathcal{W}^1(\mathbb{R}^3)} = 2\pi \int_{\mathbb{R}^+\times\mathbb{R}} \Bigg(\frac{\breve{w}_\theta^2}{\big(1+(r^2+z^2)^{\frac{1}{2}}\big)^2} + \frac{\breve{w}_\theta^2}{r^2} + \Big(\frac{\partial \breve{w}_\theta}{\partial r}\Big)^2 + \Big(\frac{\partial \breve{w}_\theta}{\partial z}\Big)^2 \Bigg)\, r\, dr\, dz.$$

But we have

$$\int_{\mathbb{R}^+\times\mathbb{R}} \frac{\breve{w}_\theta^2\, r}{\big(1+(r^2+z^2)^{\frac{1}{2}}\big)^2}\, dr\, dz \le \int_{\mathbb{R}^+\times\mathbb{R}} \frac{\breve{w}_\theta^2}{r}\, dr\, dz.$$

This proves the result. □

Applying this result, we introduce the space

$$\breve{\mathcal{W}} := \Big\{ \psi : \mathbb{R}^+\times \mathbb{R} \to \mathbb{C};\ \frac{1}{r}\,\psi,\ \frac{\partial \psi}{\partial r},\ \frac{\partial \psi}{\partial z} \in \breve{\mathcal{L}}^2(\mathbb{R}^+\times \mathbb{R}) \Big\},$$

with its natural norm

$$\|\psi\|_{\breve{\mathcal{W}}} := \Bigg(\Big\|\frac{1}{r}\,\psi\Big\|^2_{\breve{\mathcal{L}}^2(\mathbb{R}^+\times\mathbb{R})} + \Big\|\frac{\partial \psi}{\partial r}\Big\|^2_{\breve{\mathcal{L}}^2(\mathbb{R}^+\times\mathbb{R})} + \Big\|\frac{\partial \psi}{\partial z}\Big\|^2_{\breve{\mathcal{L}}^2(\mathbb{R}^+\times\mathbb{R})} \Bigg)^{\frac{1}{2}}.$$

We obtain the variational problem

$$\text{Find } \phi \in \breve{\mathcal{W}} \quad \text{such that} \quad \mathscr{B}_V(\phi,\psi) = \mathscr{L}_V(\psi) \qquad \forall\, \psi \in \breve{\mathcal{W}}, \tag{5.25}$$

where

$$\mathscr{B}_V(\phi,\psi) := \int_{\mathbb{R}^+\times\mathbb{R}} \mu^{-1}\Big(\frac{1}{r^2}\phi\overline{\psi} + \frac{\partial \phi}{\partial r}\frac{\partial \overline{\psi}}{\partial r} + \frac{\partial \phi}{\partial z}\frac{\partial \overline{\psi}}{\partial z}\Big) r\, dr\, dz + \mathrm{i}\omega \int_{\breve{\Omega}} \sigma \phi\, \overline{\psi}\, r\, dr\, dz,$$

$$\mathscr{L}_V(\psi) := \frac{V}{2\pi} \int_{\breve{\Omega}_1} \sigma \overline{\psi}\, dr\, dz.$$

We deduce the following

Theorem 5.3.3. *Problem* (5.25) *has a unique solution. Moreover, this solution has a null trace on the axis* $r = 0$.

Proof. The sesquilinear form $\mathscr{B}_V$ is continuous and coercive on $\breve{\mathcal{W}}$ since we have from (2.7) and (2.11),

$$|\mathscr{B}_V(\phi,\psi)| \le C\,\|\phi\|_{\breve{\mathcal{W}}}\,\|\psi\|_{\breve{\mathcal{W}}}.$$

Furthermore, again from (2.7),

$$\left|\mathscr{B}_V(\psi,\psi)\right| \ge \left|\mathrm{Re}\left(\mathscr{B}_V(\psi,\psi)\right)\right| \ge \mu_M^{-1}\,\|\psi\|^2_{\breve{\mathcal{W}}}.$$

The antilinear form $\mathscr{L}_V$ can be bounded as follows using (2.11) and the Cauchy–Schwarz inequality:

$$\begin{aligned} |\mathscr{L}_V(\psi)| &\le \frac{VR^{\frac{1}{2}}}{2\pi}\int_{\breve{\Omega}} |\psi|\,r^{-\frac{1}{2}}\,dr\,dz \\ &\le \frac{VR^{\frac{1}{2}}}{2\pi}|\breve{\Omega}|^{\frac{1}{2}}\Big(\int_{\breve{\Omega}} |\psi|^2\frac{1}{r}\,dr\,dz\Big)^{\frac{1}{2}} \\ &\le C\,\|\psi\|_{\breve{\mathcal{W}}}, \end{aligned}$$

where R is the upper bound of r in $\breve{\Omega}$.

The Lax-Milgram theorem (Theorem 1.2.1) yields the existence and uniqueness of a solution. □

5.3.2 A Prescribed Total Current Model

The current intensity in the setup is defined in the inductor by the expression

$$I := \int_{\breve{\Omega}_1} J_\theta\,dr\,dz = \int_{\breve{\Omega}_1} \sigma\,E_\theta\,dr\,dz.$$

In the inductor $\breve{\Omega}_1$, the current is given, using (5.15), by

$$I = \int_{\breve{\Omega}_1} \sigma\left(\frac{C_1}{r} - \mathrm{i}\omega\phi\right)dr\,dz,$$

from which we deduce

$$C_1 = \frac{I + \mathrm{i}\omega\int_{\breve{\Omega}_1}\sigma\phi\,dr\,dz}{\int_{\breve{\Omega}_1}\sigma\,r^{-1}\,dr\,dz}.$$

The value $C_2 = 0$ is already obtained from (5.18). Equation (5.8) can then be written,

$$-\frac{\partial}{\partial r}\Big(\frac{\mu^{-1}}{r}\frac{\partial}{\partial r}(r\phi)\Big) - \frac{\partial}{\partial z}\Big(\mu^{-1}\frac{\partial \phi}{\partial z}\Big) + \mathrm{i}\omega\,\sigma\,(\phi - \chi(\phi)) = \delta \qquad \text{in } \breve{\Omega}, \quad (5.26)$$

where $\chi = \chi_k$ in $\breve{\Omega}_k$, with

$$\chi_1(\phi) = \frac{1}{r}\frac{\int_{\breve{\Omega}_1} \sigma\phi\,dr\,dz}{\int_{\breve{\Omega}_1} \sigma\,r^{-1}\,dr\,dz}, \quad \chi_2(\phi) = 0$$

and

$$\delta = \begin{cases} \dfrac{\sigma I}{r\int_{\breve{\Omega}_1} \sigma\,r^{-1}\,dr\,dz} & \text{in } \breve{\Omega}_1, \\ 0 & \text{in } \breve{\Omega}_2. \end{cases}$$

In order to obtain a variational formulation of (5.26), (5.9)–(5.12), we multiply equations by $r\,\psi$ where $\psi \in \breve{\mathcal{W}}$, integrate over $\mathbb{R}^2$ and use the Green formula. We obtain the problem

$$\text{Find } \phi \in \breve{\mathcal{W}} \text{ such that} \quad \mathscr{B}_I(\phi, \psi) = \mathscr{L}_I(\psi) \qquad \forall\,\psi \in \breve{\mathcal{W}}, \qquad (5.27)$$

where

$$\mathscr{B}_I(\phi, \psi) = \int_{\mathbb{R}^+\times\mathbb{R}} \mu^{-1}\Big(\frac{1}{r^2}\phi\overline{\psi} + \frac{\partial\phi}{\partial r}\frac{\partial\overline{\psi}}{\partial r} + \frac{\partial\phi}{\partial z}\frac{\partial\overline{\psi}}{\partial z}\Big)\,r\,dr\,dz$$
$$+\,\mathrm{i}\omega\int_{\breve{\Omega}} \sigma(\phi - \chi(\phi))\,\overline{\psi}\,r\,dr\,dz,$$
$$\mathscr{L}_I(\psi) = \int_{\breve{\Omega}} \delta\overline{\psi}\,r\,dr\,dz = I\,\frac{\int_{\breve{\Omega}_1} \sigma\psi\,dr\,dz}{\int_{\breve{\Omega}_1} \sigma r^{-1}\,dr\,dz}.$$

We have the following result:

Theorem 5.3.4. *Problem* (5.27) *has a unique solution* $\phi \in \mathcal{V}$.

Proof. We use the Lax–Milgram theorem (Theorem 1.2.1). Clearly, the forms $\mathscr{B}_I$ and $\mathscr{L}_I$ are continuous on $\breve{\mathcal{W}}$. Moreover, we have for all $\phi, \psi \in \breve{\mathcal{W}}$,

$$\int_{\breve{\Omega}} \sigma(\phi - \chi(\phi))\,\overline{\psi}\,r\,dr\,dz = \int_{\breve{\Omega}_1} \sigma(\phi - \chi_1(\phi))\,(\overline{\psi} - \chi_1(\overline{\psi}))\,r\,dr\,dz$$
$$+ \int_{\breve{\Omega}_2} \sigma\phi\,\overline{\psi}\,r\,dr\,dz.$$

Therefore

$$\begin{aligned}\operatorname{Re}\left(\mathscr{B}_I(\psi,\psi)\right) = & \int_{\mathbb{R}_+\times\mathbb{R}} \mu^{-1}\Big(\frac{1}{r}\Big|\frac{\partial(r\psi)}{\partial r}\Big|^2 + r\,\Big|\frac{\partial\psi}{\partial z}\Big|^2\Big)\,dr\,dz \\ & + \omega \int_{\breve{\Omega}_1} \sigma\,|\psi - \chi_1(\psi)|^2\, r\,dr\,dz \\ & + \omega \int_{\breve{\Omega}_2} \sigma\,|\psi|^2\, r\,dr\,dz.\end{aligned}$$

Using (2.7), (2.11) and Theorem 5.3.1, we obtain

$$\left|\mathscr{B}_I(\psi,\psi)\right| \ge \left|\operatorname{Re}\left(\mathscr{B}_I(\psi,\psi)\right)\right| \ge \mu_M^{-1}\,|\psi|^2_{\breve{\mathcal{W}}}.$$

The form $\mathscr{B}_I$ is thus coercive and existence and uniqueness follow. □

5.3.3 A Boundary Integral Formulation

Problems (5.27) and (5.25) are not well suited for numerical solution. We shall then proceed as we did for the two-dimensional cartesian case. For the sake of simplicity we shall treat the voltage model (5.25) only.

Let us first remark that (5.21) implies

$$\Delta f = 0 \qquad \text{in } \Omega_{\text{ext}},$$

where

$$f(\boldsymbol{x}) = \phi(r,z)\sin\theta \quad \text{with } \boldsymbol{x} = \boldsymbol{\Phi}(r,\theta,z),$$

and Δ is the Laplace operator considered in the cartesian plane Orz. According to Theorem 1.3.7, and extending f by 0 to Ω, we have

$$\frac{1}{2}f^+(\boldsymbol{x}) = -\int_\Gamma \frac{\partial f^+}{\partial n}(\boldsymbol{y})\,G(\boldsymbol{x},\boldsymbol{y})\,ds(\boldsymbol{y}) + \int_\Gamma f^+(\boldsymbol{y})\frac{\partial G}{\partial \boldsymbol{n}_y}(\boldsymbol{x},\boldsymbol{y})\,ds(\boldsymbol{y})$$

for all $\boldsymbol{x} \in \Gamma$. We recall that f^+ is the external part of f (restriction to Ω_{ext}). Remark that

$$\nabla f(\boldsymbol{x}) = \frac{\partial\phi^+}{\partial r}\sin\theta\,\boldsymbol{e}_r + \frac{1}{r}\phi^+\cos\theta\,\boldsymbol{e}_\theta + \frac{\partial\phi^+}{\partial z}\sin\theta\,\boldsymbol{e}_z$$

and

$$\frac{\partial f^+}{\partial n} = \frac{\partial\phi^+}{\partial\breve{n}}\sin\theta \qquad \text{on } \Gamma,$$

where ϕ^+ denotes the function ϕ in the domain $\breve{\Omega}_{\text{ext}}$. Since ϕ is continuous in $\mathbb{R}^3$, we have

$$\frac{1}{2}\phi(r,z)\sin\theta$$
$$= -\int_0^{2\pi}\Big(\int_{\breve{\Gamma}} \frac{\partial \phi^+}{\partial \breve{n}}(r',z')\,\sin\theta'\,\breve{G}((r,\theta,z);(r',\theta',z'))\,r'\,ds(r',z')\Big)\,d\theta'$$
$$+\int_0^{2\pi}\Big(\int_{\breve{\Gamma}} \phi(r',z')\,\sin\theta'\,\breve{G}_{\breve{\boldsymbol{n}}_{(r',z')}}((r,\theta,z);(r',\theta',z'))\,r'\,ds(r',z')\Big)\,d\theta',$$

where $\breve{\boldsymbol{n}}_{(r',z')} = n_r(r',z')\,\boldsymbol{e}_r + n_z(r',z')\,\boldsymbol{e}_z$ and $ds(\boldsymbol{y}) = r'\,ds(r',z')\,d\theta'$ (Here $ds(r',z')$ is the length element on Γ in the plane Orz for the euclidean norm). By choosing $\theta = \frac{\pi}{2}$ we obtain

$$\frac{1}{2}\phi(r,z) = -\int_{\breve{\Gamma}} \frac{\partial \phi^+}{\partial \breve{\boldsymbol{n}}}(r',z')\Big(\int_0^{2\pi} \ell((r,z),(r',z'),\theta')\,\sin\theta'\,d\theta'\Big)\,r'\,ds(r',z')$$
$$+\int_{\breve{\Gamma}} \phi^+(r',z')\Big(\int_0^{2\pi} m((r,z),(r',z'),\theta')\,\sin\theta'\,d\theta'\Big)\,r'\,ds(r',z')$$

where

$$\ell((r,z),(r',z'),\theta') = \breve{G}((r,\frac{\pi}{2},z),(r',\theta',z'))$$
$$= \frac{1}{4\pi}\frac{1}{\left(r^2+r'^2-2rr'\sin\theta'+(z-z')^2\right)^{\frac{1}{2}}}$$

and

$$m((r,z),(r',z'),\theta') = \frac{\partial\breve{G}}{\partial\breve{\boldsymbol{n}}_{(r',z')}}((r,\frac{\pi}{2},z),(r',\theta',z'))$$
$$= \frac{1}{4\pi}\frac{-r'\cos\theta' n_r(r',z')+(z-z')n_z(r',z')}{\left(r^2+r'^2-2rr'\sin\theta'+(z-z')^2\right)^{\frac{1}{2}}}.$$

Setting

$$g((r,z),(r',z')) = \int_0^{2\pi} \ell((r,z),(r',z'),\theta')\,\sin\theta'\,d\theta',$$
$$g_n((r,z),(r',z')) = \int_0^{2\pi} m((r,z),(r',z'),\theta')\,\sin\theta'\,d\theta',$$

we obtain

$$\frac{1}{2}\phi(r,z) = -\int_{\breve{\Gamma}} \xi(r',z')\, g((r,z),(r',z'))\, r'\, ds(r',z')$$
$$+ \int_{\breve{\Gamma}} \phi(r',z')\, g_n((r,z),(r',z'))\, r'\, ds(r',z') \tag{5.28}$$

for $(r,z) \in \breve{\Gamma}$, where

$$\xi(r',z') := \frac{\partial \phi^+}{\partial \breve{\boldsymbol{n}}}(r',z') \qquad (r',z') \in \breve{\Gamma}.$$

Remark 5.3.2. The function g can be expressed by

$$g((r,z),(r'z')) = \frac{1}{4\pi} \frac{1}{(r^2 + r'^2 + (z-z')^2)^{\frac{1}{2}}} \int_0^{2\pi} \frac{\sin\theta'}{(1-\lambda \sin\theta')^{\frac{1}{2}}}\, d\theta', \tag{5.29}$$

where

$$\lambda = \lambda((r,z),(r',z')) = \frac{2rr'}{r^2 + r'^2 + (z-z')^2}.$$

As mentioned in [146], the function λ ranges between 0 and 1 and reaches 1 only when $r = r'$ and $z = z'$. In this last case, the integral (5.29) is singular and $g((r,z),(r',z'))$ is not defined.

An analogous remark can be made for the function g_n. However (5.28) makes sense and its numerical integration has to be carefully handled (see [146]).

Equation (5.28) provides a relation between the trace of ϕ and its exterior normal derivative that can be used to couple the inner equation (5.19) with the integral representation.

We have from the interface conditions (5.22) and (5.23),

$$0 = \left[\frac{1}{\mu r}\frac{\partial(r\phi)}{\partial r} n_r + \frac{1}{\mu}\frac{\partial \phi}{\partial z} n_z\right]_{\breve{\Gamma}_k}$$
$$= \left[\mu^{-1}\left(\frac{\partial \phi}{\partial r} n_r + \frac{\partial \phi}{\partial z} n_z\right)\right]_{\breve{\Gamma}_k} + \frac{1}{r}[\mu^{-1}]_{\breve{\Gamma}_k} \phi\, n_r$$
$$= \left[\mu^{-1}\frac{\partial \phi}{\partial \breve{n}}\right]_{\breve{\Gamma}_k} + \frac{1}{r}[\mu^{-1}]_{\breve{\Gamma}_k} \phi\, n_r,$$

for $k = 1, 2$. It follows that

$$\frac{1}{\mu_0}\xi + \frac{1}{\mu_0 r}\phi\, n_r = \frac{1}{\mu_0}\frac{\partial \phi^+}{\partial \breve{n}} + \frac{1}{\mu_0 r}\phi\, n_r = \frac{1}{\mu}\frac{\partial \phi^-}{\partial \breve{n}} + \frac{1}{\mu r}\phi\, n_r. \tag{5.30}$$

Multiplying (5.19) by $r\overline{\psi}$, where $\psi \in \mathcal{X}$,

$$\mathcal{X} := \{\psi : \breve{\Omega} \to \mathbb{C};\ \frac{1}{r}\psi, \frac{\partial \psi}{\partial r}, \frac{\partial \psi}{\partial z} \in \breve{\mathcal{L}}^2(\breve{\Omega})\},$$

and integrating by parts, we obtain after using (5.30):

$$\begin{aligned}
-\int_{\breve{\Omega}} \Big(\frac{\partial}{\partial r}\Big(\frac{\mu^{-1}}{r}\frac{\partial}{\partial r}(r\phi)\Big) + \frac{\partial}{\partial z}\Big(\mu^{-1}\frac{\partial \phi}{\partial z}\Big)\Big) r\overline{\psi}\,dr\,dz + \mathrm{i}\omega \int_{\breve{\Omega}} \sigma\phi\overline{\psi}\,r\,dr\,dz \\
= \int_{\breve{\Omega}} \Big(\frac{1}{\mu r}\frac{\partial}{\partial r}(r\phi)\,\frac{\partial}{\partial r}(r\overline{\psi}) + \frac{r}{\mu}\frac{\partial \phi}{\partial z}\frac{\partial \overline{\psi}}{\partial z}\Big)\,dr\,dz + \mathrm{i}\omega \int_{\breve{\Omega}} \sigma\phi\overline{\psi}\,r\,dr\,dz \\
- \int_{\breve{\Gamma}} \Big(\frac{1}{\mu r}\frac{\partial}{\partial r}(r\phi^-)\,n_r + \frac{1}{\mu}\frac{\partial \phi^-}{\partial z}\,n_z\Big)\overline{\psi}\,r\,ds(r,z) \\
= \int_{\breve{\Omega}} \Big(\frac{1}{\mu r}\frac{\partial}{\partial r}(r\phi)\,\frac{\partial}{\partial r}(r\overline{\psi}) + \frac{r}{\mu}\frac{\partial \phi}{\partial z}\frac{\partial \overline{\psi}}{\partial z}\Big)\,dr\,dz + \mathrm{i}\omega \int_{\breve{\Omega}} \sigma\phi\overline{\psi}\,r\,dr\,dz \\
- \int_{\breve{\Gamma}} \mu_0^{-1}\xi\overline{\psi}\,r\,ds(r,z).
\end{aligned}$$

Let us then define the sesquilinear forms:

$$\begin{aligned}
\breve{\mathscr{B}}_V(\phi,\psi) &:= \int_{\breve{\Omega}} \Big(\frac{1}{\mu r}\frac{\partial}{\partial r}(r\phi)\,\frac{\partial}{\partial r}(r\overline{\psi}) + \frac{r}{\mu}\frac{\partial \phi}{\partial z}\frac{\partial \overline{\psi}}{\partial z}\Big)\,dr\,dz \\
&\quad + \mathrm{i}\omega \int_{\breve{\Omega}} \sigma\phi\overline{\psi}\,r\,dr\,dz - \int_{\breve{\Gamma}} \mu_0^{-1}\phi\overline{\psi}n_r\,ds(r,z), \\
\mathscr{C}(\xi,\psi) &:= \int_{\breve{\Gamma}} \xi\overline{\psi}\,r\,ds(r,z),
\end{aligned}$$

we obtain from (5.25)

$$\breve{\mathscr{B}}_V(\phi,\psi) - \mathscr{C}(\xi,\psi) = \mu_0\mathscr{L}_V(\psi) \qquad \forall\,\psi \in \mathcal{X}. \tag{5.31}$$

On the other hand, multiplying (5.28) by $r\overline{\eta}$ where η is a function defined on $\breve{\Gamma}$, and integrating over $\breve{\Gamma}$, we obtain

$$\begin{aligned}
\overline{\mathscr{C}}(\eta,\phi) = -2\int_{\breve{\Gamma}} \Big(\int_{\breve{\Gamma}} g((r,z),(r',z'))\xi(r',z')\,r'\,ds(r',z')\Big)\overline{\eta}(r,z)\,r\,ds(r,z) \\
-2\int_{\breve{\Gamma}} \Big(\int_{\breve{\Gamma}} g_n((r,z),(r',z'))\phi(r',z')\,r'\,ds(r',z')\Big)\overline{\eta}(r,z)\,r\,ds(r,z).
\end{aligned} \tag{5.32}$$

Let us define

$$\mathscr{D}(\xi,\eta) := 2\int_{\breve{\Gamma}} \Big(\int_{\breve{\Gamma}} g((r,z),(r',z'))\, \xi(r',z')\, r'\, ds(r',z') \Big) \overline{\eta}(r,z)\, r\, ds(r,z),$$

and

$$\mathscr{K}(\phi,\eta) = 2\int_{\breve{\Gamma}} \Big(\int_{\breve{\Gamma}} g_n((r,z),(r',z'))\phi(r',z')\, r'\, ds(r',z') \Big) \overline{\eta}(r,z)\, r\, ds(r,z).$$

We obtain from (5.32)

$$\overline{\mathscr{C}}(\eta,\phi) + \mathscr{K}(\phi,\eta) + \mathscr{D}(\xi,\eta) = 0. \tag{5.33}$$

We complete this setting by giving the function space in which we seek ξ: We define $\mathcal{Q}$ as the dual space of $\mathcal{X}_\Gamma$ where

$$\mathcal{X}_\Gamma := \{\psi_{|\breve{\Gamma}};\ \psi \in \mathcal{X}\}.$$

The final variational problem reads then:

$$\begin{cases} \text{Find } (\phi,\xi) \in \mathcal{X} \times \mathcal{Q} \text{ such that:} \\ \breve{\mathscr{B}}_V(\phi,\psi) - \mathscr{C}(\xi,\psi) = \mu_0 \mathscr{L}_V(\psi) & \forall\, \psi \in \mathcal{X}, \\ \overline{\mathscr{C}}(\eta,\phi) + \mathscr{K}(\phi,\eta) + \mathscr{D}(\xi,\eta) = 0 & \forall\, \eta \in \mathcal{Q}. \end{cases} \tag{5.34}$$

It is worth noting that coupling a variational formulation in the interior domain with an integral representation on the boundary gives here very similar settings to the two-dimensional transversal case (Sect. 3.3). The main difference concerns the integral representation itself and its numerical evaluation when this topic is involved.

Chapter 6
Eddy Current Models with Thin Inductors

We investigate in this chapter, the derivation of eddy current models when some specific geometric properties of the conductors are considered. More precisely, in practice, eddy current devices generally involve two types of conductors:

– "Thick" conductors in which eddy currents are induced. These are generally the conductors in which "treatment" takes place (heating, melting, stirring, ...).
– "Thin" conductors that carry current from a power or voltage generator. These ones are generally made of highly conducting material (e.g. Copper) and consist in thin wires or coils.

Combined situations where the treated workpieces are thin conductors are involved can however be considered. The presented results cover these situations without major changes. In other words, in an eddy current process, the conductors may exhibit different space scales. This situation may be at the origin of difficulties for numerical solution. A standard treatment would require the use of fine meshing for the thin conductors and this dramatically increases the cost of numerical simulations in terms of computer time. On the other hand, one is generally not specifically interested in the electric and magnetic fields in the regions involved by this treatment. For these reasons, it is natural to resort to asymptotic approximations in order to simplify equations in the thin devices. We shall see hereafter that these ones reduce, in the present case of time harmonic fields, to algebraic equations in the inductors. It is even noticeable that, in some situations, the derivation of the limit models results in well known circuit equations.

Let us outline the main advantages of considering such simplified models:

1. As we shall see, the obtained models are rather less expensive in terms of computational time than the complete ones. Equations in the inductors are reduced indeed to algebraic equations rather than partial differential ones.
2. The obtained problems are better conditioned that the original ones. In fact, numerical solution of problems with various space scales requires using meshes with variable sizes which can lead to ill conditioning.

R. Touzani and J. Rappaz, *Mathematical Models for Eddy Currents and Magnetostatics: With Selected Applications*, Scientific Computation, DOI 10.1007/978-94-007-0202-8_6,

3. As far as optimal shape of inductors is involved (see Chap. 8), the limit models involve only a finite number of inductor geometrical descriptors (length, area, curvature) which considerably simplifies the setting of the optimization problem.

Our goal is to show how standard techniques of asymptotic expansions enable justifying the obtained models. For this, we reconsider throughout this chapter some of the models presented in previous chapters and prove the derivation of the limit problems. In the cases where proofs are rather technical and delicate we skip the details and mention the obtained results with references to papers where the proofs are given in detail.

All the results quoted in this chapter are borrowed from the papers by Touzani [169, 170] and Amirat–Touzani [8–10]. Let us also indicate that, since the material presented in this chapter is still in progress, some of the mentioned results will be given in their formal form without proof, i.e. they are to be considered as conjectures that are planned for future research.

In the following, we shall use the same classification of the studied models as in Chaps. 3 and 4. For the sake of simplicity, we shall consider, throughout this chapter conductors with constant conductivities and magnetic permeabilities. More precisely, we assume

$$\sigma_{|\Omega_k} = \text{Const.} \qquad \forall\, k, \tag{6.1}$$

$$\mu = \mu_0 \qquad \text{in } \mathbb{R}^3. \tag{6.2}$$

These assumptions simplify the obtention of limit models but can be removed without major difficulty.

6.1 The Two-Dimensional Solenoidal Model

We recall that this model, described in Sect. 3.2, consists in considering fields of the form

$$\boldsymbol{J}(x_1, x_2, x_3) = J_1(x_1, x_2)\, \boldsymbol{e}_1 + J_2(x_1, x_2)\, \boldsymbol{e}_2,$$

$$\boldsymbol{H}(x_1, x_2, x_3) = H(x_1, x_2)\, \boldsymbol{e}_3.$$

Let us consider the typical two-conductor example depicted in Fig. 6.1 where an annular inductor that surrounds a "thick" conductor $\Omega_2 = \Lambda_2 \times \mathbb{R}$. The inductor will be defined more precisely as follows: Let γ_1^- denote a closed curve of class $\mathcal{C}^2$ parameterized by its curvilinear abscissa by the function

$$\boldsymbol{X} : s \in [0, \ell_1) \mapsto \boldsymbol{X}(s) = (X_1(s), X_2(s)) \in \mathbb{R}^2,$$

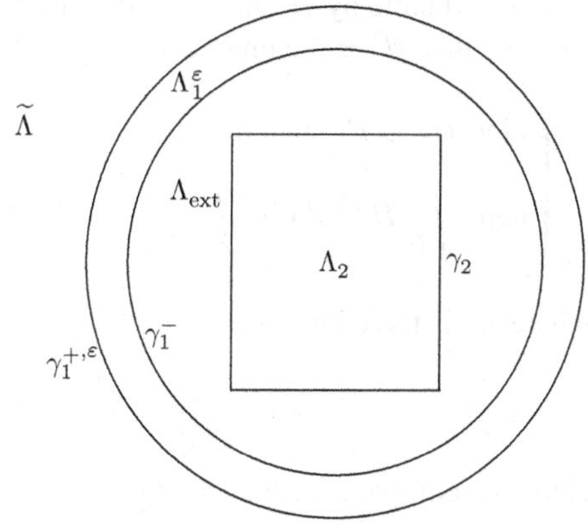

Fig. 6.1 A typical configuration of the conductors for the solenoidal model

where ℓ_1 is the length of the curve γ_1^-. We assume the properties

$$\boldsymbol{X}(0) = \boldsymbol{X}(\ell_1),\ \boldsymbol{X}'(0) = \boldsymbol{X}'(\ell_1). \tag{6.3}$$

Note that due to the choice of this parameterization, we have $|\boldsymbol{X}'(s)| = 1$ for $0 \leq s \leq \ell_1$. We denote by $\boldsymbol{n}(s)$ and $\boldsymbol{t}(s)$ the unit normal and tangent vectors to γ_1^- respectively given by

$$\boldsymbol{n}(s) = (X_2'(s), -X_1'(s)),\ \boldsymbol{t}(s) = \boldsymbol{X}'(s) \qquad 0 \leq s < \ell_1.$$

We define the mapping

$$\boldsymbol{F}_\varepsilon(s, \xi) := \boldsymbol{X}(s) + \varepsilon\, \xi\, \boldsymbol{n}(s) \qquad 0 \leq s < \ell_1,\ 0 < \xi < 1.$$

The description of the inductor is well defined if its "thickness" ε is small enough. The jacobian of the mapping $\boldsymbol{F}_\varepsilon$ is given by

$$J_\varepsilon(s, \xi) = \varepsilon \left|1 + \varepsilon\, \xi\, \boldsymbol{t}(s) \cdot \boldsymbol{n}'(s)\right| \qquad (s, \xi) \in [0, \ell_1) \times (0, 1).$$

Note that, owing to (6.3), there exists $\varepsilon_0 > 0$ such that if $\varepsilon \leq \varepsilon_0$:

$$\varepsilon \leq J_\varepsilon(s, \xi) \leq C_1\, \varepsilon \qquad \forall\ (s, \xi) \in [0, \ell_1) \times (0, 1), \tag{6.4}$$

where the constant C_1 is independent of ε.

The inductor domain is then defined by $\Lambda_1^\varepsilon := \boldsymbol{F}_\varepsilon([0, \ell_1) \times (0, 1))$. The exterior boundary of the inductor is given by $\gamma_1^{+,\varepsilon} := \boldsymbol{F}_\varepsilon([0, \ell_1) \times \{1\})$. The inner free space, which is independent of ε, is the bounded domain Λ_{ext} such that γ_1^- is the boundary of $\Lambda_{ext} \cup \overline{\Lambda}_2$. We also set $\hat{\Lambda}^\varepsilon := \overline{\Lambda}_2 \cup \overline{\Lambda}_{ext} \cup \Lambda_1^\varepsilon$.

Let us denote by H^ε the x_3–component of the magnetic field $\boldsymbol{H}$. It is proven in Sect. 3.2 that H^ε is the unique solution of the variational problem:

$$\begin{cases} \text{Find } H^\varepsilon \in \mathcal{H}^\varepsilon \text{ such that} \\ \mathrm{i}\omega\mu_0 \displaystyle\int_{\hat{\Lambda}^\varepsilon} H^\varepsilon \overline{\phi}\, d\boldsymbol{x} + \int_{\Lambda_1^\varepsilon \cup \Lambda_2} \sigma^{-1} \nabla H^\varepsilon \cdot \nabla \overline{\phi}\, d\boldsymbol{x} = V \overline{\phi}_{|\Lambda_{\text{ext}}} \quad \forall\, \phi \in \mathcal{H}^\varepsilon, \end{cases} \tag{6.5}$$

where $\mathcal{H}^\varepsilon$ is the Hilbert space

$$\mathcal{H}^\varepsilon = \{\phi \in \mathcal{H}_0^1(\hat{\Lambda}^\varepsilon);\ \phi_{|\Lambda_{\text{ext}}} = \text{Const.}\}.$$

Here, the complex constant V stands for the current voltage.

Our purpose is to study the limit of the solution H^ε when $\varepsilon \to 0$. A careful analysis (see [170]) shows that, in order to obtain a "reasonable" limit problem, it is necessary to assume that the electric conductivity in Λ_1^ε is large enough. More precisely, we assume that

$$\frac{1}{\sigma} = \varepsilon\,\alpha + \mathcal{O}(\varepsilon^2) \qquad \text{in } \Lambda_1^\varepsilon, \tag{6.6}$$

where α is a positive real number. In the following we shall omit the remainder $\mathcal{O}(\varepsilon^2)$ for the sake of simplicity. This assumption roughly means that while the inductor gets thin, its resistivity goes to zero. This is in some sense necessary to maintain the same total current intensity in it.

Let us define the following Hilbert space in which lies the limit solution

$$\mathcal{H}^0 := \{\phi \in \mathcal{H}^1(\Lambda_2);\ \phi_{|\gamma_2} = \text{Const.}\},$$

endowed with the norm

$$\|\phi\|_{\mathcal{H}^0} := \left(\|\nabla\phi\|^2_{\mathcal{L}^2(\Lambda_2)} + |\phi_{|\gamma_2}|^2\right)^{\frac{1}{2}}.$$

Let us furthermore define the following problem:

$$\begin{cases} \text{Find } H \in \mathcal{H}^0 \text{ such that} \\ \mathrm{i}\omega\mu_0 \displaystyle\int_{\Lambda_2} H\overline{\phi}\, d\boldsymbol{x} + \sigma^{-1} \int_{\Lambda_2} \nabla H \cdot \nabla\overline{\phi}\, d\boldsymbol{x} \\ \qquad + (\mathrm{i}\omega\mu_0\, |\Lambda_{\text{ext}}| + \alpha\,\ell_1)\, H_{|\gamma_2} \overline{\phi}_{|\gamma_2} = V\overline{\phi}_{|\gamma_2} \qquad \forall\, \phi \in \mathcal{H}_0, \end{cases} \tag{6.7}$$

where ℓ_1 is the length of the curve γ_1^- and where we recall that $|\Lambda_{\text{ext}}|$ is the area of Λ_{ext}.

Theorem 6.1.1. *Problem* (6.7) *possesses a unique solution.*

Proof. Let us define:

$$\begin{aligned}\mathscr{B}(\phi,\psi) &:= \mathrm{i}\omega\mu_0 \int_{\Lambda_2} \phi\overline{\psi}\,dx + \sigma^{-1}\int_{\Lambda_2} \nabla\phi\cdot\nabla\overline{\psi}\,dx \\ &\quad + \left(\mathrm{i}\omega\mu_0\,|\Lambda_{\mathrm{ext}}| + \alpha\,\ell_1\right)\phi_{|\gamma_2}\overline{\psi}_{|\gamma_2},\\ \mathscr{L}(\psi) &:= V\overline{\psi}_{|\gamma_2}.\end{aligned}$$

Existence and uniqueness can be proved by using the Lax–Milgram theorem (Theorem 1.2.1). The continuity of the sesquilinear form $\mathscr{B}$ and the antilinear form $\mathscr{L}$ are easy to prove. The coercivity is established by using the Poincaré–Friedrichs inequality (1.5) as follows

$$\mathrm{Re}\left(\mathscr{B}(\phi,\phi)\right) = \sigma^{-1}\int_{\Lambda_2} |\nabla\phi|^2\,dx + \alpha\ell_1|\phi_{|\gamma_2}|^2.$$

Then

$$|\mathscr{B}(\phi,\phi)| \ge \left|\mathrm{Re}\,\mathscr{B}(\phi,\phi)\right| \ge C\,\|\phi\|^2_{\mathcal{H}^1(\Lambda_2)}.$$

Therefore, the Lax–Milgram theorem applies. □

We have the following result proven in [170]:

Theorem 6.1.2. *Under the hypothesis* (6.6), *the function* H^ε *converges in* $\mathcal{H}^1(\Lambda_2)$ *toward the unique solution* H *of the variational problem* (6.7).

Proof. Choosing $\phi = H^\varepsilon$ in (6.5) we obtain

$$\begin{aligned}&\mathrm{i}\omega\mu_0\int_{\Lambda_2\cup\Lambda_1^\varepsilon} |H^\varepsilon|^2\,dx + \sigma^{-1}\int_{\Lambda_2} |\nabla H^\varepsilon|^2\,dx + \varepsilon\alpha\int_{\Lambda_1^\varepsilon} |\nabla H^\varepsilon|^2\,dx\\ &\quad + \mathrm{i}\omega\mu_0|\Lambda'|\,|H^\varepsilon_{|\gamma_2}|^2 = V\overline{H}^\varepsilon_{|\gamma_2}.\end{aligned}$$

Therefore

$$\|H^\varepsilon\|^2_{\mathcal{L}^2(\Lambda_2)} + \|H^\varepsilon\|^2_{\mathcal{L}^2(\Lambda_1^\varepsilon)} + \|\nabla H^\varepsilon\|^2_{\boldsymbol{\mathcal{L}}^2(\Lambda_2)} + \varepsilon\,\|\nabla H^\varepsilon\|^2_{\boldsymbol{\mathcal{L}}^2(\Lambda_1^\varepsilon)} + |H^\varepsilon_{|\gamma_2}|^2 \le C_1\,|V|^2.$$

This yields the estimates:

$$\begin{aligned}&\|H^\varepsilon\|_{\mathcal{L}^2(\Lambda_2)} + \|H^\varepsilon\|_{\mathcal{L}^2(\Lambda_1^\varepsilon)} + \|\nabla H^\varepsilon\|_{\boldsymbol{\mathcal{L}}^2(\Lambda_2)} + \varepsilon^{\frac{1}{2}}\,\|\nabla H^\varepsilon\|_{\boldsymbol{\mathcal{L}}^2(\Lambda_1^\varepsilon)}\\ &\quad + |H^\varepsilon_{|\gamma_2}| \le C_2.\end{aligned} \tag{6.8}$$

From this we deduce the following convergence, when $\varepsilon \to 0$, for a subsequence of (H^ε), still denoted by (H^ε):

$$\begin{aligned} H^\varepsilon &\to H && \text{in } \mathcal{L}^2(\Lambda_2), \\ \nabla H^\varepsilon &\rightharpoonup \nabla H && \text{in } \boldsymbol{\mathcal{L}}^2(\Lambda_2) \text{ weakly}, \\ H^\varepsilon_{|\gamma_2} &\to H_{|\gamma_2} && \text{in } \mathcal{L}^2(\Lambda_2). \end{aligned}$$

We obtain then the limits when $\varepsilon \to 0$ for any $\phi \in \mathcal{H}^0$:

$$\begin{aligned} &\int_{\Lambda_2} H^\varepsilon \overline{\phi}\, d\boldsymbol{x} \to \int_{\Lambda_2} H \overline{\phi}\, d\boldsymbol{x}, \\ &\int_{\Lambda_2} \nabla H^\varepsilon \cdot \nabla \overline{\phi}\, d\boldsymbol{x} \to \int_{\Lambda_2} \nabla H \cdot \nabla \overline{\phi}\, d\boldsymbol{x}, \\ &H^\varepsilon_{|\gamma_2} \overline{\phi}_{|\gamma_2} \to H_{|\gamma_2} \overline{\phi}_{|\gamma_2}. \end{aligned}$$

In addition, we have by the Lebesgue convergence theorem:

$$\lim_{\varepsilon \to 0} \int_{\Lambda_1^\varepsilon} H^\varepsilon \overline{\phi}\, d\boldsymbol{x} = 0 \qquad \forall\, \phi \in \mathcal{H}^0.$$

Taking the limit in the term involving Λ_1^ε is more delicate.

Let us define, for a function $\phi \in \mathcal{H}^1(\Lambda_1^\varepsilon)$, the function $\hat{\phi}(s,\xi) := \phi(\boldsymbol{F}_\varepsilon(s,\xi))$ for $(s,\xi) \in [0,\ell_1) \times (0,1)$. A straightforward calculation gives

$$\int_{\Lambda_1^\varepsilon} \nabla\phi \cdot \nabla\overline{\psi}\, d\boldsymbol{x} = \int_0^1 d\xi \int_0^{\ell_1} \frac{1}{J_\varepsilon} \Big(\varepsilon^2 \frac{\partial \hat{\phi}}{\partial s} \frac{\partial \overline{\hat{\psi}}}{\partial s} + \delta_\varepsilon \frac{\partial \hat{\phi}}{\partial \xi} \frac{\partial \overline{\hat{\psi}}}{\partial \xi} \Big)\, ds,$$

where $\delta_\varepsilon := |\boldsymbol{X}' + \varepsilon\, \xi\, \boldsymbol{n}'|^2$. Therefore the estimate (6.8) implies

$$\varepsilon \left\| \frac{\partial \hat{H}^\varepsilon}{\partial s} \right\|_{\mathcal{L}^2([0,\ell_1)\times(0,1))} + \left\| \frac{\partial \hat{H}^\varepsilon}{\partial \xi} \right\|_{\mathcal{L}^2([0,\ell_1)\times(0,1))} \le C.$$

This implies that, up to an extraction of a subsequence,

$$\frac{\partial \hat{H}^\varepsilon}{\partial \xi} \rightharpoonup \frac{\partial \hat{H}}{\partial \xi} \qquad \text{in } \mathcal{L}^2([0,\ell_1) \times (0,1)) \text{ weakly},$$

and we then have the limits, when $\varepsilon \to 0$:

$$\alpha\varepsilon^3 \int_0^1 d\xi \int_0^{\ell_1} \frac{1}{J_\varepsilon} \frac{\partial \hat{H}^\varepsilon}{\partial s} \frac{\partial \overline{\hat{\phi}}}{\partial s}\, ds \to 0,$$

$$\alpha\varepsilon \int_0^1 d\xi \int_0^{\ell_1} \frac{\delta_\varepsilon}{J_\varepsilon} \frac{\partial \hat{H}^\varepsilon}{\partial \xi} \frac{\partial \overline{\hat{\phi}}}{\partial \xi}\, ds = \alpha \int_0^1 d\xi \int_0^{\ell_1} \frac{|\boldsymbol{X}' + \varepsilon\,\xi\,\boldsymbol{n}'|^2}{|1 + \varepsilon\,\xi\,\boldsymbol{t}\cdot\boldsymbol{n}'|} \frac{\partial \hat{H}^\varepsilon}{\partial \xi} \frac{\partial \overline{\hat{\phi}}}{\partial \xi}\, ds$$

$$\to \alpha \int_0^1 d\xi \int_0^{\ell_1} \frac{\partial \hat{H}}{\partial \xi} \frac{\partial \overline{\hat{\phi}}}{\partial \xi}\, ds.$$

Let us now select a particular test function v. We choose $\hat{\phi}(s,\xi) = \phi_{|\gamma_2}(1-\xi)$ and then obtain

$$\alpha\varepsilon \int_0^1 d\xi \int_0^{\ell_1} \frac{\delta_\varepsilon}{J_\varepsilon} \frac{\partial \hat{H}^\varepsilon}{\partial \xi} \frac{\partial \overline{\hat{\phi}}}{\partial \xi}\, ds \to -\alpha\overline{\phi}_{|\gamma_2}\ell_1 \int_0^1 \frac{\partial \hat{H}}{\partial \xi}\, d\xi = \alpha\ell_1 H_{|\gamma_2}\overline{\phi}_{|\gamma_2}.$$

Finally, the uniqueness of solutions of (6.7) implies that the whole sequence (H^ε) converges to H in $\mathcal{H}^1(\Lambda_2)$ weakly. □

It is possible to formulate the limit problem (6.7) like (3.32). For this, Let $\tilde{H} = H/H_{|\gamma_2}$. Problem (6.7) can be interpreted as the following boundary value problem:

$$\begin{cases} -\sigma^{-1}\Delta\tilde{H} + \mathrm{i}\omega\mu_0\tilde{H} = 0 & \text{in } \Lambda_2, \\ \tilde{H} = 1 & \text{on } \gamma_2. \end{cases} \tag{6.9}$$

A simple calculation using the Green formula leads to the identity:

$$H = \frac{V\tilde{H}}{\delta}, \quad \text{with } \delta = \sigma^{-1}\int_{\gamma_2} \frac{\partial \tilde{H}}{\partial n}\, ds + \mathrm{i}\omega\mu_0|\Lambda_{\text{ext}}| + \alpha\ell_1.$$

An interesting case is the one where we deal with only one inductor $\Omega_1^\varepsilon = \Lambda_1^\varepsilon \times \mathbb{R}$ and no workpiece. The obtained limit problem reduces in this case to the algebraic equation:

$$(\mathrm{i}\omega\mu_0\,|\Lambda_{\text{ext}}| + \alpha\ell_1)\, H_{|\gamma_2} = V. \tag{6.10}$$

We have indeed from (6.9) and since, in this case, $\sigma = 0$ in Λ_2,

$$\Delta\tilde{H} = 0 \quad \text{in } \Lambda_2.$$

Using the boundary condition in (6.9) we deduce that $\tilde{H} \equiv 1$ in Λ_2. Thus $\dfrac{\partial \tilde{H}}{\partial n} = 0$ on γ_2 and we have:

$$V = \delta H_{|\gamma_2} = H_{|\gamma_2}(\mathrm{i}\omega\mu_0|\Lambda_{\text{ext}}| + \alpha\ell_1).$$

The relation (6.10) is nothing else but the well known *Kirchhoff equation* in electrodynamics for a RL (Resistance – Self Inductance) circuit (see [139] for instance). To see this, one must consider a unit length of depth of the inductor (in the x_3–direction). Defining the resistance and the inductance as usual in electrotechnics respectively by

$$R = \alpha \ell_1, \ L = \mu_0 \, |\Lambda_{\text{ext}}|,$$

recalling that α stands for the ratio between resistivity (inverse of conductivity) and the size of the inductor, we obtain the Kirchhoff circuit equation:

$$(\mathrm{i}\omega L + R)\, H_{|\gamma_2} = V. \tag{6.11}$$

Note that $H_{|\gamma_2}$ is the magnetic field value for a unit depth which is also the current intensity. This observation suggests that the presented limit process has yielded a coupling between a field equation and a circuit equation.

6.2 The Two-Dimensional Transversal Model

Let us now turn to the 2-D transversal model described in Sect. 3.3. The geometrical set up of the conductors consists (see Fig. 6.2) in one conductor $\Omega_1 = \Lambda_1 \times \mathbb{R}$ and two inductors $\Omega_k = \Lambda_k \times \mathbb{R}$, $k = 2, 3$ that we assume thin. Let us recall that in fact Ω_2 and Ω_3 (and then Λ_2 and Λ_3) represent one unique inductor with two branches that can be seen as "linked at the infinity". We denote by γ_k the boundary of Λ_k and set $\gamma = \gamma_1 \cup \gamma_2 \cup \gamma_3$. This assumption is implemented as follows: We consider that $\Lambda_k = z_k + \varepsilon\, \hat{\Lambda}_k$, for $k = 2, 3$ where z_2 and z_3 are two points in $\mathbb{R}^2$ that do not lie in $\overline{\Lambda}_1$, $\varepsilon \ll 1$ is a small parameter and $\hat{\Lambda}_2$, $\hat{\Lambda}_3$ are two domains in $\mathbb{R}^2$. For this reason, the sets Λ_k will be referred to as Λ_k^ε for $k = 2, 3$, where

$$\Lambda_k^\varepsilon := \{\boldsymbol{x} = z_k + \varepsilon\, \hat{\boldsymbol{x}}_k,\ \hat{\boldsymbol{x}}_k \in \hat{\Lambda}_k\}.$$

Clearly, the domains Λ_2^ε and Λ_3^ε degenerate to z_2 and z_3 respectively when $\varepsilon \to 0$. Furthermore, for ε small enough, the domains $\overline{\Lambda}_2^\varepsilon$, $\overline{\Lambda}_3^\varepsilon$ do not intersect $\overline{\Lambda}_1$.

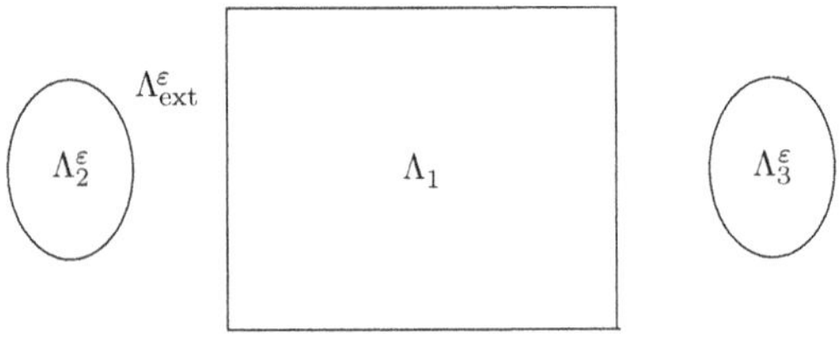

Fig. 6.2 A typical configuration of the conductors for the transversal model

Let us now recall the eddy current problem for such configurations. The potential A defined by (3.36), and denoted here by A^ε, is solution of the set of equations (see (3.38)–(3.41)), by using assumptions (6.1)–(6.2):

$$-\Delta A^\varepsilon + \mathrm{i}\omega\mu_0\sigma(A^\varepsilon - \tilde{A}_1) = 0 \qquad \text{in } \Lambda_1, \tag{6.12}$$

$$-\Delta A^\varepsilon + \mathrm{i}\omega\mu_0\sigma(A^\varepsilon - \tilde{A}_2^\varepsilon) = \frac{\mu_0 I}{|\Lambda_2^\varepsilon|} \qquad \text{in } \Lambda_2^\varepsilon, \tag{6.13}$$

$$-\Delta A^\varepsilon + \mathrm{i}\omega\mu_0\sigma(A^\varepsilon - \tilde{A}_3^\varepsilon) = -\frac{\mu_0 I}{|\Lambda_3^\varepsilon|} \qquad \text{in } \Lambda_3^\varepsilon, \tag{6.14}$$

$$\Delta A^\varepsilon = 0 \qquad \text{in } \Lambda_{\text{ext}}^\varepsilon, \tag{6.15}$$

$$\tilde{A}_1 = 0, \tag{6.16}$$

$$[A^\varepsilon] = \left[\frac{\partial A^\varepsilon}{\partial n}\right] = 0 \qquad \text{on } \gamma, \tag{6.17}$$

$$A^\varepsilon(\boldsymbol{x}) = \beta + \mathcal{O}(|\boldsymbol{x}|^{-1}) \qquad |\boldsymbol{x}| \to \infty, \tag{6.18}$$

where the complex number I stands for the total current intensity flowing in the inductor, and the symbol $\tilde{}$ means for a function ψ:

$$\tilde{\psi}_1 := \frac{1}{|\Lambda_1|}\int_{\Lambda_1} \psi \, d\boldsymbol{x},$$

$$\tilde{\psi}_k^\varepsilon := \frac{1}{|\Lambda_k^\varepsilon|}\int_{\Lambda_k^\varepsilon} \psi \, d\boldsymbol{x} \qquad k = 2, 3.$$

Let us define the space

$$\mathcal{V} := \{\phi \in \mathcal{W}^1(\mathbb{R}^2);\ \tilde{\phi}_1 = 0\}.$$

It is noteworthy that on $\mathcal{V}$, the semi-norm $|\cdot|_{\mathcal{W}^1(\Lambda_1)}$ is a norm on $\mathcal{V}$ which is equivalent to $\|\cdot\|_{\mathcal{W}^1(\Lambda_1)}$ (see for this [149]). We have for Problem (6.12)–(6.18) the variational formulation:

$$\begin{cases} \text{Find } A^\varepsilon \in \mathcal{V} \text{ such that} \\ \displaystyle\int_{\mathbb{R}^2} \nabla A^\varepsilon \cdot \nabla\overline{\phi}\, d\boldsymbol{x} + \mathrm{i}\omega\mu_0 \int_{\Lambda_1} \sigma A^\varepsilon \overline{\phi}\, d\boldsymbol{x} \\ \displaystyle\quad + \mathrm{i}\omega\mu_0 \sum_{k=2}^{3} \int_{\Lambda_k^\varepsilon} \sigma(A^\varepsilon - \tilde{A}_k^\varepsilon)\overline{\phi}\, d\boldsymbol{x} = \mu_0\, I(\tilde{\phi}_2^\varepsilon - \tilde{\phi}_3^\varepsilon) \qquad \forall\, \phi \in \mathcal{V}. \end{cases} \tag{6.19}$$

Thanks to Theorem 3.3.2, Problem (6.19) has a unique solution.

We are interested in the limit of A^ε when $\varepsilon \to 0$. In the following, we prove, in a precise sense, that the limit problem is given by:

$$-\Delta A + \mathrm{i}\omega\mu_0\sigma\, A = 0 \qquad \text{in } \Lambda_1, \tag{6.20}$$

$$-\Delta A = \mu_0 I\,(\delta_{z_1} - \delta_{z_2}) \qquad \text{in } \mathbb{R}^2 \setminus \overline{\Lambda}_1, \tag{6.21}$$

$$[A] = \left[\frac{\partial A}{\partial n}\right] = 0 \qquad \text{on } \gamma_1, \tag{6.22}$$

$$A(\boldsymbol{x}) = \beta + \mathcal{O}(|\boldsymbol{x}|^{-1}) \qquad |\boldsymbol{x}| \to \infty, \tag{6.23}$$

where δ_{z_k} is the Dirac delta concentrated at z_k.

We shall, in the following, present the proof given in (Amirat–Touzani [10]). For this, we first prove an existence and uniqueness result for (6.20)–(6.23). Let us define the space

$$\mathcal{L}^2_\varrho(\mathbb{R}^2) := \{v : \mathbb{R}^2 \to \mathbb{C};\ \varrho v \in \mathcal{L}^2(\mathbb{R}^2)\},$$

where

$$\varrho(\boldsymbol{x}) = \frac{1}{(1+|\boldsymbol{x}|)\,\ln(2+|\boldsymbol{x}|)} \qquad \boldsymbol{x} \in \mathbb{R}^2.$$

Theorem 6.2.1. *Problem* (6.20)–(6.23) *has a unique weak solution in* $\mathcal{L}^2_\varrho(\mathbb{R}^2)$.

Proof. Let us define the function

$$\hat{A}(\boldsymbol{x}) := \mu_0 I\,(G(\boldsymbol{x},z_1) - G(\boldsymbol{x},z_2)) = \frac{\mu_0 I}{2\pi}\ln\left(\frac{|\boldsymbol{x}-z_2|}{|\boldsymbol{x}-z_1|}\right) \qquad \boldsymbol{x} \in \mathbb{R}^2.$$

Using the properties of the Green function G we check that

$$-\Delta\hat{A} = \mu_0 I(\delta_{z_1} - \delta_{z_2}) \quad \text{and } \hat{A} \in \mathcal{L}^2_\varrho(\mathbb{R}^2).$$

Defining $\tilde{A} = A - \hat{A}$, we obtain the problem

$$\begin{cases} -\Delta\tilde{A} + \mathrm{i}\omega\mu_0\sigma\tilde{A} = \mathrm{i}\omega\mu_0\sigma\hat{A} & \text{in } \Lambda_1, \\ \Delta\tilde{A} = 0 & \text{in } \mathbb{R}^2 \setminus \overline{\Lambda}_1, \\ [\tilde{A}] = \left[\dfrac{\partial\tilde{A}}{\partial n}\right] = 0 & \text{on } \gamma_1, \\ \tilde{A}(\boldsymbol{x}) = \beta + \mathcal{O}(|\boldsymbol{x}|^{-1}) & |\boldsymbol{x}| \to \infty. \end{cases}$$

Therefore, we look for a function $\tilde{A} \in \mathcal{W}^1(\mathbb{R}^2)$ such that for all $\varphi \in \mathcal{W}^1(\mathbb{R}^2)$

$$\int_{\mathbb{R}^2} \nabla \tilde{A} \cdot \nabla \overline{\varphi} \, d\boldsymbol{x} + \mathrm{i}\omega\mu_0 \int_{\Lambda_1} \sigma \tilde{A} \overline{\varphi} \, d\boldsymbol{x} = \mathrm{i}\omega\mu_0 \int_{\Lambda_1} \sigma \hat{A} \overline{\varphi} \, d\boldsymbol{x}. \tag{6.24}$$

Using the Lax–Milgram theorem (Theorem 1.2.1), we conclude by the existence and uniqueness of a solution of (6.24).

Let A_1 and A_2 denote two weak solutions in $\mathcal{L}^2_\varrho(\mathbb{R}^2)$ of (6.20)–(6.23). We have for the difference $A = A_1 - A_2$, the equation

$$\int_{\mathbb{R}^2} A \, (-\Delta\overline{\varphi} + \mathrm{i}\omega\mu_0\chi_1\sigma\overline{\varphi}) \, d\boldsymbol{x} = 0 \qquad \forall \, \varphi \in \mathscr{D}(\mathbb{R}^2),$$

where χ_1 is the characteristic function of the domain Λ_1, (i.e., $\chi_1 = 1$ on Λ_1, 0 elsewhere). Now, if $\psi \in \mathcal{L}^2_\varrho(\mathbb{R}^2)$, there exists a unique $\varphi \in \mathcal{W}^1(\mathbb{R}^2)$ that satisfies

$$-\Delta\overline{\varphi} + \mathrm{i}\omega\mu_0\sigma\chi_1\overline{\varphi} = \varrho^2\overline{\psi} \qquad \text{in } \mathbb{R}^2.$$

It follows that

$$\int_{\mathbb{R}^2} \varrho^2 A \overline{\psi} \, d\boldsymbol{x} = 0 \qquad \forall \, \psi \in \mathcal{L}^2_\varrho(\mathbb{R}^2),$$

By choosing $\psi = A$, we obtain $A = 0$, which implies $A_1 = A_2$ and uniqueness follows. □

It is noteworthy that since (6.20)–(6.23) cannot have a solution in $\mathcal{W}^1(\mathbb{R}^2)$, the right-hand side of (6.21) being a combination of Dirac masses, then the convergence of A^ε to A cannot take place in this space. The following result shows that this one holds in $\mathcal{L}^2$ and gives a bound for the error. In (Amirat–Touzani [10]), a stronger convergence result is proven.

Theorem 6.2.2. *The potential A^ε converges in $\mathcal{L}^2_\varrho(\mathbb{R}^2)$ toward the unique solution A of* (6.20)–(6.23)*. Moreover, there exists a constant C independent of ε such that the following error bound holds*

$$\|\varrho(A^\varepsilon - A)\|_{L^2(\mathbb{R}^2)} \le C \, \varepsilon^\alpha,$$

for any $0 < \alpha < \frac{1}{2}$.

The proof of this result makes use of standard duality techniques due to Lions and Magenes ([124], p. 177). We decompose the proof in some separate steps for the sake of clearness. We start by deriving a dual formulation and then prove some preliminary result on this formulation before passing to the limit.

Let us multiply Eqs. (6.12)–(6.15) by a function $\varphi \in \mathcal{V} \cap \mathcal{H}^2_{\mathrm{loc}}(\mathbb{R}^2)$, and integrate them over their respective domains, we have

$$\int_{\mathbb{R}^2} \nabla A^\varepsilon \cdot \nabla \overline{\varphi}\, d\boldsymbol{x} + \mathrm{i}\omega\mu_0 \sum_{k=2}^{3} \int_{\Lambda_k^\varepsilon} \sigma(A^\varepsilon - \tilde{A}_k^\varepsilon)\, \overline{\varphi}\, d\boldsymbol{x}$$
$$+ \mathrm{i}\omega\mu_0 \int_{\Lambda_1} \sigma A^\varepsilon \overline{\varphi}\, d\boldsymbol{x} = \mu_0 I(\overline{\tilde{\varphi}_2^\varepsilon} - \overline{\tilde{\varphi}_3^\varepsilon}).$$

Using the Green formula and the identity

$$\int_{\Lambda_k^\varepsilon} (A^\varepsilon - \tilde{A}_k^\varepsilon)\overline{\varphi}\, d\boldsymbol{x} = \int_{\Lambda_k^\varepsilon} A^\varepsilon(\overline{\varphi} - \overline{\tilde{\varphi}_k^\varepsilon})\, d\boldsymbol{x} = \int_{\Lambda_k^\varepsilon} (A^\varepsilon - \tilde{A}_k^\varepsilon)(\overline{\varphi} - \overline{\tilde{\varphi}_k^\varepsilon})\, d\boldsymbol{x},$$

we get

$$\int_{\mathbb{R}^2} A^\varepsilon\Big(- \Delta\overline{\varphi}\, d\boldsymbol{x} + \mathrm{i}\omega\mu_0\sigma \sum_{k=1}^{3} \chi_k^\varepsilon(\overline{\varphi} - \overline{\tilde{\varphi}_k^\varepsilon})\Big)\, d\boldsymbol{x} = \mu_0 I\, (\overline{\tilde{\varphi}_2^\varepsilon} - \overline{\tilde{\varphi}_3^\varepsilon}),$$

where χ_k^ε is the characteristic function of Λ_k^ε. Let ψ denote a function in $\mathcal{L}_\varrho^2(\mathbb{R}^2)$ and let $\varphi^\varepsilon \in \mathcal{V}$ be the solution of

$$- \Delta\varphi^\varepsilon + \mathrm{i}\omega\mu_0\sigma \sum_{k=2}^{3} \chi_k^\varepsilon(\varphi^\varepsilon - \tilde{\varphi}^\varepsilon) + \mathrm{i}\omega\mu_0\sigma\chi_1\varphi^\varepsilon = \varrho^2\psi \qquad \text{in } \mathbb{R}^2. \tag{6.25}$$

Then we have

$$\int_{\mathbb{R}^2} \varrho^2 A^\varepsilon \overline{\psi}\, d\boldsymbol{x} = \mu_0 I(\overline{\tilde{\varphi}_2^\varepsilon} - \overline{\tilde{\varphi}_3^\varepsilon}). \tag{6.26}$$

In the following, we prove that the sequence (φ^ε) converges to a function $\varphi \in \mathcal{V}$ that solves the equation:

$$\int_{\mathbb{R}^2} \nabla\varphi \cdot \nabla\overline{v}\, d\boldsymbol{x} + \mathrm{i}\omega\mu_0 \int_{\Lambda_1} \sigma\varphi\,\overline{v}\, d\boldsymbol{x}$$
$$+ \mathrm{i}\omega\mu_0 \sum_{i=2}^{3} \int_{\Lambda_i} (\varphi - \tilde{\varphi})\overline{v}\, d\boldsymbol{x} = \int_{\mathbb{R}^2} \varrho^2\psi\overline{v}\, d\boldsymbol{x} \qquad \forall\, v \in \mathcal{V}. \tag{6.27}$$

Note that by using classical regularity results for elliptic equations (see Dautray–Lions [62] for instance), we have $\varphi \in \mathcal{H}_{\text{loc}}^2(\mathbb{R}^2)$. This result will then be used to take the limit $\varepsilon \to 0$ in the formulation (6.26).

Lemma 6.2.1. *The sequence* (φ^ε) *converges toward* φ *in* $\mathcal{H}^2(B)$ *for any ball of* $\mathbb{R}^2$ *containing* Λ^ε. *Moreover we have the error bound:*

$$\|\varphi^\varepsilon - \varphi\|_{\mathcal{H}^2(B)} \le C\varepsilon \, \|\varrho\psi\|_{\mathcal{L}^2(\mathbb{R}^2)}, \tag{6.28}$$

where the constant C *is independent of* ε.

Proof. Multiplying (6.25) by $\overline{\varphi}^\varepsilon$, integrating and using the Green formula, we obtain

$$\int_{\mathbb{R}^2} |\nabla\varphi^\varepsilon|^2 \, d\boldsymbol{x} + i\omega\mu_0 \sum_{k=2}^{3} \int_{\Lambda_k^\varepsilon} \sigma|\varphi^\varepsilon - \tilde{\varphi}_k^\varepsilon|^2 \, d\boldsymbol{x} + i\omega\mu_0 \int_{\Lambda_1} \sigma|\varphi^\varepsilon|^2 \, d\boldsymbol{x} = \int_{\mathbb{R}^2} \varrho^2 \overline{\psi} \varphi^\varepsilon \, d\boldsymbol{x}.$$

From the bound (2.11) we deduce

$$\int_{\mathbb{R}^2} |\nabla\varphi^\varepsilon|^2 \, d\boldsymbol{x} + \sum_{k=2}^{3} \int_{\Lambda_k^\varepsilon} |\varphi^\varepsilon - \tilde{\varphi}^\varepsilon|^2 \, d\boldsymbol{x} + \int_{\Lambda_1} |\varphi^\varepsilon|^2 \, d\boldsymbol{x} \le C \left| \int_{\mathbb{R}^2} \varrho^2 \overline{\psi} \varphi^\varepsilon \, d\boldsymbol{x} \right|.$$

Using the Cauchy–Schwarz inequality we deduce

$$\int_{\mathbb{R}^2} |\nabla\varphi^\varepsilon|^2 \, d\boldsymbol{x} \le C \, \|\varrho\psi\|_{\mathcal{L}^2(\mathbb{R}^2)} \|\varrho\varphi^\varepsilon\|_{\mathcal{L}^2(\mathbb{R}^2)}.$$

Then

$$\left(\int_{\mathbb{R}^2} |\nabla\varphi^\varepsilon|^2 \, d\boldsymbol{x} \right)^{\frac{1}{2}} \le C \, \|\varrho\psi\|_{\mathcal{L}^2(\mathbb{R}^2)}.$$

To derive $\mathcal{L}^2$–estimates we first use the equivalence of norms (see [149]) to obtain

$$\begin{aligned} \|\varphi^\varepsilon\|_{\mathcal{L}^2(\Lambda_1)} &\le C_1 \, \|\nabla\varphi^\varepsilon\|_{\mathcal{L}^2(\mathbb{R}^2)} \\ &\le C_2 \, \|\varrho\psi\|_{\mathcal{L}^2(\mathbb{R}^2)}. \end{aligned} \tag{6.29}$$

On the other hand, using the Poincaré–Wirtinger inequality (see [41], p. 194) and a scaling argument, we obtain for $k = 2, 3$:

$$\begin{aligned} \|\varphi^\varepsilon - \tilde{\varphi}_k^\varepsilon\|_{\mathcal{L}^2(\Lambda_k^\varepsilon)} &\le C_3\varepsilon \, \|\nabla\varphi^\varepsilon\|_{\mathcal{L}^2(\Lambda_k^\varepsilon)} \\ &\le C_4\varepsilon \, \|\varrho\psi\|_{\mathcal{L}^2(\mathbb{R}^2)}. \end{aligned} \tag{6.30}$$

This proves also that (φ^ε) is bounded in $\mathcal{H}^1(B)$ for any bounded open set B of $\mathbb{R}^2$.

Let us prove the $\mathcal{H}^2$–estimate. Using standard regularity results for elliptic equations (see [92], p. 183 for instance), we obtain for any ball B of $\mathbb{R}^2$ containing Λ^ε, and any regular domain D containing $\overline{B}$, by using (6.25), (6.29) and (6.30),

$$\begin{aligned}\|\varphi^\varepsilon\|_{\mathcal{H}^2(B)} &\le C_1\Big(\|\varphi^\varepsilon\|_{\mathcal{H}^1(D)} + \|\varrho^2\psi\|_{\mathcal{L}^2(D)} + \|\varphi^\varepsilon\|_{\mathcal{L}^2(\Lambda_1)} \\ &\qquad + \sum_{k=2}^{3} \|\varphi^\varepsilon - \tilde{\varphi}_k^\varepsilon\|_{\mathcal{L}^2(\Lambda_k^\varepsilon)}\Big) \\ &\le C_2\, \|\varrho\psi\|_{\mathcal{L}^2(\mathbb{R}^2)}. \end{aligned} \tag{6.31}$$

The estimates (6.29)–(6.31) enable concluding that a subsequence of (φ^ε) converges toward φ weakly in $\mathcal{H}^2(B)$ for any ball B of $\mathbb{R}^2$. Let us characterize the limit.

Let $\phi^\varepsilon = \varphi^\varepsilon - \varphi \in \mathcal{V} \cap \mathcal{H}^2_{\text{loc}}(\mathbb{R}^2)$. We have the variational equation

$$\int_{\mathbb{R}^2} \nabla\phi^\varepsilon \cdot \nabla\overline{v}\, d\boldsymbol{x} + \mathrm{i}\omega\mu_0 \int_{\Lambda_1} \sigma\phi^\varepsilon\overline{v}\, d\boldsymbol{x} + \mathrm{i}\omega\mu_0 \sum_{k=2}^{3} \int_{\Lambda_k^\varepsilon} \sigma(\varphi^\varepsilon - \tilde{\varphi}_k^\varepsilon)\,\overline{v}\, d\boldsymbol{x} = 0 \quad \forall\, v \in \mathcal{V}.$$

Choosing $v = \phi^\varepsilon$, we obtain

$$\int_{\mathbb{R}^2} |\nabla\phi^\varepsilon|^2\, d\boldsymbol{x} + \mathrm{i}\omega\mu_0 \int_{\Lambda_1} \sigma|\phi^\varepsilon|^2\, d\boldsymbol{x} = -\mathrm{i}\omega\mu_0 \sum_{k=2}^{3} \int_{\Lambda_k^\varepsilon} \sigma(\varphi^\varepsilon - \tilde{\varphi}_k^\varepsilon)\overline{\phi}^\varepsilon\, d\boldsymbol{x}.$$

Then using the estimates (6.30), (2.11) and the Cauchy–Schwarz inequality, we have

$$\begin{aligned}\int_{\mathbb{R}^2} |\nabla\phi^\varepsilon|^2\, d\boldsymbol{x} + \int_{\Lambda_1} |\phi^\varepsilon|^2\, d\boldsymbol{x} &\le C_1 \sum_{k=2}^{3} \|\varphi^\varepsilon - \tilde{\varphi}_k^\varepsilon\|_{\mathcal{L}^2(\Lambda_k^\varepsilon)}\, \|\phi^\varepsilon\|_{\mathcal{L}^2(\Lambda_k^\varepsilon)} \\ &\le C_2\varepsilon\, \|\varrho\psi\|_{L^2(\mathbb{R}^2)}\, \|\nabla\phi^\varepsilon\|_{\boldsymbol{\mathcal{L}}^2(\mathbb{R}^2)}.\end{aligned}$$

Therefore, by (1.46), we have the bounds

$$\|\nabla\phi^\varepsilon\|_{\boldsymbol{\mathcal{L}}^2(\mathbb{R}^2)} \le C_2\varepsilon\, \|\varrho\psi\|_{\mathcal{L}^2(\mathbb{R}^2)}, \tag{6.32}$$

$$\|\phi^\varepsilon\|_{\mathcal{L}^2(\Lambda_1)} \le C_3\, \|\nabla\phi^\varepsilon\|_{\boldsymbol{\mathcal{L}}^2(\mathbb{R}^2)} \le C_4\,\varepsilon\, \|\varrho\psi\|_{\mathcal{L}^2(\mathbb{R}^2)}. \tag{6.33}$$

The sequence (φ^ε) converges then to φ strongly in $\mathcal{W}^1(\mathbb{R}^2)$. This proves the limit problem (6.27).

We next have from (1.46) and (6.32), for $k = 2, 3$,

$$\begin{aligned}\|\phi^\varepsilon\|_{\mathcal{L}^2(\Lambda_k^\varepsilon)} &\le C_1\, \|\varrho\phi^\varepsilon\|_{\mathcal{L}^2(\Lambda_k^\varepsilon)} \\ &\le C_1\, \|\varrho\phi^\varepsilon\|_{\mathcal{L}^2(\mathbb{R}^2)} \\ &\le C_2\, \|\nabla\phi^\varepsilon\|_{\boldsymbol{\mathcal{L}}^2(\mathbb{R}^2)} \\ &\le C_3\varepsilon\, \|\varrho\psi\|_{\mathcal{L}^2(\mathbb{R}^2)}.\end{aligned}$$

The $\mathcal{H}^2$–estimate is handled in the following way: By subtracting (6.25) from (6.27) written in its strong form, we obtain

$$-\Delta\phi^\varepsilon = -\mathrm{i}\omega\mu_0\sigma\chi_0\phi^\varepsilon - \mathrm{i}\omega\mu_0\sigma\sum_{k=2}^{3}\chi_k^\varepsilon(\varphi^\varepsilon - \tilde{\varphi}_k^\varepsilon) \qquad \text{in } \mathbb{R}^2.$$

Using (6.30), (6.33) and classical regularity results for elliptic problems (see [92], p. 183 for instance), we get

$$\begin{aligned}\|\phi^\varepsilon\|_{\mathcal{H}^2(B)} &\le C_1\Big(\|\phi^\varepsilon\|_{\mathcal{H}^1(D)} + \|\phi^\varepsilon\|_{\mathcal{L}^2(\Lambda_1)} + \sum_{k=2}^{3}\|\varphi^\varepsilon - \tilde{\varphi}_k^\varepsilon\|_{\mathcal{L}^2(\Lambda_k^\varepsilon)}\Big)\\ &\le C_2\Big(\varepsilon\,\|\nabla\phi^\varepsilon\|_{\mathcal{L}^2(\mathbb{R}^2)} + \|\phi^\varepsilon\|_{\mathcal{L}^2(\Lambda_1)} + \varepsilon\,\|\varrho\psi\|_{\mathcal{L}^2(\mathbb{R}^2)}\Big)\\ &\le C_3\,\varepsilon\,\|\varrho\psi\|_{\mathcal{L}^2(\mathbb{R}^2)},\end{aligned}$$

for all compact subsets B of $\mathbb{R}^2$ and all regular domains D that contain $\overline{B}$. Note that the constant C_3 depends actually on B. □

We can now derive the convergence result for A^ε. Consider the problem (6.26) and the following one, for $\psi \in \mathcal{L}^2_\varrho(\mathbb{R}^2)$,

$$\int_{\mathbb{R}^2}\varrho^2 A\,\overline{\psi}\,d\boldsymbol{x} = \mu I\,(\overline{\varphi}(z_1) - \overline{\varphi}(z_2)). \tag{6.34}$$

where φ is the solution of Problem (6.27). Then

$$\begin{aligned}\int_{\mathbb{R}^2}\varrho^2(A^\varepsilon - A)\overline{\psi}\,d\boldsymbol{x} &= \mu I\Big(\frac{1}{|\Lambda_2^\varepsilon|}\int_{\Lambda_2^\varepsilon}\varphi^\varepsilon\,d\boldsymbol{x} - \varphi(z_2)\Big)\\ &- \mu I\Big(\frac{1}{|\Lambda_3^\varepsilon|}\int_{\Lambda_3^\varepsilon}\varphi^\varepsilon\,d\boldsymbol{x} - \varphi(z_3)\Big).\end{aligned} \tag{6.35}$$

Since $\varphi \in \mathcal{H}^2(B) \subset C^{0,\alpha}(\overline{B})$ for all α with $0 < \alpha < 1$ (see [41] for instance) and all compact subsets B of $\mathbb{R}^2$, where $C^{0,\alpha}(\overline{B})$ is the space of Hölder continuous functions with exponent α on $\overline{B}$, we have for $k = 2, 3$,

$$\begin{aligned}\Big|\frac{1}{|\Lambda_k^\varepsilon|}\int_{\Lambda_k^\varepsilon}\varphi(\boldsymbol{x})\,d\boldsymbol{x} - \varphi(z_k)\Big| &\le \frac{1}{|\Lambda_k^\varepsilon|}\int_{\Lambda_{k\varepsilon}}|\varphi(\boldsymbol{x}) - \varphi(z_k)|\,d\boldsymbol{x}\\ &\le C\,\frac{1}{|\Lambda_k^\varepsilon|}\int_{\Lambda_k^\varepsilon}|\boldsymbol{x} - z_k|^\alpha\,d\boldsymbol{x}\\ &\le C\,\varepsilon^\alpha.\end{aligned} \tag{6.36}$$

Furthermore, we have from (6.31), the imbedding $\mathcal{H}^2(B) \subset \mathcal{C}^0(\overline{B})$ and the mean value theorem,

$$
\begin{aligned}
\frac{1}{|\Lambda_k^\varepsilon|}\Big|\int_{\Lambda_k^\varepsilon} (\varphi^\varepsilon - \varphi)\, d\boldsymbol{x}\Big| &\le C_1\, \|\varphi^\varepsilon - \varphi\|_{\mathcal{C}^0(B)} \\
&\le C_2\, \|\varphi^\varepsilon - \varphi\|_{\mathcal{H}^2(B)} \\
&\le C_3\varepsilon\, \|\varrho\psi\|_{\mathcal{L}^2(\mathbb{R}^2)}.
\end{aligned} \tag{6.37}
$$

Recalling (6.35) and using (6.36), (6.37), we get

$$
\lim_{\varepsilon\to 0}\int_{\mathbb{R}^2} (A^\varepsilon - A)\, \varrho^2\overline{\psi}\, d\boldsymbol{x} = 0 \qquad \forall\ \psi \in \mathcal{L}^2_\varrho(\mathbb{R}^2).
$$

The sequence (A^ε) converges then weakly to A in $\mathcal{L}^2_\varrho(\mathbb{R}^2)$. To obtain the strong convergence of A^ε, we choose $\psi = (A^\varepsilon - A) \in \mathcal{L}^2_\varrho(\mathbb{R}^2)$ in (6.35). We have by using again (6.36), (6.37),

$$
\begin{aligned}
\|\varrho(A^\varepsilon - A)\|^2_{\mathcal{L}^2(\mathbb{R}^2)} &\le \mu I \sum_{k=2}^{3}\Big|\int_{\Lambda_k^\varepsilon} (\varphi^\varepsilon - \varphi)\, d\boldsymbol{x}\Big| + \mu I \sum_{k=2}^{3}\Big|\frac{1}{|\Lambda_k^\varepsilon|}\int_{\Lambda_k^\varepsilon} \varphi\, d\boldsymbol{x} - \varphi(z_k)\Big| \\
&\le C_4\, \varepsilon + C_5\, \varepsilon^\alpha \le C\, \varepsilon^\alpha.
\end{aligned}
$$

Although no convergence in $\mathcal{H}^1$–spaces can be obtained, the convergence result of Theorem 6.2.2 is rather weak. We quote here a sharper convergence result obtained in [10].

Theorem 6.2.3. *We have when $\varepsilon \to 0$, the convergence*

$$
\begin{aligned}
A^\varepsilon &\to A && \text{in } \mathcal{L}^p(B), \\
\nabla A^\varepsilon &\rightharpoonup \nabla A && \text{in } \boldsymbol{\mathcal{L}}^p(B) \text{ weakly},
\end{aligned}
$$

for all $p \in [1, 2)$, and all bounded sets B of $\mathbb{R}^2$.

Theorems 6.2.2 and 6.2.3 show that the limit problem is a singular one in the sense that it is an elliptic problem with a measure (linear combination of Dirac masses) as right-hand side.

Problem (6.20)–(6.23) can be written in an equivalent formulation that is more adapted to numerical approximation. We start, for this, by considering the Green function (1.21) that satisfies

$$
-\Delta G(z_k, \cdot) = \delta_{z_k} \qquad k = 1, 2.
$$

We recall that the function G is given in 2-D by

$$
G(\boldsymbol{x}, \boldsymbol{y}) = -\frac{1}{2\pi} \ln |\boldsymbol{x} - \boldsymbol{y}|.
$$

Let us consider the following problem:

$$-\Delta \hat{A} = \mu_0 I \, (\delta_{z_1} - \delta_{z_2}) \qquad \text{in } \mathbb{R}^2,$$

the solution of which is given by

$$\hat{A}(\boldsymbol{x}) = \frac{\mu_0 I}{2\pi} \ln \frac{|\boldsymbol{x} - z_2|}{|\boldsymbol{x} - z_1|}.$$

The solution of Problem (6.20)–(6.23) is then given by $A = \tilde{A} + \hat{A}$ where $\tilde{A}$ is the solution of the problem:

$$- \Delta \tilde{A} + \mathrm{i}\omega\mu_0\sigma \tilde{A} = -\mathrm{i}\omega\mu_0\sigma \hat{A} \qquad \text{in } \Lambda_1, \tag{6.38}$$

$$\Delta \tilde{A} = 0 \qquad \text{in } \mathbb{R}^2 \setminus \overline{\Lambda}_1, \tag{6.39}$$

$$[\tilde{A}] = \left[\frac{\partial \tilde{A}}{\partial n}\right] = 0 \qquad \text{on } \gamma_1, \tag{6.40}$$

$$\tilde{A}(\boldsymbol{x}) = \beta + \mathcal{O}(|\boldsymbol{x}|^{-1}) \qquad |\boldsymbol{x}| \to \infty. \tag{6.41}$$

Problem (6.38)–(6.41) is an elliptic problem in $\mathbb{R}^2$ that possesses a unique solution in $\mathcal{W}^1(\mathbb{R}^2)$ since that right-hand side term does not exhibit any singularity. Solving (6.38)–(6.41) has at least two advantages over the primary model (6.12)–(6.18):

1. No computation is to be performed in the inductor domains Λ_2^ε, Λ_3^ε any more.
2. The position of the inductors in a setup can be modified without dramatically penalizing the computational cost in the sense that this modification has an effect only in the right-hand side of (6.38)–(6.41).

The last issue has a great importance in the optimization of eddy current processes, mainly when we are interested in optimal positions of the inductors.

6.3 Three-Dimensional Models

Handling thin inductors in three-dimensional models exhibits more mathematical difficulties. As only partial results were obtained in this area, we shall give them hereafter and mention formal results that need to be mathematically justified. In the following, we start by considering a model problem in three-dimensional geometries. This problem concerns the computation of the potential $\boldsymbol{A}$ when the current density $\boldsymbol{J}$ is known. We shall show that, the case of a thin wire, the limit problem results in the so-called Biot–Savart law.

An interesting problem that illustrates the mathematical difficulty of treating such situations is the computation of the self inductance of a (non simply connected) inductor. We shall, for this problem, retrieve the formal result given in [116].

6.3.1 *The Biot–Savart Law*

Let us consider a conductor $\Omega \subset \mathbb{R}^3$ in which a current of density $\boldsymbol{J}$ flows and consider the vector potential $\boldsymbol{A}$ given by (2.12) and (2.13). Let us recall (2.19)–(2.21) by using hypothesis (6.2):

$$-\Delta \boldsymbol{A} = \mu_0 \boldsymbol{J} \qquad \text{in } \mathbb{R}^3, \tag{6.42}$$

$$\operatorname{div} \boldsymbol{A} = 0 \qquad \text{in } \mathbb{R}^3, \tag{6.43}$$

$$|\boldsymbol{A}(\boldsymbol{x})| = \mathcal{O}(|\boldsymbol{x}|^{-1}) \qquad |\boldsymbol{x}| \to \infty. \tag{6.44}$$

Let us first notice that since $\operatorname{div} \boldsymbol{J} = 0$ (see (2.70)), then, if $u = \operatorname{div} \boldsymbol{A}$, we have by taking the divergence of (6.42),

$$-\Delta u = 0 \qquad \text{in } \mathbb{R}^3.$$

If we look for smooth solution, i.e., that satisfy

$$\operatorname{div} \boldsymbol{u}(\boldsymbol{x}) \to 0 \qquad \text{when } |\boldsymbol{x}| \to \infty,$$

then, we obtain by uniqueness $\boldsymbol{u} = 0$. This means that (6.43) is redundant, and we are led to the solution of problem:

$$\begin{cases} -\Delta \phi = f & \text{in } \mathbb{R}^3, \\ \phi(\boldsymbol{x}) = \mathcal{O}(|\boldsymbol{x}|^{-1}) & |\boldsymbol{x}| \to \infty, \end{cases} \tag{6.45}$$

where ϕ is any component of the potential $\boldsymbol{A}$ and f is the corresponding component of $\mu_0 \boldsymbol{J}$. It is also important to note that f vanishes outside the conductors.

We consider here the case where the conductor Ω is a "thin wire" aligned with the x_3–axis in two distinct situations:

1. The case where the "wire" Ω is infinite, that is $\Omega = \Lambda \times \mathbb{R}$, where Λ is a given bounded domain in $\mathbb{R}^2$.
2. The case where Ω is bounded, that is

$$\Omega := \{(\tilde{\boldsymbol{x}}, x_3) \in \mathbb{R}^3;\ \tilde{\boldsymbol{x}} \in \Lambda(x_3),\ 0 < x_3 < \ell\},$$

where $\Lambda(x_3)$ is a given bounded domain in $\mathbb{R}^2$ for each $0 < x_3 < \ell$, $\ell > 0$.

In both cases we shall admit that the domain Λ_ε is defined by $\Lambda_\varepsilon = \varepsilon\Lambda$ for all $0 < \varepsilon \ll 1$. In order to define the right scales for the potential problem and then to obtain the convenient limit problem, we shall replace in the following (6.45) by

$$-\Delta\phi^\varepsilon = \frac{1}{|\Lambda_\varepsilon|} f^\varepsilon \qquad \text{in } \mathbb{R}^3,$$

where $f^\varepsilon(\tilde{\boldsymbol{x}}, x_3) = f(\tilde{\boldsymbol{x}}/\varepsilon, x_3)$, $\tilde{\boldsymbol{x}} \in \mathbb{R}^2$, $x_3 \in \mathbb{R}$. The choice of a right-hand side that varies rapidly within the x_3–variable means that the support of the current density reduces within the inductor size. The division of this density by the section area ensures that the total current intensity remains constant when the inductor becomes thinner.

Let us rewrite Problem (6.45) in the present context. We have

$$\begin{cases} -\Delta\phi^\varepsilon = \dfrac{1}{|\Lambda_\varepsilon|} f^\varepsilon & \text{in } \mathbb{R}^3, \\ \phi^\varepsilon(\boldsymbol{x}) = \mathcal{O}(|\boldsymbol{x}|^{-1}) & |\boldsymbol{x}| \to \infty, \end{cases} \tag{6.46}$$

where the support of f^ε is Λ_ε. To treat this problem, we proceed as in [159] by taking the Fourier Transform in the x_3–variable. Let us define, for a function $g \in \mathcal{L}^1(\mathbb{R}^3)$, the function

$$\hat{g}(\tilde{x}, \xi) := \int_{\mathbb{R}} g(\tilde{\boldsymbol{x}}, x_3)\, e^{-2\pi i x_3 \xi}\, dx_3 \qquad \tilde{\boldsymbol{x}} \in \mathbb{R}^2,\ \xi \in \mathbb{R}.$$

Using classical rules of the Fourier transform, Problem (6.46) becomes

$$\begin{cases} -\tilde{\Delta}\hat{\phi}^\varepsilon + \xi^2\hat{\phi}^\varepsilon = \dfrac{1}{|\Lambda_\varepsilon|} \hat{f}^\varepsilon & \text{in } \mathbb{R}^2, \\ \hat{\phi}^\varepsilon(\tilde{\boldsymbol{x}}, \xi) = \mathcal{O}(|\tilde{\boldsymbol{x}}|^{-1}) & |\tilde{\boldsymbol{x}}| \to \infty, \end{cases}$$

for $\xi \in \mathbb{R}$, where $\tilde{\Delta}$ is the Laplace operator with respect to the variables $\tilde{\boldsymbol{x}} = (x_1, x_2)$. Let us mention the identity (using $|\Lambda_\varepsilon| = \varepsilon^2|\Lambda|$):

$$\frac{1}{|\Lambda^\varepsilon|}\int_{\mathbb{R}^2} \hat{f}^\varepsilon(\tilde{\boldsymbol{x}}, \xi)\hat{\psi}(\tilde{\boldsymbol{x}})\, d\tilde{\boldsymbol{x}} = \frac{1}{|\Lambda|}\int_{\mathbb{R}^2} \hat{f}(\tilde{\boldsymbol{x}}, \xi)\hat{\psi}(\varepsilon\tilde{\boldsymbol{x}})\, d\tilde{\boldsymbol{x}},$$

Let $\tilde{\nabla}$ is the 2–D gradient with respect to the variables $\tilde{\boldsymbol{x}} = (x_1, x_2)$. Then we have for a.e. $\xi \in \mathbb{R}$ the variational formulation:

$$\begin{cases} \text{Find } \hat{\phi}^{\varepsilon}(\cdot,\xi) \in \mathcal{W}^1(\mathbb{R}^2) \text{ such that} \\ \displaystyle\int_{\mathbb{R}^2} \tilde{\nabla}\hat{\phi}^{\varepsilon}(\cdot,\xi)\cdot\tilde{\nabla}\hat{\psi}\,d\tilde{x} + \int_{\mathbb{R}^2} \xi^2\hat{\phi}^{\varepsilon}\hat{\psi}\,d\tilde{x} \\ \qquad\qquad = \displaystyle\frac{1}{|\Lambda|}\int_{\mathbb{R}^2} \hat{f}(\tilde{\boldsymbol{x}},\xi)\hat{\psi}(\varepsilon\tilde{x})\,d\tilde{x} \quad \forall\,\hat{\psi}\in\mathcal{W}^1(\mathbb{R}^2). \end{cases}$$

Let us take the limit $\varepsilon \to 0$ in this problem. This is done classically by taking $\hat{\psi} = \hat{\phi}^{\varepsilon}(\cdot,\xi)$ for $\xi \in \mathbb{R}$. We obtain

$$\int_{\mathbb{R}^2} (|\tilde{\nabla}\hat{\phi}^{\varepsilon}(\cdot,\xi)|^2 + \xi^2(\hat{\phi}^{\varepsilon}(\cdot,\xi))^2)\,d\tilde{x} = \frac{1}{|\Lambda|}\int_{\Lambda} \hat{f}(\tilde{\boldsymbol{x}},\xi)\hat{\phi}^{\varepsilon}(\varepsilon\tilde{\boldsymbol{x}},\xi)\,d\tilde{\boldsymbol{x}}.$$

Using the Cauchy–Schwarz inequality and Inequality (1.46), we get

$$\|\tilde{\nabla}\hat{\phi}^{\varepsilon}(\cdot,\xi)\|_{\mathcal{L}^2(\mathbb{R}^2)} + |\xi|\,\|\hat{\phi}^{\varepsilon}(\cdot,\xi)\|_{\mathcal{L}^2(\Lambda)} \le C\,\|\hat{f}(\cdot,\xi)\|_{\mathcal{L}^2(\Lambda)} \qquad \text{for a.e. } \xi\in\mathbb{R}.$$

We obtain the limit problem:

$$\begin{cases} -\tilde{\Delta}\hat{u}(\cdot,\xi) + \xi^2\hat{u}(\cdot,\xi) = \hat{f}_{\Lambda}(\xi)\,\delta & \text{in } \mathbb{R}^2, \\ \hat{u}(\tilde{\boldsymbol{x}}) = \mathcal{O}(|\tilde{\boldsymbol{x}}|^{-1}) & |\tilde{\boldsymbol{x}}| \to \infty, \end{cases} \tag{6.47}$$

where

$$\hat{f}_{\Lambda}(\xi) := \frac{1}{|\Lambda|}\int_{\Lambda} \hat{f}(\tilde{\boldsymbol{x}},\xi)\,d\tilde{\boldsymbol{x}},$$

and δ is the Dirac mass concentrated at $\tilde{\boldsymbol{x}} = 0$. The solution of (6.47) is given explicitly by (see [159]):

$$\hat{u}(\tilde{\boldsymbol{x}},\xi) = \hat{f}_{\Lambda}(\xi)\,K_0(|\xi|\,|\tilde{\boldsymbol{x}}|) \qquad \tilde{\boldsymbol{x}}\in\mathbb{R}^2,\ \xi\in\mathbb{R},$$

where K_0 is the modified Bessel function of order 0 (see [2]). The limit u is therefore given by computing the inverse Fourier transform:

$$u(\tilde{x},x_3) = \int_{\mathbb{R}} \hat{f}_{\Lambda}(\xi)\,K_0(|\xi|\,|\tilde{\boldsymbol{x}}|)\,e^{2\pi i x_3\xi}\,d\xi \qquad \tilde{\boldsymbol{x}}\in\mathbb{R}^2,\ x_3\in\mathbb{R}.$$

We can also write the limit potential:

$$\boldsymbol{A}(\tilde{x},x_3) = \int_{\mathbb{R}} \hat{\boldsymbol{J}}_{\Lambda}(\xi)\,K_0(|\xi|\,|\tilde{\boldsymbol{x}}|)\,e^{2\pi i x_3\xi}\,d\xi \qquad \tilde{\boldsymbol{x}}\in\mathbb{R}^2,\ x_3\in\mathbb{R}.$$

6.3.2 Self Inductance of a Thin Inductor

We have shown in Chap. 4 how the self inductance appears as input data in some three-dimensional models. Assume now that we are in presence of a unique conductor Ω that is doubly connected, i.e. there are surfaces S and Σ such that the domains

$$\Omega \setminus S \quad \text{and} \quad \Omega_{\text{ext}} \setminus \Sigma = (\mathbb{R}^3 \setminus \overline{\Omega}) \setminus \Sigma$$

are simply connected. More precisely, the inductor can be assumed toroidal. Here again, in order to obtain a limit model for thin inductor, it is necessary to describe precisely how the inductor geometry depends on its thickness. Due to the heaviness of the proof, we have chosen to mention the main result and let the interested reader consulting the reference [8] that contains the detailed proof.

In order to describe the dependency of the inductor on its thickness small parameter, we consider a closed smooth curve γ in $\mathbb{R}^3$ that is parameterized by its curvilinear abcissa using a $\mathcal{C}^2$–function $\boldsymbol{X} : [0, \ell_\gamma] \to \mathbb{R}^3$ that satisfies

$$\boldsymbol{X}(0) = \boldsymbol{X}(\ell_\gamma), \ \boldsymbol{X}'(0) = \boldsymbol{X}'(\ell_\gamma), \tag{6.48}$$

where ℓ_γ is the length of γ. We denote by the triple $(\boldsymbol{t}(s), \boldsymbol{\nu}(s), \boldsymbol{b}(s))$ for $s \in [0, \ell_\gamma)$ the Serret–Frénet coordinates at the point $\boldsymbol{X}(s)$, i.e. $\boldsymbol{t}(s)$, $\boldsymbol{\nu}(s)$, $\boldsymbol{b}(s)$ stand respectively for the unit tangent vector to γ, the principal normal and the binormal given by:

$$\boldsymbol{t}(s) = \boldsymbol{X}'(s), \ \boldsymbol{\nu}(s) = \frac{\boldsymbol{t}'(s)}{|\boldsymbol{t}'(s)|}, \ \boldsymbol{b}(s) = \boldsymbol{t}(s) \times \boldsymbol{\nu}(s), \quad s \in [0, \ell_\gamma).$$

Let us define the mapping $\boldsymbol{F}_\varepsilon : [0, \ell_\gamma) \times (0, 1) \times (0, 2\pi] \to \mathbb{R}^3$ by

$$\boldsymbol{F}_\varepsilon(s, \xi, \theta) := \boldsymbol{X}(s) + \varepsilon\, \xi\, (\cos\theta\, \boldsymbol{\nu}(s) + \sin\theta\, \boldsymbol{b}(s)).$$

We define the conductor Ω^ε by

$$\Omega^\varepsilon := \boldsymbol{F}_\varepsilon([0, \ell_\gamma) \times (0, 1) \times [0, 2\pi)), \tag{6.49}$$

and its boundary by

$$\Gamma^\varepsilon := \boldsymbol{F}_\varepsilon([0, \ell_\gamma) \times \{1\} \times [0, 2\pi)).$$

The exterior domain is given by $\breve{\Omega}^\varepsilon_{\text{ext}} := \mathbb{R}^3 \setminus \overline{\Omega}^\varepsilon$. It is important to note that the cut surface Σ can be chosen independent of the parameter ε.

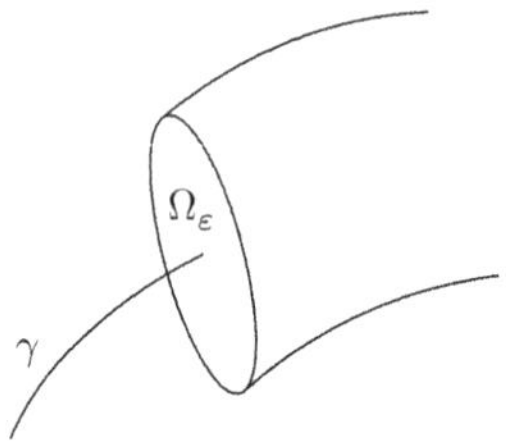

Fig. 6.3 A cut in the inductor

We recall that the inductance is given by

$$L^\varepsilon := \mu_0 \int_\Sigma \frac{\partial p^\varepsilon}{\partial n}\, ds = \mu_0 \int_{\breve{\Omega}^\varepsilon_{\mathrm{ext}}\setminus\Sigma} |\nabla p^\varepsilon|^2\, d\boldsymbol{x}, \tag{6.50}$$

where p^ε is the unique solution to the problem:

$$\begin{cases} \Delta p^\varepsilon = 0 & \text{in } \breve{\Omega}^\varepsilon_{\mathrm{ext}} \setminus \Sigma, \\ \dfrac{\partial p^\varepsilon}{\partial n^\varepsilon} = 0 & \text{on } \Gamma^\varepsilon, \\ [p^\varepsilon]_\Sigma = 1, & \\ \left[\dfrac{\partial p^\varepsilon}{\partial n}\right]_\Sigma = 0, & \\ p^\varepsilon(\boldsymbol{x}) = \mathcal{O}(|\boldsymbol{x}|^{-1}) & |\boldsymbol{x}| \to \infty. \end{cases} \tag{6.51}$$

Above $\boldsymbol{n}^\varepsilon$ is the outward unit normal to Γ^ε. A closer look to the definition (6.50) shows that L^ε might tend to the infinity when $\varepsilon \to 0$. In fact, when ε vanishes the domain $\breve{\Omega}^\varepsilon_{\mathrm{ext}} \setminus \Sigma$ tends to the domain $\mathbb{R}^3 \setminus \Sigma$ and then (6.51) formally tends to an elliptic problem with a discontinuous trace, the discontinuity appearing on the boundary of Σ. It is well known that a function on the boundary that involves a jump cannot lie in the space $\mathcal{H}^{\frac{1}{2}}$ (see for instance [92]). In addition, in [116], using rough arguments, the authors claim that L^ε behaves like $\ln \varepsilon$ for small values of ε. Our goal is to prove this result by giving more precisely the singular term and the remainder in an asymptotic expansion in function of ε.

Theorem 6.3.1. *We have the expansion*

$$L^\varepsilon = -\frac{\mu_0\, \ell_\gamma}{2\pi} \ln \varepsilon + \mathcal{O}(1) \qquad \textit{when } \varepsilon \to 0.$$

This result justifies the approximation given in Landau–Lifshitz [116]. As it was mentioned before, the complete proof of this result with an additional term in the expansion can be found in [8].

6.3.3 The Complete 3-D Model

Let us end this chapter with some words about the derivation of a complete model for the three-dimensional case. As this was shown in Sect. 6.1, the use of an $\boldsymbol{H}$–model is necessary to obtain as a limit problem a circuit Kirchhoff equation like (6.11). Unfortunately, a rigorous derivation remains an open problem for the three-dimensional model.

Let us however give some preliminary settings for interested reader. The starting point is the $\boldsymbol{H}$–model presented in Sect. 4.2. Here again, we consider first the case of a unique toroidal conductor Ω^ε. The domain Ω^ε is defined by (6.49). We recall the first formulation of the $\boldsymbol{H}$–model given by (4.24), i.e.

$$\begin{cases} \text{Find } \boldsymbol{H}^\varepsilon \in \mathcal{H}^\varepsilon \text{ such that} \\ \mathrm{i}\omega \displaystyle\int_{\mathbb{R}^3} \mu \boldsymbol{H}^\varepsilon \cdot \overline{\boldsymbol{w}}\, d\boldsymbol{x} + \int_{\Omega^\varepsilon} \sigma^{-1}\, \mathbf{curl}\, \boldsymbol{H}^\varepsilon \cdot \mathbf{curl}\, \overline{\boldsymbol{w}}\, d\boldsymbol{x} \\ \qquad\qquad = V \displaystyle\int_{S^\varepsilon} \mathrm{curl}_{S^\varepsilon} \overline{\boldsymbol{w}}\, ds \qquad \forall\, \boldsymbol{w} \in \mathcal{H}^\varepsilon, \end{cases} \tag{6.52}$$

where

$$\mathcal{H}^\varepsilon := \{\boldsymbol{v} \in \mathcal{H}(\mathbf{curl}, \mathbb{R}^3);\ \mathbf{curl}\, \boldsymbol{v} = 0 \text{ in } \Omega^\varepsilon_{\mathrm{ext}}\},$$
$$\Omega^\varepsilon_{\mathrm{ext}} := \mathbb{R}^3 \setminus \overline{\Omega}^\varepsilon.$$

Note that we have used the superscript ε as we did it throughout this chapter to mention the dependency of the various fields and space on the thickness of the inductor.

Using the Green formula we can see that the right-hand side in the variational formulation (6.52) can be also written

$$\int_{S^\varepsilon} \mathrm{curl}_{S^\varepsilon} \overline{\boldsymbol{w}}\, ds = \int_{\Omega^\varepsilon \setminus S^\varepsilon} \nabla \varphi^\varepsilon \cdot \mathbf{curl}\, \overline{\boldsymbol{w}}\, d\boldsymbol{x},$$

where φ^ε is a solution of (1.17). It is proven in [11] that we have the estimate

$$\left| \int_{\Omega^\varepsilon \setminus S^\varepsilon} \nabla \varphi^\varepsilon \cdot \mathbf{curl}\, \overline{\boldsymbol{w}}\, d\boldsymbol{x} \right| \le C\varepsilon\, \|\, \mathbf{curl}\, \boldsymbol{w} \|_{\mathcal{L}^2(\Omega^\varepsilon)}, \tag{6.53}$$

where the constant C is independent of ε.

A first estimate for the magnetic field is then classically obtained by choosing $\boldsymbol{w} = \boldsymbol{H}^\varepsilon$ in (6.52): We have from (6.53) and the bounds (2.7), (2.11),

$$\|\boldsymbol{H}^\varepsilon\|_{\mathcal{L}^2(\mathbb{R}^3)} + \|\, \mathbf{curl}\, \boldsymbol{H}^\varepsilon \|_{\mathcal{L}^2(\Omega^\varepsilon)} \le C\varepsilon. \tag{6.54}$$

Therefore, the magnetic field $\boldsymbol{H}^\varepsilon$ tends to 0 when $\varepsilon \to 0$ in $\mathcal{L}^2(\mathbb{R}^3)$.

Further information on the limit can be obtained if we write $\boldsymbol{H}^\varepsilon = \varepsilon \boldsymbol{H}_1^\varepsilon$ and look for the limit of $\boldsymbol{H}_1^\varepsilon$.

As we have mentioned it, our conjecture is that, like in the 2-D case, the limit problem should be given by

$$(\mathrm{i}\omega L + R)\, I = V.$$

Here L is the self inductance of the inductor and R is its resistance defined by $R = \sigma^{-1}\ell_\gamma$. Furthermore, I is the total current flowing in the inductor, defined by

$$I = \lim_{\varepsilon \to 0} \int_{S^\varepsilon} \mathbf{curl}\, \boldsymbol{H}^\varepsilon \cdot \boldsymbol{n}\, ds.$$

Chapter 7
Numerical Methods

This chapter is devoted to the numerical solution of various problems we have derived in the previous chapters. Our goal is to define some numerical methods that can be used to approximate the solutions of the presented problems and give their main properties.

7.1 Introduction and Main Notations

Numerical solution of eddy current problems present specific difficulties due to the fact that most of the problems we have derived involve both partial differential equations in bounded domains and integral representations on the boundaries of these domains. This suggests the use of numerical methods that couple classical techniques (finite differences, finite elements, finite volumes, ...) for partial differential equations with numerical schemes for boundary integral equations (collocation, finite elements, ...). For its flexibility and popularity, we focus our presentation on the finite element method. In addition, we shall often restrict the presentation to the lowest order methods.

We give in the sequel a general abstract framework for the finite element method for elliptic problems. This is done for the sake of completeness but does not constitute a general introduction to finite element methods. A more detailed presentation can be found in more specialized literature: The references [38, 40, 50] provide theoretical foundations of the method while more practical aspects can be found in [106, 168] for instance. More specific applications of the method to the various problems we have presented are then presented:

1. Standard $\mathcal{H}^1$–finite element method for two-dimensional eddy current equations in the conductors like those presented in Chap. 3.
2. $\boldsymbol{\mathcal{H}}(\mathbf{curl})$–finite element methods for three-dimensional eddy current equations as in Chap. 4.

R. Touzani and J. Rappaz, *Mathematical Models for Eddy Currents and Magnetostatics: With Selected Applications*, Scientific Computation, DOI 10.1007/978-94-007-0202-8_7,

3. $\mathcal{H}(\mathrm{div})$–finite element methods for three-dimensional models using the current density as presented in Sect. 4.17 and for two-dimensional problems arising in axisymmetric formulations obtained in Chap. 5.
4. Finite element methods for integral equations that represent the solution of harmonic problems in the free space. These are generally referred to as *Boundary Element methods*.

Once all these elements are presented, the remaining of this chapter is devoted to the numerical approximation of the models coupling interior and exterior problems, i.e. the coupling of finite element and boundary element methods.

Let us consider an abstract variational formulation as follows: Assume we are given a complex Hilbert space $\mathcal{V}$ and the variational problem:

$$\text{Find } u \in \mathcal{V} \quad \text{such that} \quad \mathscr{B}(u, v) = \mathscr{L}(v) \qquad \forall\, v \in \mathcal{V}, \tag{7.1}$$

where $\mathscr{B}$ is a sesquilinear, continuous and coercive form on $\mathcal{V} \times \mathcal{V}$ and $\mathscr{L}$ is an antilinear continuous form on $\mathcal{V}$ so that, owing to the Lax–Milgram theorem (Theorem 1.2.1) (7.1) admits a unique solution. Let $\mathcal{V}_h$ stand for a finite–dimensional subspace of $\mathcal{V}$ where h is a parameter that tends to 0 as the dimension of $\mathcal{V}_h$ tends to the infinity in such a way that

$$\overline{\bigcup_h \mathcal{V}_h} = \mathcal{V}.$$

The discrete problem is defined by

$$\text{Find } u_h \in \mathcal{V}_h \quad \text{such that} \qquad \mathscr{B}(u_h, v) = \mathscr{L}(v) \qquad \forall\, v \in \mathcal{V}_h. \tag{7.2}$$

Classically, u_h is referred to as *Galerkin approximation* of u.

Since $\mathcal{V}_h$ is a closed subspace of $\mathcal{V}$, being of finite dimension, it follows that (7.2) admits a unique solution. Moreover we have obviously

$$\mathscr{B}(u - u_h, v) = 0 \qquad \forall\, v \in \mathcal{V}_h.$$

A direct consequence of this is that

$$\mathscr{B}(u - u_h, u - u_h) = \mathscr{B}(u - u_h, u - v) \qquad \forall\, v \in \mathcal{V}_h,$$

and then the coercivity and the continuity of $\mathscr{B}$ imply

$$\|u - u_h\|_{\mathcal{V}} \le C \inf_{v \in \mathcal{V}_h} \|u - v\|_{\mathcal{V}}, \tag{7.3}$$

where C is a positive constant that depends on the constants of continuity and coercivity of $\mathscr{B}$ only.

The inequality (7.3) is basic in finite element theory (see [38,40,50] for instance). It means in particular that the convergence rate of the method can be obtained by using basic approximation theory in numerical analysis. The quantity $\|u - v\|_{\mathcal{V}}$ at the right-hand side of (7.3) depends indeed on the subspace $\mathcal{V}_h$ and u.

As a consequence of (7.3), an error estimate for a given finite element method can be obtained by constructing a projection $\pi_h u$ of u on $\mathcal{V}_h$ and bounding the error between u and $\pi_h u$.

Let $(\phi_i)_{i=1}^N$ denote a basis of the space $\mathcal{V}_h$ and consider an expansion of u_h in this basis. We have

$$u_h = \sum_{j=1}^{N} u_j \phi_j. \tag{7.4}$$

Replacing in (7.2) and taking $v = \phi_i$, we get

$$\sum_{j=1}^{N} \mathscr{B}(\phi_j, \phi_i)\, u_j = \mathscr{L}(\phi_i) \qquad 1 \le i \le N.$$

Denoting by $\mathbf{A}$ the matrix with entries $A_{ij} = \mathscr{B}(\phi_j, \phi_i)$, by $\mathbf{b}$ the vector with components $b_i = \mathscr{L}(\phi_i)$ and by $\mathbf{u}$ the vector $\mathbf{u} = (u_i)$ we obtain the linear system

$$\mathbf{A}\,\mathbf{u} = \mathbf{b}. \tag{7.5}$$

Theoretical aspects of the finite element method can be summarized in the following steps:

1. To build up a finite–dimensional subspace $\mathcal{V}_h$ of $\mathcal{V}$ with basis functions (ϕ_i) having local support. This property ensures, when this is possible, sparsity of the resulting matrix $\mathbf{A}$.
2. When the evaluation of the matrix and right-hand side vector entries is impossible or too costly, one generally resorts to numerical integration. In this case, (7.3) is no more valid and must be adapted. This topic is described in [50] for instance.
3. To evaluate an upper bound of the right-hand side in (7.3) in order to obtain an error estimate. In the applications, when u is a smooth function, one looks for an estimate of a projection $\pi_h u$ of u on $\mathcal{V}_h$ to obtain

   $$\|u - u_h\| \le C\, \|u - \pi_h u\|.$$

   This projection is generally obtained from an interpolation or a local projection of u.

Note that the matrix structure (symmetry, sparseness, …) depends strongly on the chosen numerical method and even on the choice of the basis of the space $\mathcal{V}_h$. In addition, each choice of the basis (ϕ_i) gives a particular interpretation of the coefficients u_i in the expansion (7.4).

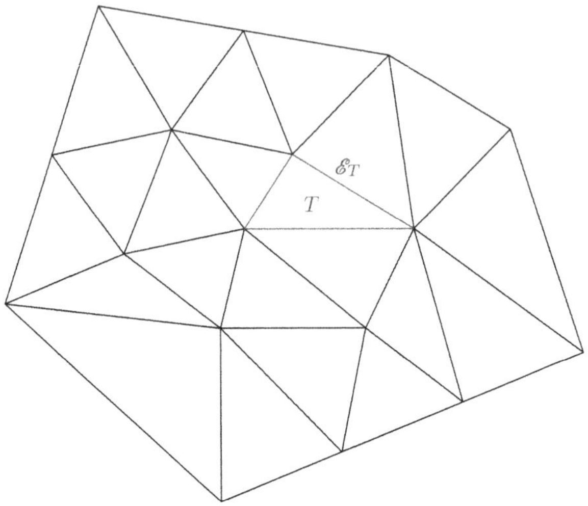

Fig. 7.1 A typical finite element mesh

Remark 7.1.1. Since all eddy current problems we have considered so far use complex valued functions, as a consequence of using time–harmonic modelling, we present hereafter the model problems in complex variables. However, basis functions can be taken real valued.

Let us, before going into the exploration of finite element methods, define a mesh of the considered domains. For clarity we distinguish hereafter 2-D and 3-D situations.

For the 2-D case, let Λ denote a polygonal domain of $\mathbb{R}^2$ with boundary γ. We define a finite element mesh of Λ as a collection of triangles $T \in \mathscr{T}_h$, where h is the maximal edge length (see Fig. 7.1), that satisfies the following hypotheses:

- The union of triangles, considered as closed sets is the closure of Λ.
- The intersection of any pair of triangles, considered as closed sets of $\mathbb{R}^2$ is either empty, or a vertex or an edge common to these elements.
- For each $T \in \mathscr{T}_h$, if r_T and h_T stand respectively for the diameter of the inscribed and the circumscribed circle of T, then there exists a constant $\beta > 0$, independent of h such that

$$\inf_{T \in \mathscr{T}_h} \frac{r_T}{h_T} \geq \beta. \tag{7.6}$$

For such a mesh, the following additional notations will be needed:

- $\mathscr{E}_h$ is the set of edges of the mesh $\mathscr{T}_h$. Among these elements $e \in \mathscr{E}_h$ we can consider the set of boundary edges $\mathscr{E}_h^\gamma$ which forms a mesh of the boundary, and the set of internal edges $\mathscr{E}_h^{\text{int}}$ (of course $\mathscr{E}_h = \mathscr{E}_h^{\text{int}} \cup \mathscr{E}_h^\gamma$).

- For an edge $e \in \mathscr{E}_h$, $\mathscr{T}_e$ will stand for the set of the pair of elements (of $\mathscr{T}_h$) that share e.
- For an element $T \in \mathscr{T}_h$, $\mathscr{E}_T$ will stand for the set of edges of T.
- For an element $T \in \mathscr{T}_h$ and an edge e of T ($e \in \mathscr{E}_T$), the vector $\boldsymbol{n}_e^T$ will stand for the outward unit normal to T on e.
- Since we shall sometimes deal with multiple domains Λ_k, we shall denote by $\mathscr{T}_h^k$ the mesh of the k–th domain and by $\mathscr{T}_h$ the union of theses meshes.
- The set of all mesh nodes will be denoted by $\mathscr{N}_h$ and the set of boundary nodes by $\mathscr{N}_h^\gamma$. It is also sometimes useful to use the set of node labels, respectively $\mathscr{I}_h$ and $\mathscr{I}_h^\gamma$.
- The set $\mathbb{P}_r(X)$ (or simply $\mathbb{P}_r$) will stand for the space of polynomials of degree $\leq r$ on the set $X \subset \mathbb{R}^2$, i.e. of the form

$$\sum_{\substack{i,j=0\\ i+j\leq r}}^{r} \alpha_{ij} x_1^i x_2^j \qquad \boldsymbol{x} = (x_1, x_2) \in X,\ \alpha_{ij} \in \mathbb{C}.$$

In the three-dimensional case, we consider a polyhedral domain Ω of $\mathbb{R}^3$ with boundary Γ. A *mesh* of this domain is defined as its subdivision into tetrahedra T that will be sometimes referred to as *elements*. The set of all the elements will be called $\mathscr{T}_h$ where h is the maximal edge length. Analogous hypotheses to the 2-D case are assumed, i.e.

- The union of tetrahedra, considered as closed sets is the closure of Ω.
- The intersection of any pair of tetrahedra, considered as closed sets of $\mathbb{R}^3$ is either empty, a vertex, an edge or a face common to these elements.
- For each $T \in \mathscr{T}_h$, if r_T and h_T stand respectively for diameter of the inscribed and circumscribed sphere of T, then there exists a constant $\beta > 0$, independent of h such that

$$\inf_{T\in\mathscr{T}_h} \frac{r_T}{h_T} \geq \beta. \tag{7.7}$$

The same additional notations hold for the 3-D case, and we add the set of faces of tetrahedra that will be denoted by $\mathscr{F}_h$, whereas internal (*resp.* boundary) faces belong the set $\mathscr{F}_h^{\text{int}}$ (*resp.* $\mathscr{F}_h^\Gamma$). Analogous definitions are given to the sets $\mathscr{T}_f$, $\mathscr{F}_T$, ... Finally, $\mathbb{P}_r(X)$ (or $\mathbb{P}_r$) will stand for the set of polynomials of the form

$$\sum_{\substack{i,j,k=0\\ i+j+k\leq r}}^{r} \alpha_{ijk} x_1^i x_2^j x_3^k \qquad \boldsymbol{x} = (x_1, x_2, x_3) \in X,\ \alpha_{ijk} \in \mathbb{C}.$$

Remark 7.1.2. We have considered the case of polyhedral domains. In the general case of a domain with smooth boundary is generally handled by considering its approximation by polyhedral domains. The presentation of the methods remains the same excepting that the domains Λ and Ω are replaced by their respective approximations Λ_h and Ω_h but the analysis is more difficult since in this case, we have to introduce the approximation of the boundary of Λ or Ω by the one of Λ_h or Ω_h.

Remark 7.1.3. It is sometimes convenient to index the vectors or matrices that have the same size as the number of elements by these elements rather than by their labels. For example, for a vector $\boldsymbol{b} \in \mathbb{R}^N$ where N is the total number of elements of a mesh $\mathscr{T}_h$, issued for instance from an elementwise constant approximation, we will sometimes refer to its entries as b_T, $T \in \mathscr{T}_h$. The same convention will also hold for faces or edges.

7.2 Standard ($\mathcal{H}^1$) Finite Element Methods

Let us start with the most popular finite element method. We mean by standard finite element methods, those who may be applied to boundary value problems that result in variational formulations like (7.1) where $\mathcal{V}$ is a space that satisfies $\mathcal{H}_0^1(X) \subset \mathcal{V} \subset \mathcal{H}^1(X)$, where X is a polygonal (*resp.* polyhedral) domain in $\mathbb{R}^2$ (*resp.* $\mathbb{R}^3$). In view of our applications, a typical example for which this method is applied is the 2-D solenoidal model presented in Sect. 3.2.

7.2.1 A Finite Element Method for the 2-D Solenoidal Model

Let us consider (3.31). We assume that the domains Λ_1 and Λ_2 are polygonal. This assumption is not fundamental but simplifies the setting of the finite element method. Our purpose is to present the simplest finite element method to discretize (3.31).

A finite–dimensional approximation of the space $\mathcal{H}$ in (3.31) is defined by

$$\mathcal{H}_h := \{\phi \in C^0(\overline{\Lambda});\ \phi_{|\Lambda_{\text{ext}}} = \text{Const.},\ \phi_{|T} \in \mathbb{P}_1(T)\ \forall\, T \in \mathscr{T}_h^1 \cup \mathscr{T}_h^2,\\ \phi = 0 \text{ on } \partial\hat{\Lambda}\}.$$

Let then $(\boldsymbol{a}_i)$ denote the set of vertices of all triangles of $\mathscr{T}_h$. For a function $u \in \mathcal{H}$ such that $u_{|\Lambda_k} \in \mathcal{H}^2(\Lambda_k)$, $k = 1, 2$ we define $\pi_h u$ as the $\mathcal{H}_h$–interpolant of u, i.e.

$$\begin{aligned} &\pi_h u \in \mathcal{H}_h, \\ &\pi_h u(\boldsymbol{a}_i) = u(\boldsymbol{a}_i) \quad \text{for all vertices } \boldsymbol{a}_i. \end{aligned} \tag{7.8}$$

We have then the following interpolation error bound (see [50] for instance):

Theorem 7.2.1. *Let u denote a function in $\mathcal{H}$ such that $u_{|\Lambda_k} \in \mathcal{H}^2(\Lambda_k)$ for $k = 1, 2$ and let $\pi_h u$ denote its interpolant defined by* (7.8)*. Then there exists a constant C, independent of h, such that*

$$\|u - \pi_h u\|_{\mathcal{H}^1(\Lambda)} \leq Ch\left(\|u\|_{\mathcal{H}^2(\Lambda_1)} + \|u\|_{\mathcal{H}^2(\Lambda_2)}\right).$$

The discrete problem is defined by the variational formulation:

$$\begin{cases} \text{Find } H_h \in \mathcal{H}_h \text{ such that} \\ \displaystyle\int_\Lambda \sigma^{-1} \nabla H_h \cdot \nabla \overline{\phi}\, d\boldsymbol{x} + \mathrm{i}\omega \int_{\hat{\Lambda}} \mu H_h \overline{\phi}\, d\boldsymbol{x} = V\overline{\phi}_{|\Lambda_{\text{ext}}} \qquad \forall\, \phi \in \mathcal{H}_h. \end{cases} \tag{7.9}$$

As for (3.31), Problem (7.9) possesses a unique solution in $\mathcal{H}_h$. Moreover, if $H_{|\Lambda_k} \in \mathcal{H}^2(\Lambda_k)$ for $k = 1, 2$, then Theorem 7.2.1 with (7.3) imply:

Theorem 7.2.2. *There exists a constant C independent of h such that, for $k = 1, 2$,*

$$\|H - H_h\|_{\mathcal{H}^1(\Lambda_k)} \leq Ch\big(\|H\|_{\mathcal{H}^2(\Lambda_1)} + \|H\|_{\mathcal{H}^2(\Lambda_2)}\big).$$

It can be noticed that the formulation (7.9) couples the conductors Λ_1 and Λ_2 although the coupling reduces to the magnetic field constant value in the region Λ_{ext}. This coupling can be removed by using the procedure introduced in Remark 3.2.3. To present the numerical version of this procedure, we define the spaces in which lie the normalized magnetic fields:

$$\mathcal{H}_h^k := \{\phi \in C^0(\overline{\Lambda}_k);\ \phi_{|T} \in \mathbb{P}_1(T)\ \forall\, T \in \mathscr{T}_h^k\},\ k = 1, 2.$$

and the spaces of test functions:

$$\mathcal{H}_{h0}^k := \{\phi \in \mathcal{H}_h^k;\ \phi = 0 \text{ on } \gamma_k\},\ k = 1, 2.$$

We next define the following variational problems:

$$\begin{cases} \text{Find } \hat{H}_h^1 \in \mathcal{H}_h^1 \text{ such that:} \\ \displaystyle\int_{\Lambda_1} \sigma^{-1} \nabla \hat{H}_h^1 \cdot \nabla \overline{\phi}\, d\boldsymbol{x} + \mathrm{i}\omega \int_{\Lambda_1} \mu \hat{H}_h^1 \overline{\phi}\, d\boldsymbol{x} = 0 & \forall\, \phi \in \mathcal{H}_{h0}^1, \\ \hat{H}_h^1 = 1 & \text{on } \gamma_1^-, \\ \hat{H}_h^1 = 0 & \text{on } \gamma_1^+. \end{cases} \tag{7.10}$$

$$\begin{cases} \text{Find } \hat{H}_h^2 \in \mathcal{H}_h^2 \text{ such that:} & \\ \displaystyle\int_{\Lambda_2} \sigma^{-1} \nabla \hat{H}_h^2 \cdot \nabla \overline{\phi}\, dx + \mathrm{i}\omega \int_{\Lambda_2} \mu \hat{H}_h^2 \overline{\phi}\, dx = 0 & \forall\, \phi \in \mathcal{H}_{h0}^2, \\ \hat{H}_h^2 = 1 & \text{on } \gamma_2. \end{cases} \tag{7.11}$$

We consider the identity (3.32) that gives the value H_0 of H on the boundary $\gamma_1^- \cup \gamma_2$. Considering now the numerical approximation, the use of the same finite element method as for (7.9) gives the approximate magnetic field $\hat{H}$. It remains then to give a discrete version of (3.32). For this, it is well known that a standard treatment of the normal derivative (i.e., by calculating the gradient on elements that contain boundary edges and taking their normal component), results in poor accuracy. Instead, we propose the following approach to calculate H_0: Let $\mathscr{E}_h^1$ stand for the set of edges on the boundary γ_1^-. For any $e \in \mathscr{E}_h^1$, we denote by T_e the triangle that owns this edge. We also denote by $\phi_e \in \mathcal{H}_h^1$ the function that equals 1 on e and 0 on the remaining vertex of T_e.

Let us now consider (3.23) and let $\hat{H}_1 = H/H_0$, we have the boundary value problem

$$\begin{cases} -\operatorname{div}(\sigma^{-1} \nabla \hat{H}_1) + \mathrm{i}\omega\mu \hat{H}_1 = 0 & \text{in } \Lambda_1, \\ \hat{H}_1 = 1 & \text{on } \gamma_1^-, \\ \hat{H}_1 = 0 & \text{on } \gamma_1^+. \end{cases} \tag{7.12}$$

Multiplying the first equation of (7.12) by a function $\phi \in \mathcal{H}^1(\Lambda_1)$ with $\phi = 0$ on γ_1^+ and using the Green formula, we get

$$\int_{\gamma_1^-} \sigma^{-1} \frac{\partial \hat{H}_1}{\partial n} \overline{\phi}\, ds = \mathrm{i}\omega \int_{\Lambda_1} \mu \hat{H}_1 \overline{\phi}\, dx + \int_{\Lambda_1} \sigma^{-1} \nabla \hat{H}_1 \cdot \nabla \overline{\phi}\, dx.$$

Now choosing $\phi = 1$ on γ_1^- and $\phi = 0$ on the nodes that do not lie on γ_1^- we have

$$\int_{\gamma_1^-} \sigma^{-1} \frac{\partial \hat{H}_1}{\partial n}\, ds = \mathrm{i}\omega \int_{\Lambda_1} \mu \hat{H}_1 \overline{\phi}\, dx + \int_{\Lambda_1} \sigma^{-1} \nabla \hat{H}_1 \cdot \nabla \overline{\phi}\, dx.$$

The discrete version of this equation is given by

$$\int_{\gamma_1^-} \sigma^{-1} \frac{\partial \tilde{H}_h^1}{\partial n}\, ds = \mathrm{i}\omega \int_{\Lambda_1} \mu \tilde{H}_1 \overline{\phi}\, dx + \int_{\Lambda_1} \sigma^{-1} \nabla \tilde{H}_1 \cdot \nabla \overline{\phi}\, dx.$$

Hence we can define an approximation of (3.32) by

$$\frac{V}{H_{0h}} = \mathrm{i}\omega \int_{\Lambda} \mu \hat{H}_h\, dx + \mathrm{i}\omega\mu_0 |\Lambda_{\text{ext}}| + \int_{\Lambda_1} \sigma^{-1} \nabla \hat{H}_h \cdot \nabla \overline{\phi}\, dx$$

where ϕ is a function in $\mathcal{H}_h^1$ such that $\phi = 1$ on γ_1^- and $\phi = 0$ on any node that does not belong to γ_1^-.

Consequently, the actual approximate magnetic field can be defined by

$$H_h^1 := \hat{H}_h^1 H_{0h}, \quad H_h^2 := \hat{H}_h^2 H_{0h}.$$

Considering the Total Current model defined in (3.22), the numerical solution by the finite element method is straightforward and does not need further development.

Remark 7.2.1. Classically, the resulting linear system of equations is obtained by choosing the *canonical Lagrange basis* of the space $\mathcal{H}_{h0}^k$, i.e. such that

$$\phi_i(\boldsymbol{a}_j) = \delta_{ij} \qquad i, j \in \mathcal{N}_h.$$

As a result, the matrix of the linear system is sparse since the support of any basis function ϕ_i is reduced to the union of the triangles that share the node $\boldsymbol{a}_i$. □

7.2.2 Finite Elements for the Axisymmetric Model

Let us present a finite element method for the numerical solution of a variant of (5.8)–(5.12) that consists in considering it in a bounded domain. The numerical approximation of (5.8)–(5.12) will be given later when coupled interior/exterior problems will be investigated. Let us then define the intermediate model problem:

$$\begin{cases} -\dfrac{\partial}{\partial r}\Big(\dfrac{\mu^{-1}}{r}\dfrac{\partial}{\partial r}(ru)\Big) - \dfrac{\partial}{\partial z}\Big(\mu^{-1}\dfrac{\partial u}{\partial z}\Big) + \mathrm{i}\omega\sigma u = f & \text{in } \Lambda, \\ u = 0 & \text{on } \gamma, \end{cases} \tag{7.13}$$

where

$$\Lambda := \{(r,z) \in \mathbb{R}^+ \times \mathbb{R};\ 0 < r < \Phi(z),\ 0 < z < L\},$$
$$\gamma := \{(r,z) \in \partial\Lambda;\ r > 0\},$$

where the function Φ is positive, smooth and defined on $[0, L]$ and $\partial\Lambda$ stands for the boundary of Λ. In (7.13), μ and σ are positive functions that satisfy conditions (2.7) and (2.11) respectively, and $\omega \in \mathbb{R}$.

Remark 7.2.2. To interpret (7.13) as an axisymmetric version of a 3-D problem, one has to consider the following boundary value problem:

$$\begin{cases} \mathbf{curl}\,(\mu^{-1}\,\mathbf{curl}\,\boldsymbol{v}) + \mathrm{i}\omega\sigma\boldsymbol{v} = \boldsymbol{g} & \text{in } \Omega, \\ \operatorname{div}\boldsymbol{v} = 0 & \text{in } \Omega, \\ \boldsymbol{v} = 0 & \text{on } \Gamma, \end{cases}$$

where $\boldsymbol{g} = f(r,z)\,\boldsymbol{e}_\theta$,

$$\Omega = \{\boldsymbol{x} = (r\cos\theta, r\sin\theta, z);\ (r,z) \in \Lambda,\ 0 \le \theta < 2\pi\},$$

and Γ is its boundary. Writing $\boldsymbol{v}$ in cylindrical coordinates (see Chap. 5)

$$\breve{\boldsymbol{v}} = \breve{v}_r\,\boldsymbol{e}_r + \breve{v}_\theta\,\boldsymbol{e}_\theta + \breve{v}_z\,\boldsymbol{e}_z$$

and looking for axisymmetric solutions with the property $\breve{v}_r = \breve{v}_z = 0$, it can be shown that $u = \breve{v}_\theta$ is a solution to (7.13).

Problem (7.13) differs from (7.10) and (7.11) essentially by the fact that this one invokes a singularity at the line $r = 0$. For this we have to choose an adequate functional setting and numerical approximation to obtain a well posed variational formulation. Following the formulation (5.25), we define the space

$$\mathcal{W} := \Big\{v : \Lambda \to \mathbb{C};\ \frac{v}{r},\ \frac{\partial v}{\partial r},\ \frac{\partial v}{\partial z} \in \breve{\mathcal{L}}^2(\Lambda),\ v = 0 \text{ on } \gamma\Big\},$$

where we recall that

$$\breve{\mathcal{L}}^2(\Lambda) = \{v : \Lambda \to \mathbb{C};\ \int_\Lambda |v(r,z)|^2\, r\,dr\,dz < \infty\}.$$

Using Theorem 5.3.2, this space can be equipped with the norm:

$$\|v\|_{\mathcal{W}} := \left(\|r^{-1}v\|^2_{\breve{\mathcal{L}}^2(\Lambda)} + \left\|\frac{\partial v}{\partial r}\right\|^2_{\breve{\mathcal{L}}^2(\Lambda)} + \left\|\frac{\partial v}{\partial z}\right\|^2_{\breve{\mathcal{L}}^2(\Lambda)}\right)^{\frac{1}{2}}.$$

A variational formulation of (7.13) is given by:

$$\begin{cases} \text{Find } u \in \mathcal{W} \text{ such that} \\ \displaystyle\int_\Lambda \mu^{-1}\Big(\frac{1}{r}u\overline{v} + r\,\frac{\partial u}{\partial r}\frac{\partial \overline{v}}{\partial r} + r\,\frac{\partial u}{\partial z}\frac{\partial \overline{v}}{\partial z}\Big)\,dr\,dz \\ \displaystyle\qquad + \mathrm{i}\omega \int_\Lambda \sigma u\overline{v}\,r\,dr\,dz = \int_\Lambda f\overline{v}\,r\,dr\,dz \qquad \forall\, v \in \mathcal{W}. \end{cases} \tag{7.14}$$

Remark 7.2.3. According to Theorem 5.3.2, the solution u of (7.14) experiences the property:

$$u(0,z) = 0 \qquad \forall\, z \in (0,L). \tag{7.15}$$

In order to discretize (7.14), we define the space

$$\mathcal{W}_h = \{v \in \mathcal{C}^0(\overline{\Lambda});\ v_{|T} \in \mathbb{P}_1(T)\ \forall\, T \in \mathcal{T}_h,\ v = 0 \text{ on } \gamma,\\ v(0,z) = 0 \text{ for } z \in [0,L]\}.$$

Let us first notice that without the boundary condition at $r = 0$, the space $\mathcal{W}_h$ would not be included in $\mathcal{W}$ (see Remark 7.2.3). Prescribing (7.15) as a boundary condition ensures that $\mathcal{W}_h$ is a subspace of $\mathcal{W}$.

We can now formulate the finite element approximation of (7.14):

$$\begin{cases} \text{Find } u_h \in \mathcal{W}_h \text{ such that} \\ \displaystyle\int_\Lambda \mu^{-1}\Big(\frac{1}{r}u_h\overline{v} + r\frac{\partial u_h}{\partial r}\frac{\partial \overline{v}}{\partial r} + r\,\frac{\partial u_h}{\partial z}\frac{\partial \overline{v}}{\partial z}\Big)\,dr\,dz \\ \displaystyle\qquad + \mathrm{i}\omega \int_\Lambda \sigma u_h \overline{v}\, r\,dr\,dz = \int_\Lambda f\overline{v}\, r\,dr\,dz \qquad \forall\, v \in \mathcal{W}_h. \end{cases} \tag{7.16}$$

We have the following convergence result (see [122] for instance):

Theorem 7.2.3. *Assume that the unique solution of* (7.14) *has the regularity property* $u \in \breve{\mathcal{H}}^2(\Lambda)$ *where*

$$\breve{\mathcal{H}}^2(\Lambda) = \Big\{v \in \mathcal{D}'(\Lambda);\ \frac{\partial^{i+j} v}{\partial r^i\,\partial z^j} \in \breve{\mathcal{L}}^2(\Lambda),\ 0 \le i+j \le 2\Big\}.$$

Then there exists a constant C*, independent of* h *and* u*, such that*

$$\|u - u_h\|_{\mathcal{W}} \le C h\, \|u\|_{\breve{\mathcal{H}}^2(\Lambda)}.$$

Proof. We define the sesquilinear and antilinear forms:

$$\mathcal{B}(u,v) = \int_\Lambda \mu^{-1}\Big(\frac{1}{r}u\overline{v} + r\frac{\partial u}{\partial r}\frac{\partial \overline{v}}{\partial r} + r\,\frac{\partial u}{\partial z}\frac{\partial \overline{v}}{\partial z}\Big)\,dr\,dz + \mathrm{i}\omega \int_\Lambda \sigma u \overline{v}\, r\,dr\,dz,$$

$$\mathcal{L}(v) = \int_\Lambda f\overline{v}\, r\,dr\,dz.$$

Since $\mathcal{B}$ is continuous and coercive and $\mathcal{L}$ is continuous, we obtain the result by using Theorem 7.2.1. □

Remark 7.2.4. In order to achieve the numerical solution of (5.8)–(5.12), a coupling condition with the exterior problem must be added. This issue is treated in the following sections.

7.3 Finite Elements in $\mathcal{H}(\mathbf{curl})$–Spaces

Most electromagnetic problems in the three-dimensional case are defined in $\mathcal{H}(\mathbf{curl}, \cdot)$–spaces. For their numerical treatment a stable approximation must fulfill the regularity of these spaces. In other words tangential components of the sought vector fields must be continuous across element boundaries.

Let us consider the following model problem:

$$\begin{cases} \mathbf{curl}\,(\sigma^{-1}\,\mathbf{curl}\,\boldsymbol{u}) + \mathrm{i}\omega\mu\boldsymbol{u} = \boldsymbol{f} & \text{in } \Omega, \\ \boldsymbol{u} \times \boldsymbol{n} = 0 & \text{on } \Gamma, \end{cases} \tag{7.17}$$

where Ω is a bounded domain in $\mathbb{R}^3$ with boundary Γ, $\omega \in \mathbb{R}$, μ and σ are positive functions that satisfy conditions (2.7) and (2.11) respectively, and $\boldsymbol{f} \in \mathcal{L}^2(\Omega)$. Denoting by $\mathcal{H}_0(\mathbf{curl}, \Omega)$ the space

$$\mathcal{H}_0(\mathbf{curl}, \Omega) := \{\boldsymbol{v} \in \mathcal{H}(\mathbf{curl}, \Omega);\ \boldsymbol{v} \times \boldsymbol{n} = 0 \text{ on } \Gamma\},$$

endowed with the norm of $\mathcal{H}(\mathbf{curl}, \Omega)$, we have for (7.17) the variational formulation:

$$\begin{cases} \text{Find } \boldsymbol{u} \in \mathcal{H}_0(\mathbf{curl}, \Omega) \text{ such that} \\ \displaystyle\int_\Omega (\sigma^{-1}\,\mathbf{curl}\,\boldsymbol{u} \cdot \mathbf{curl}\,\overline{\boldsymbol{v}} + \mathrm{i}\omega\mu\boldsymbol{u} \cdot \overline{\boldsymbol{v}})\,d\boldsymbol{x} = \int_\Omega \boldsymbol{f} \cdot \overline{\boldsymbol{v}}\,d\boldsymbol{x} \\ \qquad\qquad\qquad\qquad \forall\ \boldsymbol{v} \in \mathcal{H}_0(\mathbf{curl}, \Omega). \end{cases} \tag{7.18}$$

The construction of a finite–dimensional subspace of $\mathcal{W}_h$ of $\mathcal{H}(\mathbf{curl}, \Omega)$ will be made through the construction of its basis functions.

For a tetrahedron $T \in \mathscr{T}_h$ let $\mathscr{E}_T$ stand for the set of the six edges of T. Along each edge $e \in \mathscr{E}_h$, we define a unit tangent vector $\boldsymbol{t}_e$ to e with arbitrary orientation. We then associate to the edge e the function $\boldsymbol{\xi}_e$ defined by:

$$\boldsymbol{\xi}_e(\boldsymbol{x}) = \boldsymbol{b}_T + \boldsymbol{c}_T \times \boldsymbol{x}, \quad \boldsymbol{b}_T, \boldsymbol{c}_T \in \mathbb{C}^3, \quad \forall\ \boldsymbol{x} \in T, \quad \forall\ T \in \mathscr{T}_h,$$

$$\int_{e'} \boldsymbol{\xi}_e \cdot \boldsymbol{t}_{e'}\,d\ell = \delta_{ee'} \qquad \forall\ e' \in \mathscr{E}_h.$$

A more practical form of the basis functions $(\boldsymbol{\xi}_e)$ can be derived as follows: For a tetrahedron T with vertices $\boldsymbol{a}_1$, $\boldsymbol{a}_2$, $\boldsymbol{a}_3$ and $\boldsymbol{a}_4$ such that $\boldsymbol{a}_1$ and $\boldsymbol{a}_2$ are the ends of an edge $e \in \mathscr{E}_T$, the tangent vector $\boldsymbol{t}_e$ being directed from $\boldsymbol{a}_1$ to $\boldsymbol{a}_2$, we denote by $\phi_i \in \mathbb{P}_1(T)$ the Lagrange basis of degree 1 on T, i.e. such that $\phi_i(\boldsymbol{a}_j) = \delta_{ij}$ for $1 \le i, j \le 4$. It can be shown (see [136] for instance) that for $\boldsymbol{x} \in T$,

$$\boldsymbol{\xi}_e(\boldsymbol{x}) = \phi_2(\boldsymbol{x})\nabla\phi_1 - \phi_1(\boldsymbol{x})\nabla\phi_2. \tag{7.19}$$

Basic properties of the functions $\boldsymbol{\xi}_e$ can be summarized in the following:

1. We have $\mathbf{curl}\,\boldsymbol{\xi}_e \in \mathcal{L}^2(\Omega)$.
2. The support of the function $\boldsymbol{\xi}_e$ is the union of all tetrahedra that share the edge e.
3. The inner product $\boldsymbol{\xi}_e \cdot \boldsymbol{t}_{e'}$ is constant on the edge $e' \in \mathscr{E}_T$.
4. The jump of the trace $\boldsymbol{\xi}_e \times \boldsymbol{n}$ is null on any face common to two tetrahedra.

Owing to these properties it is easy to see that $\boldsymbol{\xi}_e \in \mathcal{H}(\mathbf{curl}, \Omega)$ for all $e \in \mathscr{E}_h$, and $\mathcal{W}_h$ is simply defined as the space spanned by the family of vector functions $(\boldsymbol{\xi}_e)_{e \in \mathscr{E}_h}$.

Remark 7.3.1. Let $\boldsymbol{w} \in \mathcal{W}_h$, and let us consider its expansion in the basis $\boldsymbol{\xi}_e$:

$$\boldsymbol{w}(\boldsymbol{x}) = \sum_{e' \in \mathscr{E}_h} \alpha_{e'} \boldsymbol{\xi}_{e'}(\boldsymbol{x}) \qquad \boldsymbol{x} \in \Omega.$$

The circulation of $\boldsymbol{w}$ on any edge $e \in \mathscr{E}_h$ is given by

$$\int_e \boldsymbol{w} \cdot \boldsymbol{t}\, d\ell = \sum_{e' \in \mathscr{E}_h} \alpha_{e'} \int_e \boldsymbol{\xi}_{e'} \cdot \boldsymbol{t}\, d\ell = \alpha_e.$$

Therefore the coefficients of the expansion of the vector field $\boldsymbol{w}$ in the basis $(\boldsymbol{w}_e)$ are the circulations along mesh edges.

The construction, as well as the following convergence result for this finite element are due to Nédélec [136, 137].

Theorem 7.3.1. *Let $\boldsymbol{u} \in \mathcal{H}^1(\Omega)$ with $\mathbf{curl}\,\boldsymbol{u} \in \mathcal{H}^1(\Omega)$ and let $\boldsymbol{\pi}_h \boldsymbol{u} \in \mathcal{W}_h$ be defined by*

$$\int_e \boldsymbol{\pi}_h \boldsymbol{u} \cdot \boldsymbol{t}_e\, d\ell = \int_e \boldsymbol{u} \cdot \boldsymbol{t}_e\, d\ell \qquad \forall\, e \in \mathscr{E}_h.$$

Then, there is a constant C, independent of h, such that

$$\|\boldsymbol{u} - \boldsymbol{\pi}_h \boldsymbol{u}\|_{\mathcal{L}^2(\Omega)} + \|\,\mathbf{curl}\,(\boldsymbol{u} - \boldsymbol{\pi}_h \boldsymbol{u})\|_{\mathcal{L}^2(\Omega)} \le Ch\big(\|\boldsymbol{u}\|_{\mathcal{L}^2(\Omega)} + \|\,\mathbf{curl}\,\boldsymbol{u}\|_{\mathcal{H}^1(\Omega)}\big).$$

Let us now define the space

$$\mathcal{W}_h^0 := \{\boldsymbol{v} \in \mathcal{W}_h;\ \boldsymbol{v} \times \boldsymbol{n} = 0 \text{ on } \Gamma\}.$$

The finite element approximation of (7.18) is given by the variational problem:

$$\begin{cases} \text{Find } \boldsymbol{u}_h \in \mathcal{W}_h^0 \text{ such that} \\ \displaystyle\int_\Omega (\sigma^{-1}\, \mathbf{curl}\, \boldsymbol{u}_h \cdot \mathbf{curl}\, \overline{\boldsymbol{v}} + \mathrm{i}\omega\mu\, \boldsymbol{u}_h \cdot \overline{\boldsymbol{v}})\, d\boldsymbol{x} = \int_\Omega \boldsymbol{f} \cdot \overline{\boldsymbol{v}}\, d\boldsymbol{x} \\ \qquad\qquad\qquad\qquad \forall\, \boldsymbol{v} \in \mathcal{W}_h^0, \end{cases} \tag{7.20}$$

which possesses a unique solution thanks to the Lax–Milgram theorem (Theorem 1.2.1).

We have the following convergence theorem:

Theorem 7.3.2. *Assume that the solution of* (7.18) *fulfills the regularity properties:*

$$\boldsymbol{u} \in \mathcal{H}^1(\Omega), \ \mathbf{curl}\, \boldsymbol{u} \in \mathcal{H}^1(\Omega).$$

Then there exists a constant C*, independent of* h *and* $\boldsymbol{u}$ *such that*

$$\|\boldsymbol{u} - \boldsymbol{u}_h\|_{\mathcal{L}^2(\Omega)} + \|\,\mathbf{curl}\,(\boldsymbol{u} - \boldsymbol{u}_h)\|_{\mathcal{L}^2(\Omega)} \leq Ch\big(\|\boldsymbol{u}\|_{\mathcal{H}^1(\Omega)} + \|\,\mathbf{curl}\,\boldsymbol{u}\|_{\mathcal{H}^1(\Omega)}\big).$$

Proof. We first note that (7.18) and (7.20) can be written in the variational forms:

$$\mathscr{B}(\boldsymbol{u}, \boldsymbol{v}) = \mathscr{L}(\boldsymbol{v}) \qquad \forall\, \boldsymbol{v} \in \mathcal{H}_0(\mathbf{curl}, \Omega),$$

$$\mathscr{B}(\boldsymbol{u}_h, \boldsymbol{v}) = \mathscr{L}(\boldsymbol{v}) \qquad \forall\, \boldsymbol{v} \in \mathcal{W}_h^0,$$

where

$$\mathscr{B}(\boldsymbol{u}, \boldsymbol{v}) = \int_\Omega \sigma^{-1}\, \mathbf{curl}\, \boldsymbol{u} \cdot \mathbf{curl}\, \overline{\boldsymbol{v}}\, dx + \mathrm{i}\omega \int_\Omega \mu\, \boldsymbol{u} \cdot \overline{\boldsymbol{v}}\, dx,$$

$$\mathscr{L}(\boldsymbol{v}) = \int_\Omega \boldsymbol{f} \cdot \overline{\boldsymbol{v}}\, dx.$$

We note that owing to (2.11) and (2.7), we have the existence of a positive constant C such that

$$\big|\mathscr{B}(\boldsymbol{v}, \boldsymbol{v})\big| \geq C\,\big(\|\boldsymbol{v}\|^2_{\mathcal{L}^2(\Omega)} + \|\,\mathbf{curl}\,\boldsymbol{v}\|^2_{\mathcal{L}^2(\Omega)}\big),$$

which proves that $\mathscr{B}$ is coercive. Using Theorem 7.3.1 and the abstract estimate (7.3) we deduce the error bound. □

Remark 7.3.2. In view of considering the numerical solution of a coupled interior/exterior problem, instead of imposing $\boldsymbol{u} \times \boldsymbol{n} = 0$ on Γ, a numerical procedure that couples the finite element method defined above with an integral representation can be envisaged. This will be given later in this chapter.

7.4 Finite Elements in $\mathcal{H}(\mathrm{div})$–Spaces

Another type of functional setting that can be used for the formulation of some eddy current models we have defined is the use of $\mathcal{H}(\mathrm{div})$–spaces. We have indeed in Chap. 5, and more precisely in (5.5) a mathematical setting that imposes this use.

Furthermore, 3-D models based on the current density (4.17) impose using these spaces.

Let us consider the following model problem in $\mathbb{R}^3$:

$$\begin{cases} -\nabla(\sigma^{-1}\operatorname{div}\boldsymbol{u}) + \mathrm{i}\omega\mu\boldsymbol{u} = \boldsymbol{f} & \text{in } \Omega, \\ \boldsymbol{u}\cdot\boldsymbol{n} = 0 & \text{on } \Gamma, \end{cases} \tag{7.21}$$

where Ω is a bounded domain in $\mathbb{R}^3$ with boundary Γ, $\omega > 0$, σ and μ are functions that fulfill conditions (2.11) and (2.7) respectively, and $\boldsymbol{f} \in \mathcal{L}^2(\Omega)$. Let us define the space

$$\mathcal{H}_0(\mathrm{div},\Omega) := \{\boldsymbol{v} \in \mathcal{H}(\mathrm{div},\Omega);\ \boldsymbol{v}\cdot\boldsymbol{n} = 0 \text{ on } \Gamma\},$$

endowed with the norm of $\mathcal{H}(\mathrm{div},\Omega)$. We deduce for (7.21) the variational formulation:

$$\begin{cases} \text{Find } \boldsymbol{u} \in \mathcal{H}_0(\mathrm{div},\Omega) \text{ such that} \\ \displaystyle\int_\Omega \left(\sigma^{-1}\operatorname{div}\boldsymbol{u}\,\operatorname{div}\overline{\boldsymbol{v}} + \mathrm{i}\omega\mu\,\boldsymbol{u}\cdot\overline{\boldsymbol{v}}\right) d\boldsymbol{x} \\ \qquad\qquad = \displaystyle\int_\Omega \boldsymbol{f}\cdot\boldsymbol{v}\,d\boldsymbol{x} \qquad \forall\,\boldsymbol{v} \in \mathcal{H}_0(\mathrm{div},\Omega). \end{cases} \tag{7.22}$$

Using the Lax–Migram theorem (Theorem 1.2.1) we have the following result:

Theorem 7.4.1. *Problem* (7.22) *has a unique solution. Moreover, we have the estimate*

$$\|\boldsymbol{u}\|_{\mathcal{L}^2(\Omega)} + \|\operatorname{div}\boldsymbol{u}\|_{L^2(\Omega)} \le C\,\|\boldsymbol{f}\|_{\mathcal{L}^2(\Omega)}.$$

Like for the space $\mathcal{H}(\mathbf{curl},\Omega)$ we shall construct an approximation space for $\mathcal{H}(\mathrm{div},\Omega)$ by defining its basis functions. For each face of a tetrahedron $a \in \mathscr{F}_h$, we denote by $\boldsymbol{x}_a$ its barycenter. We associate to each face a the function z_a defined by

$$z_a(\boldsymbol{x}) = \boldsymbol{b}_T + c_T\boldsymbol{x} \qquad \boldsymbol{b}_T \in \mathbb{C}^3,\ c_T \in \mathbb{C}, \quad \forall\,\boldsymbol{x} \in T, \quad \forall\,T \in \mathscr{T}_h,$$

$$\int_{a'} z_a\cdot\boldsymbol{n}_{a'}\,ds = \delta_{aa'} \qquad \forall\,a' \in \mathscr{F}_h.$$

Basic properties of the functions z_a can be summarized in the following:

1. We have $\operatorname{div} z_a \in L^2(\Omega)$.
2. The support of z_a is the union of the two elements containing the face a, except in the case of a face on the boundary Γ where the support is the unique element containing this face.

3. The inner product $z_a \cdot \boldsymbol{n}_{a'}$ is constant on the face $a' \in \mathscr{F}_h$. Moreover, z_a is orthogonal to the other faces of $\mathscr{F}_h$.
4. The jump of the trace $z_a \cdot \boldsymbol{n}$ is null on any face common to two elements.

The above properties enable defining the space spanned by the functions z_a, that will be referred to as $\mathcal{V}_h$, as a subspace of $\mathcal{H}(\text{div}, \Omega)$. We define

$$\mathcal{V}_h^0 := \{\boldsymbol{w} \in \mathcal{V}_h;\ \boldsymbol{w} \cdot \boldsymbol{n} = 0 \text{ on } \Gamma\}. \tag{7.23}$$

The discrete problem is defined by the following variational formulation:

$$\begin{cases} \text{Find } \boldsymbol{u}_h \in \mathcal{V}_h^0 \text{ such that} \\ \displaystyle\int_\Omega \big(\sigma^{-1} \operatorname{div} \boldsymbol{u}_h \operatorname{div} \overline{\boldsymbol{v}}\, d\boldsymbol{x} + \mathrm{i}\omega\mu\, \boldsymbol{u}_h \cdot \overline{\boldsymbol{v}}\big)\, d\boldsymbol{x} = \int_\Omega \boldsymbol{f} \cdot \overline{\boldsymbol{v}}\, d\boldsymbol{x}\ \forall\, \boldsymbol{v} \in \mathcal{V}_h^0. \end{cases} \tag{7.24}$$

Thanks to the Lax–Milgram theorem (Theorem 1.2.1), Problem (7.24) admits a unique solution.

The construction and the following convergence result are due to Raviart–Thomas [137, 153]. This finite element is generally referred to as *The Raviart–Thomas (RT_1) element.*

Theorem 7.4.2. *Let $\boldsymbol{u} \in \mathcal{H}^1(\Omega)$ with $\operatorname{div} \boldsymbol{u} \in \mathcal{H}^1(\Omega)$ and let $\boldsymbol{\pi}_h \boldsymbol{u} \in \mathcal{V}_h^0$ be defined by*

$$\int_a \boldsymbol{\pi}_h \boldsymbol{u} \cdot \boldsymbol{n}_a\, ds = \int_a \boldsymbol{u} \cdot \boldsymbol{n}_a\, ds \qquad \forall\, a \in \mathscr{F}_h.$$

Then there exists a constant C, independent of h and $\boldsymbol{u}$ such that

$$\|\boldsymbol{u} - \boldsymbol{\pi}_h \boldsymbol{u}\|_{\mathcal{L}^2(\Omega)} + \|\operatorname{div}(\boldsymbol{u} - \boldsymbol{\pi}_h \boldsymbol{u})\|_{\mathcal{L}^2(\Omega)} \le Ch\big(\|\boldsymbol{u}\|_{\mathcal{H}^1(\Omega)} + \|\operatorname{div} \boldsymbol{u}\|_{\mathcal{H}^1(\Omega)}\big).$$

From this we easily prove the following convergence result by using Theorem 7.4.2 and (7.3):

Theorem 7.4.3. *Assume that the solution $\boldsymbol{u}$ of (7.22) fulfills the regularity property*

$$\boldsymbol{u} \in \mathcal{H}^1(\Omega),\ \operatorname{div} \boldsymbol{u} \in \mathcal{H}^1(\Omega).$$

Then there exists a constant C, independent of h and $\boldsymbol{u}$ such that

$$\|\boldsymbol{u} - \boldsymbol{u}_h\|_{\mathcal{L}^2(\Omega)} + \|\operatorname{div}(\boldsymbol{u} - \boldsymbol{u}_h)\|_{\mathcal{L}^2(\Omega)} \le Ch\big(\|\boldsymbol{u}\|_{\mathcal{H}^1(\Omega)} + \|\operatorname{div} \boldsymbol{u}\|_{\mathcal{H}^1(\Omega)}\big).$$

7.5 The Boundary Element Method for Boundary Integral Equations

We now investigate the numerical approximation of integral equations that are used in the modelling of eddy currents. We have to treat boundary integral equations that represent the solution of the harmonic equation in the exterior domain. Our aim is to give a synthetic presentation of the approximation of boundary integral equations and their coupling to partial differential equations. This coupling will be considered in view of the various eddy current problems we have analyzed so far. It is a difficult task to give a bibliographical review of the works on numerical solutions of boundary integral equations and their couplings with partial differential equations. It is maybe more instructive to consult the lecture notes of Nédélec [135], the review of Hsiao and Wendland [105], the overview paper by Hsiao [104] or the dedicated chapters in Dautray–Lions ([62], Vol. 4, pp. 359–370). For the coupling, authors have mainly considered the use of finite element equations coupled with boundary elements. For this, a lot of information can be found in e.g. [58, 87, 108, 113, 167], and many others.

To start the presentation, we shall proceed as for Chap. 1, i.e. by presenting separately the 2-D and 3-D cases.

7.5.1 *Boundary Integral Equations in* $\mathbb{R}^3$

We have seen throughout Chaps. 1 and 3–5 that two types of boundary integral equations are involved in the representation of harmonic fields in exterior domains:

- A boundary integral equation that involves the Green function as a kernel. This is referred to as a first kind integral equation.
- A boundary integral equation that involves the normal derivative of the Green function. This is a second kind integral.

Since these representations differ from each other by the used functional space, different numerical analyses are to be envisaged for them. Let us also recall (see (1.21)) that the Green function in the 3-D case is given by

$$G(\boldsymbol{x}, \boldsymbol{y}) = \frac{1}{4\pi} \frac{1}{|\boldsymbol{x} - \boldsymbol{y}|},$$

and its normal derivative in the $\boldsymbol{y}$–variable is

$$\frac{\partial}{\partial n_{\boldsymbol{y}}} G(\boldsymbol{x}, \boldsymbol{y}) = -\frac{1}{4\pi} \frac{\boldsymbol{n}(\boldsymbol{y}) \cdot (\boldsymbol{x} - \boldsymbol{y})}{|\boldsymbol{x} - \boldsymbol{y}|^3}.$$

In order to avoid considering the approximation of the surface Γ by a polyhedron, which would lead to tedious technical developments, we assume that Γ is a polyhedron. Note that this approximation is considered in ([62], Vol. 4, p. 367).

Let us recall that the domain Ω is partitioned using a finite element mesh $\mathscr{T}_h$ of tetrahedra and that the trace of this mesh on Γ is a partition $\mathscr{F}_h^\Gamma$ into triangles.

7.5.1.1 First Kind Integrals

We are concerned by the boundary integral equation presented in (1.29):

$$\frac{1}{4\pi}\int_\Gamma \frac{p(\boldsymbol{y})}{|\boldsymbol{x}-\boldsymbol{y}|}\,ds(\boldsymbol{y}) = g(\boldsymbol{x}) \qquad \boldsymbol{x}\in\Gamma, \tag{7.25}$$

where Γ is a closed polyhedral surface imbedded in $\mathbb{R}^3$ and $g \in \mathcal{H}^{\frac{1}{2}}(\Gamma)$. The solution p of this problem is sought in the space $\mathcal{H}^{-\frac{1}{2}}(\Gamma)$.

As it is already stated in (1.30), (7.25) admits the variational formulation:

$$\begin{cases} \text{Find } p \in \mathcal{H}^{-\frac{1}{2}}(\Gamma) \text{ such that} \\ \displaystyle\frac{1}{4\pi}\int_\Gamma\int_\Gamma \frac{p(\boldsymbol{y})\,\overline{q}(\boldsymbol{x})}{|\boldsymbol{x}-\boldsymbol{y}|}\,ds(\boldsymbol{y})\,ds(\boldsymbol{x}) = \int_\Gamma g\overline{q}\,ds \quad \forall\, q \in \mathcal{H}^{-\frac{1}{2}}(\Gamma), \end{cases} \tag{7.26}$$

where again the integrals stand for duality pairings between the spaces $\mathcal{H}^{\frac{1}{2}}(\Gamma)$ and $\mathcal{H}^{-\frac{1}{2}}(\Gamma)$. We recall (see [135]) that the sesquilinear form

$$\mathscr{B}(p,q) := \frac{1}{4\pi}\int_\Gamma\int_\Gamma \frac{p(\boldsymbol{y})\,\overline{q}(\boldsymbol{x})}{|\boldsymbol{x}-\boldsymbol{y}|}\,ds(\boldsymbol{y})\,ds(\boldsymbol{x})$$

is continuous and coercive in $\mathcal{H}^{-\frac{1}{2}}(\Gamma)$ and then by the Lax–Milgram theorem (Theorem 1.2.1), (7.26) possesses a unique solution.

For $r \geq 0$, we define the space

$$\mathcal{Q}_h^r := \begin{cases} \{q \in \mathcal{C}^0(\Gamma);\ q_{|a} \in \mathbb{P}_r(a),\ \forall\, a \in \mathscr{F}_h^\Gamma\} & \text{if } r > 0, \\ \{q \in \mathcal{L}^2(\Gamma);\ q_{|a} \in \mathbb{P}_0(a),\ \forall\, a \in \mathscr{F}_h^\Gamma\} & \text{if } r = 0. \end{cases}$$

In other words, we approximate the solution p either by a discontinuous piecewise constant function or by a continuous function that is polynomial of degree $\leq r$. We introduce the approximate problem:

$$\begin{cases} \text{Find } p_h \in \mathcal{Q}_h^r \text{ such that} \\ \displaystyle\frac{1}{4\pi}\int_\Gamma\int_\Gamma \frac{p_h(\boldsymbol{y})\,\overline{q}(\boldsymbol{x})}{|\boldsymbol{x}-\boldsymbol{y}|}\,ds(\boldsymbol{y})\,ds(\boldsymbol{x}) = \int_\Gamma g\,\overline{q}\,ds \quad \forall\, q \in \mathcal{Q}_h^r. \end{cases} \tag{7.27}$$

Since $\mathcal{Q}_h^r$ is a finite–dimensional subspace of $\mathcal{H}^{-\frac{1}{2}}(\Gamma)$, then (7.27) is well posed. We have the following error estimates (see [62], Vol. 4, p. 360):

Theorem 7.5.1. *Assume that $p \in \mathcal{H}^{r+1}(\Gamma)$, then we have the error estimate*

$$\|p - p_h\|_{\mathcal{H}^{-\frac{1}{2}}(\Gamma)} \leq C\, h^{r+\frac{3}{2}}\, \|p\|_{\mathcal{H}^{r+1}(\Gamma)},$$

where the constant C is independent of h, p and p_h.

Remark 7.5.1. As shown in Sect. 1.3.4, it can be useful to compute

$$u(\boldsymbol{x}) = -\frac{1}{4\pi}\int_\Gamma \frac{p(\boldsymbol{y})}{|\boldsymbol{x}-\boldsymbol{y}|}\,ds(\boldsymbol{y}) \qquad \boldsymbol{x} \in \mathbb{R}^3.$$

The function u satisfies indeed

$$\Delta u = 0 \qquad \text{in } \Omega \cup \Omega_{\text{ext}}$$

with the conditions

$$[u] = 0, \quad \left[\frac{\partial u}{\partial n}\right] = p \qquad \text{on } \Gamma.$$

Defining the approximation

$$u_h(\boldsymbol{x}) = -\frac{1}{4\pi}\int_\Gamma \frac{p_h(\boldsymbol{y})}{|\boldsymbol{x}-\boldsymbol{y}|}\,ds(\boldsymbol{y}) \qquad \boldsymbol{x} \in \mathbb{R}^3,$$

where p_h is defined by (7.27), we have the error estimate (see [62], Vol. 4, p. 360)

$$\|u - u_h\|_{\mathcal{W}^1(\mathbb{R}^3)} \leq C\, h^{r+\frac{3}{2}}\, \|p\|_{\mathcal{H}^{r+1}(\Gamma)}.$$

Let us consider some practical aspects of the method. Let $(q_k)_{k=1}^N$ denote a basis of the space $\mathcal{Q}_h^r$. We obtain the linear system

$$\mathbf{A}\mathbf{p} = \mathbf{b},$$

where the matrix $\mathbf{A} = (A_{k\ell})$ is defined by

$$A_{k\ell} := \frac{1}{4\pi}\int_\Gamma\int_\Gamma \frac{q_\ell(\boldsymbol{y})\overline{q}_k(\boldsymbol{x})}{|\boldsymbol{x}-\boldsymbol{y}|}\,ds(\boldsymbol{y})\,ds(\boldsymbol{x}), \qquad k,\ell = 1,\ldots,N,$$

and where the entries of $\mathbf{p}$ are the coefficients of p_h in the expansion

$$p_h(\boldsymbol{x}) = \sum_{\ell=1}^N p_\ell q_\ell(\boldsymbol{x}) \qquad \boldsymbol{x} \in \Gamma.$$

Let us give some remarks about the implementation of the method:

1. The matrix A is hermitian and dense. This is natural since we are solving a nonlocal problem. Consequently, numerical solution of the linear system needs particular attention since this one is rather expensive in terms of computational time.
2. Matrix entries use integrals with singular kernels. Therefore, unless the integrals are evaluated exactly, any numerical integration formula uses points near singularities which results in significant numerical discrepancies. Some works like [69–71] have explored this difficulty. It is worth noting that when the double integral involves the same triangle, the calculation can be made exactly.

We end this subsection by illustrating the simplest case consisting in piecewise constant approximations ($r = 0$). To implement the method we choose as a basis of this space the set of characteristic functions of the triangles

$$\chi_a(\boldsymbol{x}) := \begin{cases} 1 & \text{if } \boldsymbol{x} \in a, \\ 0 & \text{if not,} \end{cases} \tag{7.28}$$

for a boundary face $a \in \mathscr{F}_h^\Gamma$, i.e., p_h is defined by

$$p_h(\boldsymbol{x}) = \sum_{a \in \mathscr{F}_h^\Gamma} p_a \chi_a(\boldsymbol{x}) \qquad \boldsymbol{x} \in \Gamma.$$

Then A and b are given respectively by:

$$A_{aa'} = \frac{1}{4\pi} \int_a \int_{a'} \frac{1}{|\boldsymbol{x} - \boldsymbol{y}|} \, ds(\boldsymbol{y}) \, ds(\boldsymbol{x}) \qquad a, a' \in \mathscr{F}_h^\Gamma,$$

$$b_a = \int_a g \, ds \qquad a \in \mathscr{F}_h^\Gamma.$$

We note that well known collocation methods [15, 175] can be retrieved by adopting the following numerical integration formulae:

$$\int_a \int_{a'} \frac{1}{|\boldsymbol{x} - \boldsymbol{y}|} \, ds(\boldsymbol{y}) \, ds(\boldsymbol{x}) \approx |a| \int_{a'} \frac{1}{|\boldsymbol{x}_a - \boldsymbol{y}|} \, ds(\boldsymbol{y}),$$

$$\int_a g \, ds \approx |a| \, g(\boldsymbol{x}_a),$$

where $\boldsymbol{x}_a$ is the barycenter of the triangle a. Remark that the use of the same numerical integration scheme is not possible if $a = a'$, since this one would deal with the evaluation of singular integrands (see above remarks).

7.5.1.2 Second Kind Integrals

Let us now consider boundary integral equations of the second kind, like the one involved in Theorem 1.3.10, i.e.

$$-\frac{1}{2}\varphi(\boldsymbol{x}) + \frac{1}{4\pi}\int_\Gamma \frac{\partial}{\partial n_x}\left(\frac{1}{|\boldsymbol{x}-\boldsymbol{y}|}\right)\varphi(\boldsymbol{y})\,ds(\boldsymbol{y}) = g(\boldsymbol{x}) \qquad \boldsymbol{x} \in \Gamma, \tag{7.29}$$

where $g \in \mathcal{H}^{-\frac{1}{2}}(\Gamma)$.

We recall the existence and uniqueness (see [62], Vol. 4, p. 131) of a solution $\varphi \in \mathcal{H}^{-\frac{1}{2}}(\Gamma)$ to (7.29).

In order to derive a variational formulation of (7.29), we use the so-called solid angle identity

$$\frac{1}{2}\psi(\boldsymbol{y}) + \frac{1}{4\pi}\int_\Gamma \psi(\boldsymbol{y})\frac{\partial}{\partial n_x}\left(\frac{1}{|\boldsymbol{x}-\boldsymbol{y}|}\right)ds(\boldsymbol{x}) = 0 \qquad \forall\, \boldsymbol{y} \in \Gamma.$$

Multiplying (7.29) by $\overline{\psi}$ and integrating over Γ, we deduce

$$-\int_\Gamma \varphi\overline{\psi}\,ds + \frac{1}{4\pi}\int_\Gamma\int_\Gamma \frac{\partial}{\partial n_y}\left(\frac{1}{|\boldsymbol{x}-\boldsymbol{y}|}\right)\varphi(\boldsymbol{x})(\overline{\psi}(\boldsymbol{y}) - \overline{\psi}(\boldsymbol{x}))\,ds(\boldsymbol{y})\,ds(\boldsymbol{x}) = \int_\Gamma g\overline{\psi}\,ds.$$

This results in the variational formulation:

$$\begin{cases} \text{Find } \varphi \in \mathcal{H}^{-\frac{1}{2}}(\Gamma) \text{ such that} \\ \mathcal{B}(\varphi,\psi) = \mathcal{L}(\psi) \qquad \forall\, \psi \in \mathcal{H}^{-\frac{1}{2}}(\Gamma), \end{cases} \tag{7.30}$$

where:

$$\mathcal{B}(\varphi,\psi) := -\int_\Gamma \varphi\overline{\psi}\,ds + \frac{1}{4\pi}\int_\Gamma\int_\Gamma \frac{\partial}{\partial n_y}\left(\frac{1}{|\boldsymbol{x}-\boldsymbol{y}|}\right)\varphi(\boldsymbol{x}) (\overline{\psi}(\boldsymbol{y}) - \overline{\psi}(\boldsymbol{x}))\,ds(\boldsymbol{y})\,ds(\boldsymbol{x}),$$

$$\mathcal{L}(\psi) := \int_\Gamma g\overline{\psi}\,ds.$$

Let us define an approximation of (7.30) by

$$\begin{cases} \text{Find } \varphi_h \in \mathcal{Q}_h^r \text{ such that} \\ \mathcal{B}(\varphi_h,\psi) = \mathcal{L}(\psi) \qquad \forall\, \psi \in \mathcal{Q}_h^r. \end{cases} \tag{7.31}$$

We have the following convergence result (cf. [62], Vol. 4, p. 366):

Theorem 7.5.2. *Let* $\varphi \in \mathcal{H}^{r+1}(\Gamma)$ *denote the solution of* (7.30). *Then there exists a constant* C*, independent of* h*,* φ *and* φ_h *such that*

$$\|\varphi - \varphi_h\|_{\mathcal{L}^2(\Gamma)} \leq C h^{r+1} \|\varphi\|_{\mathcal{H}^{r+1}(\Gamma)}.$$

To end this subsection, let us investigate the simplest case of piecewise constant approximation, i.e. $r = 0$. In this case, the number of unknowns is the number of boundary faces and we obtain the linear system:

$$\mathbf{A}\boldsymbol{\varphi} = \mathbf{b},$$

where the matrix $\mathbf{A}$ and the vector $\mathbf{b}$ are defined by:

$$\begin{aligned} A_{aa'} &= -|a|\,\delta_{aa'} + \frac{1}{4\pi}\int_{a'}\int_{a} \frac{\partial}{\partial n_y}\left(\frac{1}{|\boldsymbol{x}-\boldsymbol{y}|}\right) ds(\boldsymbol{x})\,ds(\boldsymbol{y}) \\ &\quad - \frac{\delta_{aa'}}{4\pi} \sum_{a''\in\mathscr{F}_h^\Gamma} \int_{a}\int_{a''} \frac{\partial}{\partial n_y}\left(\frac{1}{|\boldsymbol{x}-\boldsymbol{y}|}\right) ds(\boldsymbol{x})\,ds(\boldsymbol{y}), \\ b_a &= \int_a g\,ds, \end{aligned}$$

for $a, a' \in \mathscr{F}_h^\Gamma$, and $\boldsymbol{\varphi}$ is the vector having as entries the values of φ_h at boundary faces.

7.5.1.3 Numerical Approximation of the Steklov–Poincaré Operator

When one deals with the numerical solution of a coupled interior–exterior problem, the exterior contribution can be represented by the use of the Steklov–Poincaré operator. This one is defined with its proper functional setting in this context in Sect. 1.3.5. A reasonable numerical discretization of the Steklov–Poincaré operator results obviously in an invertible matrix.

Let us recall from Sect. 1.3.5 that the exterior Steklov–Poincaré operator is a linear and continuous operator P from $\mathcal{H}^{\frac{1}{2}}(\Gamma) \to \mathcal{H}^{-\frac{1}{2}}(\Gamma)$, defined by

$$P = (-\frac{1}{2}I + S)\,K^{-1}$$

where

$$K : p \in \mathcal{H}^{-\frac{1}{2}}(\Gamma) \mapsto Kp \in \mathcal{H}^{\frac{1}{2}}(\Gamma)$$

is the isomorphism defined by the variational equation:

$$\int_\Gamma\int_\Gamma Kp(y)\overline{q}(x)\,ds(y)\,ds(x) := \int_\Gamma\int_\Gamma G(x,y)\,p(y)\,\overline{q}(y)\,ds(y)\,ds(x)$$

$$\forall\ p \in \mathcal{H}^{-\frac{1}{2}}(\Gamma),\ q \in \mathcal{H}^{-\frac{1}{2}}(\Gamma), \quad (7.32)$$

and

$$S : p \in \mathcal{H}^{-\frac{1}{2}}(\Gamma) \mapsto Sp \in \mathcal{H}^{\frac{1}{2}}(\Gamma)$$

is defined by

$$\int_\Gamma (Sp(x))\,\overline{\psi}(x)\,ds := \int_\Gamma p(x)\int_\Gamma \overline{\psi}(y)\frac{\partial}{\partial n_y}G(x,y)\,ds(y)\,ds(x)$$

$$\forall\ \psi \in \mathcal{H}^{\frac{1}{2}}(\Gamma),\ p \in \mathcal{H}^{-\frac{1}{2}}(\Gamma). \quad (7.33)$$

Let us now define finite–dimensional subspaces of $\mathcal{H}^{\frac{1}{2}}(\Gamma)$ and $\mathcal{H}^{-\frac{1}{2}}(\Gamma)$ respectively by:

$$\mathcal{V}_h := \{\psi \in C^0(\Gamma);\ \psi_{|a} \in \mathbb{P}_1(a)\ \forall\ a \in \mathscr{F}_h^\Gamma\},$$
$$\mathcal{Q}_h := \{q \in \mathcal{L}^2(\Gamma);\ q_{|a} \in \mathbb{P}_0(a)\ \forall\ a \in \mathscr{F}_h^\Gamma\},$$

Note that according to the choice of these spaces, the dimension of $\mathcal{V}_h$ (*resp.* $\mathcal{Q}_h$) is the number of nodes (*resp.* triangles) on the boundary Γ. Let $(\phi_i)_i$ (*resp.* $(\chi_a)_a$) stand for a basis of $\mathcal{V}_h$ (*resp.* $\mathcal{Q}_h$). More precisely, we can make for these bases the following choices:

- For $\mathcal{V}_h$ we choose the Lagrange basis functions (ϕ_i), i.e. such that $\phi_i(a_j) = \delta_{ij}$, $i, j \in \mathscr{I}_h^\Gamma$.
- For $\mathcal{Q}_h$, the natural choice for basis functions consists in characteristic functions χ_a (see (7.28)), $a \in \mathscr{F}_h^\Gamma$.

In order to derive an approximation of the operator P we shall follow the same procedure as for obtaining its representation in Sect. 1.3.5. Let φ stand for a function in $\mathcal{H}^{\frac{1}{2}}(\Gamma)$ and let φ_h denote an approximation of φ in $\mathcal{V}_h$, we have the expansion

$$\varphi_h(x) = \sum_{i\in\mathscr{I}_h^\Gamma} \varphi_i\,\phi_i(x) \qquad x \in \Gamma.$$

Consider now the problem

$$\begin{cases} \Delta u = 0 & \text{in } \Omega \cup \Omega_{\text{ext}}, \\ u = \varphi & \text{on } \Gamma, \\ u(\boldsymbol{x}) = \mathcal{O}(|\boldsymbol{x}|^{-1}) & |\boldsymbol{x}| \to \infty. \end{cases}$$

and set $p = [\frac{\partial \varphi}{\partial n}]$. According to the representation (1.25), we can define an approximation p_h of p in $\mathcal{Q}_h$ by the variational integral equation:

$$-\int_\Gamma \int_\Gamma G(\boldsymbol{x}, \boldsymbol{y})\, p_h(\boldsymbol{y})\, \overline{q}(\boldsymbol{x})\, ds(\boldsymbol{y})\, ds(\boldsymbol{x}) = \int_\Gamma \varphi_h \overline{q}\, ds \qquad \forall\, q \in \mathcal{Q}_h.$$

Choosing a basis function χ_e as a test function and using the expansion of p_h in this basis, we obtain the matrix formulation

$$\mathbf{K}\mathbf{p} = \mathbf{B}\boldsymbol{\varphi}, \tag{7.34}$$

where $\mathbf{p}$ is the vector in $\mathbb{C}^{n_q}$ with entries p_e, $\boldsymbol{\varphi}$ is the vector with entries φ_i and $\mathbf{K}$ and $\mathbf{B}$ are the matrices with coefficients:

$$\begin{aligned} K_{aa'} &= -\int_\Gamma \int_\Gamma G(\boldsymbol{x}, \boldsymbol{y})\, \chi_{a'}(\boldsymbol{y})\, \chi_a(\boldsymbol{x})\, ds(\boldsymbol{y})\, ds(\boldsymbol{x}) \\ &= -\int_a \int_{a'} G(\boldsymbol{x}, \boldsymbol{y})\, ds(\boldsymbol{y})\, ds(\boldsymbol{x}) \qquad && a, a' \in \mathcal{F}_h^\Gamma, \\ B_{aj} &= \int_\Gamma \phi_j\, \chi_a\, ds = \int_a \phi_j\, ds && j \in \mathcal{I}_h^\Gamma,\ a \in \mathcal{F}_h^\Gamma. \end{aligned}$$

Note that the matrix $\mathbf{K}$ appears here as a finite–dimensional approximation of the operator K and $\mathbf{B}$ is a projection from $\mathcal{V}_h$ on $\mathcal{Q}_h$.

To go further, we recall that the Steklov–Poincaré operator P is defined by

$$P\varphi(\boldsymbol{x}) = \frac{1}{2}\, p(\boldsymbol{x}) - \int_\Gamma \frac{\partial G}{\partial n_y}(\boldsymbol{x}, \boldsymbol{y})\, p(\boldsymbol{y})\, ds(\boldsymbol{y}) \qquad \boldsymbol{x} \in \Gamma.$$

With this we can define the approximation $P_h : \mathcal{V}_h \to \mathcal{V}_h$ by

$$\begin{aligned} \int_\Gamma P_h \varphi_h\, \overline{\psi}\, ds &= \frac{1}{2} \int_\Gamma p_h\, \overline{\psi}\, ds \\ &\quad - \int_\Gamma \int_\Gamma \frac{\partial G}{\partial n_y}(\boldsymbol{x}, \boldsymbol{y})\, p_h(\boldsymbol{y})\, \overline{\psi}(\boldsymbol{x})\, ds(\boldsymbol{y})\, ds(\boldsymbol{x}) \qquad \forall\, \psi \in \mathcal{V}_h. \end{aligned} \tag{7.35}$$

Here also we choose as a test function the basis function $\psi = \phi_i$ of $\mathcal{V}_h$ to obtain for $i \in \mathscr{I}_h^\Gamma$:

$$\int_\Gamma P_h\varphi_h\,\phi_i\,ds = \frac{1}{2}\int_\Gamma p_h\,\phi_i\,ds - \int_\Gamma\int_\Gamma \frac{\partial G}{\partial n_y}(\boldsymbol{x},\boldsymbol{y})\,p_h(\boldsymbol{y})\,\phi_i(\boldsymbol{x})\,ds(\boldsymbol{y})\,ds(\boldsymbol{x}).$$

Now, we expand φ_h and p_h in their respective bases to get eventually the linear system

$$\mathbf{P}\varphi = (\mathbf{M} + \mathbf{R})\,\mathbf{p},$$

where $\mathbf{M}$ and $\mathbf{R}$ are the matrices defined by:

$$\begin{aligned} M_{ia} &= \frac{1}{2}\int_\Gamma \chi_a\,\phi_i\,ds = \frac{1}{2}\int_a \phi_i\,ds,\\ R_{ia} &= -\int_\Gamma\int_\Gamma \frac{\partial G}{\partial n_y}(\boldsymbol{x},\boldsymbol{y})\,\chi_a(\boldsymbol{y})\,\phi_i(\boldsymbol{x})\,ds(\boldsymbol{y})\,ds(\boldsymbol{x})\\ &= -\int_\Gamma\int_a \frac{\partial G}{\partial n_y}(\boldsymbol{x},\boldsymbol{y})\,\phi_i(\boldsymbol{x})\,ds(\boldsymbol{y})\,ds(\boldsymbol{x}), \end{aligned}$$

for $i \in \mathscr{I}_h^\Gamma$, $a \in \mathscr{F}_h^\Gamma$, and $\mathbf{P}$ is the sought Steklov–Poincaré matrix. Using (7.34), we obtain

$$\mathbf{P}\varphi = (\mathbf{M} + \mathbf{R})\,\mathbf{K}^{-1}\mathbf{B}\,\varphi \qquad \forall\,\varphi = (\varphi_i)_{i\in\mathscr{I}_h^\Gamma}.$$

This implies the definition of the Steklov–Poincaré matrix:

$$\mathbf{P} = (\mathbf{M} + \mathbf{R})\,\mathbf{K}^{-1}\mathbf{B}. \tag{7.36}$$

7.5.2 Boundary Integral Equations in $\mathbb{R}^2$

Let us consider two-dimensional boundary integral equations involving as kernel the Green function. The boundary is generally considered as a union of closed arcs in $\mathbb{R}^2$. We restrict ourselves here to the case of a single closed polygonal curve γ.

We recall (see (1.50)) that the Green function in the 2-D case is given by

$$G(\boldsymbol{x},\boldsymbol{y}) = -\frac{1}{2\pi}\ln|\boldsymbol{x}-\boldsymbol{y}|,$$

and its normal derivative in the $\boldsymbol{y}$–variable is

$$\frac{\partial}{\partial n_y}G(\boldsymbol{x},\boldsymbol{y}) = -\frac{1}{2\pi}\frac{\boldsymbol{n}(\boldsymbol{y})\cdot(\boldsymbol{x}-\boldsymbol{y})}{|\boldsymbol{x}-\boldsymbol{y}|^2}.$$

Let us recall that the finite element mesh of Λ is denoted by $\mathscr{T}_h$ and that the trace of this mesh on γ is a partition of γ, denoted by $\mathscr{E}_h^\gamma$ into edges of triangles.

7.5.2.1 First Kind Integrals

Let us consider the boundary integral equation presented in (1.58):

$$\int_\gamma G(\boldsymbol{x},\boldsymbol{y})\, p(\boldsymbol{y})\, ds(\boldsymbol{y}) + \xi = g(\boldsymbol{x}) \qquad \boldsymbol{x} \in \gamma, \tag{7.37}$$

where γ is a closed regular arc imbedded in $\mathbb{R}^2$, $g \in \mathcal{H}^{\frac{1}{2}}(\gamma)$, and ξ is a constant. The solution p of this problem is required in the space

$$\tilde{\mathcal{H}}^{-\frac{1}{2}}(\gamma) := \Big\{q \in \mathcal{H}^{-\frac{1}{2}}(\gamma);\ \int_\gamma q\, ds = 0\Big\}.$$

As it is already stated in (1.59), (7.37) admits the variational formulation:

$$\begin{cases} \text{Find } p \in \tilde{\mathcal{H}}^{-\frac{1}{2}}(\gamma) \text{ such that} \\ \displaystyle\int_\gamma\int_\gamma G(\boldsymbol{x},\boldsymbol{y})\, p(\boldsymbol{y})\, \overline{q}(\boldsymbol{x})\, ds(\boldsymbol{x})\, ds(\boldsymbol{y}) = \int_\gamma g\overline{q}\, ds \quad \forall\, q \in \tilde{\mathcal{H}}^{-\frac{1}{2}}(\gamma). \end{cases} \tag{7.38}$$

We recall (see [135]) that the sesquilinear form

$$\mathscr{B}(p,q) := \int_\gamma\int_\gamma G(\boldsymbol{x},\boldsymbol{y})\, p(\boldsymbol{y})\, \overline{q}(\boldsymbol{x})\, ds(\boldsymbol{x})\, ds(\boldsymbol{y})$$

is coercive in $\tilde{\mathcal{H}}^{-\frac{1}{2}}(\gamma)$ and then by the Lax–Milgram theorem (Theorem 1.2.1) (7.38) possesses a unique solution.

In order to discretize the problem, we assume that γ is polygonal with avoiding difficulties related to regularity of the solution, like in the 3-D case. Then we consider a partitioning of the curve γ into lines (or triangle edges) $e \in \mathscr{E}_h^\gamma$ with ends $\boldsymbol{a}_i \in \mathscr{N}_h^\gamma$ ($\mathscr{N}_h^\gamma$ is the set of boundary nodes). We also denote by e_i the line joining $\boldsymbol{a}_{i-1}$ and $\boldsymbol{a}_i$ for $1 \le i \le I$ and by e_I the one with ends a_0 and a_I.

We define the spaces:

$$\mathcal{Q}_h^r := \begin{cases} \{q \in C^0(\gamma);\ q_{|e} \in \mathbb{P}_r(e),\ e \in \mathscr{E}_h^\gamma\} & \text{if } r > 0, \\ \{q \in \mathcal{L}^2(\gamma);\ q_{|e} \in \mathbb{P}_0(e),\ e \in \mathscr{E}_h^\gamma\} & \text{if } r = 0, \end{cases}$$

$$\mathcal{Q}_h^{r,0} := \Big\{q \in \mathcal{Q}_h^r;\ \int_\gamma q\, ds = 0\Big\},$$

where $r \geq 0$. In other words, we approximate the solution p by a continuous function that is piecewise polynomial of degree $\leq r$ for $r > 0$ and discontinuous piecewise constant for $r = 0$.

We introduce the approximate problem:

$$\begin{cases} \text{Find } p_h \in \mathcal{Q}_h^{r,0} \text{ such that} \\ \displaystyle\int_\gamma\int_\gamma G(\boldsymbol{x},\boldsymbol{y})\, p_h(\boldsymbol{y})\,\overline{q}(\boldsymbol{x})\,ds(\boldsymbol{x})\,ds(\boldsymbol{y}) = \int_\gamma g\,\overline{q}\,ds \qquad \forall\, q \in \mathcal{Q}_h^{r,0}. \end{cases} \tag{7.39}$$

Since $\mathcal{Q}_h^{r,0}$ is a finite–dimensional subspace $\tilde{\mathcal{H}}^{-\frac{1}{2}}(\gamma)$, then (7.39) is well posed. We have the following error estimates (see [68, 103]):

Theorem 7.5.3. *Assume that solution p has the regularity property $p \in \mathcal{H}^{r+1}(\gamma)$, then we have the error estimate*

$$\|p - p_h\|_{\mathcal{H}^{-\frac{1}{2}}(\gamma)} \leq C\, h^{r+\frac{3}{2}}\, \|p\|_{\mathcal{H}^{r+1}(\gamma)},$$

where the constant C is independent of h, p and p_h.

It is clear that it is not easy to construct an approximation that enforces the constraint in $\mathcal{Q}_h^{r,0}$. To remedy to this difficulty, a Lagrange multiplier can be introduced following [104] by substituting to (7.39) the following variational formulation:

$$\begin{cases} \text{Find } (p_h, \lambda_h) \in \mathcal{Q}_h^{r} \times \mathbb{C} \text{ such that} \\ \displaystyle\int_\gamma\int_\gamma G(\boldsymbol{x},\boldsymbol{y})\, p_h(\boldsymbol{y})\,\overline{q}(\boldsymbol{x})\,ds(\boldsymbol{x})\,ds(\boldsymbol{y}) \\ \displaystyle\qquad\qquad\qquad + \lambda_h \int_\gamma \overline{q}\,ds = \int_\gamma g\,\overline{q}\,ds \quad \forall\, q \in \mathcal{Q}_h^{r}, \\ \displaystyle\int_\gamma p_h\,ds = 0. \end{cases} \tag{7.40}$$

Let us consider now some practical aspects of the method in the case where $r = 0$. Let $(\chi_e)_{e\in\mathcal{E}_h^\gamma}$ denote the basis of the space $\mathcal{Q}_h^0$ made of characteristic functions of edges. We obtain the linear system

$$\mathbf{A}\mathbf{p} + \mathbf{c}\lambda_h = \mathbf{b},$$
$$\mathbf{c}^{\mathsf{T}}\mathbf{p} = \mathbf{0}.$$

where the matrix $\mathbf{A}$ and the vectors $\mathbf{c}$ and $\mathbf{b}$ are respectively defined by:

$$A_{e,e'} := \int_\gamma\int_\gamma G(\boldsymbol{x},\boldsymbol{y})\,\chi_e(\boldsymbol{x})\chi_{e'}(\boldsymbol{y})\,ds(\boldsymbol{x})\,ds(\boldsymbol{y}) = \int_e\int_{e'} G(\boldsymbol{x},\boldsymbol{y})\,ds(\boldsymbol{x})\,ds(\boldsymbol{y}),$$

$$c_e := \int_\gamma \chi_e \, ds = |e| \quad \text{(length of } e\text{)},$$

$$b_e := \int_\gamma g \chi_e \, ds = \int_e g \, ds,$$

for $e, e' \in \mathscr{E}_h^\gamma$, and where the entries of $\mathbf{p}$ are the coefficients of p_h in the expansion

$$p_h(\boldsymbol{x}) = \sum_{e \in \mathscr{E}_h^\gamma} p_e \chi_e(\boldsymbol{x}) \qquad \boldsymbol{x} \in \gamma.$$

Clearly, the matrix $\mathbf{A}$ is symmetric and dense. Here we have the same type of remarks than in Sect. 7.5.1.

7.5.2.2 Second Kind Integrals

Second kind integrals in the two-dimensional case can be handled by the same techniques as in Sect. 7.5.1.2. We skip here the details.

7.6 Approximation of a Domain Integral Equation

We consider in this section a formulation of 3-D eddy current problems that leads to an integral equation formulated in the domain of the conductors. The major drawback of this model is that we are faced with a nonlocal problem (and then with a dense matrix when numerical solution is involved) in the whole domain rather than on its boundary. We shall see that this difficulty can be avoided by means of an iterative procedure where the global matrix does not have to be stored.

Let us consider the three-dimensional current density model described in Sect. 4.1 with a constant magnetic permeability (i.e. $\mu = \mu_0$). We recall that in this case, the model is given by (4.20), with the variational formulation (4.21).

Remark 7.6.1. The case of a nonconstant magnetic permeability can be handled by the same method, but further developments are required. In particular, the solution of an auxiliary problem involving a scalar potential is to be supplied. We skip these details, since we are mainly interested here in the solution of a nonlocal problem in the conductors Ω.

Let us consider the variational formulation (4.17) and define a finite–dimensional approximation of the space $\boldsymbol{\mathcal{X}}$, defined by

$$\boldsymbol{\mathcal{X}}_h := \{\boldsymbol{v} \in \boldsymbol{V}_h^0;\ \operatorname{div} \boldsymbol{v} = 0 \text{ in } \Omega\},$$

where $\mathcal{V}_h^0$ is the space defined in (7.23), We have the discrete problem

$$\begin{cases} \text{Find } \boldsymbol{J}_h \in \boldsymbol{\mathcal{X}}_h \text{ such that} \\ \displaystyle\int_\Omega \Big(\mathrm{i}\omega\mu_0 \int_\Omega G(\boldsymbol{x},\boldsymbol{y})\, \boldsymbol{J}_h(\boldsymbol{y})\, d\boldsymbol{y} + \sigma^{-1} \boldsymbol{J}_h(\boldsymbol{x})\Big) \cdot \overline{\boldsymbol{v}}(\boldsymbol{x})\, d\boldsymbol{x} \\ \displaystyle\qquad\qquad = V \int_S \overline{\boldsymbol{v}} \cdot \boldsymbol{n}\, ds \qquad \forall\, \boldsymbol{v} \in \boldsymbol{\mathcal{X}}_h, \end{cases} \tag{7.41}$$

where we recall that V is the current voltage.

Let us investigate the implementation of the finite element scheme defined in (7.41). It is clear that the definition of the space $\boldsymbol{\mathcal{X}}_h$ suggests the use of $\boldsymbol{\mathcal{H}}(\mathrm{div},\cdot)$ finite elements (see Sect. 7.4) with the additional constraint of divergence free vector fields. Following Sect. 7.4, we define the basis functions

$$\boldsymbol{\xi}_e(\boldsymbol{x}) = \boldsymbol{b}_T + c_T \boldsymbol{x}, \quad \boldsymbol{b}_T \in \mathbb{C}^3,\ c_T \in \mathbb{C}, \qquad \forall\, \boldsymbol{x} \in T, \quad \forall\, T \in \mathscr{T}_h.$$

Now, the constraint $\mathrm{div}\, \boldsymbol{\xi}_e = 0$ in $T \in \mathscr{T}_h$ implies $c_T = 0$. This means that the approximate solution is a piecewise constant vector such that its normal components are continuous across the faces $a \in \mathscr{F}_h$. A convenient finite element space is then given by

$$\begin{aligned} \boldsymbol{\mathcal{X}}_(:= \{\boldsymbol{v} \in \boldsymbol{\mathcal{L}}^2(\Omega);\ \boldsymbol{v}_{|T} \in \mathbb{P}_0^3(T)\ \forall\, T \in \mathscr{T}_h,\ [\boldsymbol{v}\cdot\boldsymbol{n}]_a = 0\ \forall\, a \in \mathscr{F}_h^{\mathrm{int}}, \\ \boldsymbol{v}\cdot\boldsymbol{n} = 0 \text{ on } a\ \forall\, a \in \mathscr{F}_h^{\Gamma}\}. \end{aligned}$$

Clearly, it is not easy to enforce directly the constraint of continuous normal components across the faces. In [32], the problem is handled at the matrix level. More precisely, the author introduces a Lagrange multiplier that enables imposing the constraint. The matrix form of the variational problem is then written as

$$\begin{aligned} \mathsf{A}\mathsf{u} + \mathsf{D}^{\mathsf{T}}\mathsf{p} &= \mathsf{b}, \\ \mathsf{D}\mathsf{u} &= \mathsf{0}, \end{aligned}$$

where u is the vector whose components are the approximate constant vector $\boldsymbol{J}_h$ at elements. The second equation is the expression of the constraint of continuity of the normal components of $\boldsymbol{J}_h$ at the faces. The vector p is the Lagrange multiplier vector that results from this constraint. Note that the matrix D is sparse but A is not. In [32], the author reports good performances of an Uzawa solver that uses the Conjugate Gradient method to solve the linear system for p.

We propose here a similar procedure at the variational level by using the so-called *hybrid* formulations. To derive such a formulation, we relax the continuity of normal components and free divergence constraints by defining the spaces:

$$\tilde{\boldsymbol{\mathcal{X}}}_{(} := \{\boldsymbol{v} \in \boldsymbol{\mathcal{L}}^2(\Omega);\ \boldsymbol{v}_{|T} = \boldsymbol{b}_T + c_T\boldsymbol{x}\ \boldsymbol{b}_T \in \mathbb{C}^3,\ c_T \in \mathbb{C}\ \forall\, T \in \mathscr{T}_h\},$$

$$\boldsymbol{\mathcal{X}}_h^T := \{\boldsymbol{v} = \boldsymbol{b}_T + c_T\boldsymbol{x},\ \boldsymbol{x} \in T,\ \boldsymbol{b}_T \in \mathbb{C}^3,\ c_T \in \mathbb{C}\},$$

$$\mathcal{Q}_h := \{q \in \mathcal{L}^2(\Omega);\ q_{|T} \in \mathbb{P}_0(T)\ \forall\, T \in \mathscr{T}_h\},$$

$$\mathcal{M}_h := \{\eta \in \mathcal{L}^2(F_h);\ \eta_{|a} \in \mathbb{P}_0(a)\ \forall\, a \in \mathscr{F}_h\}.$$

We consider then the variational formulation:

Find $(\boldsymbol{J}_h, p_h, \lambda_h) \in \tilde{\boldsymbol{\mathcal{X}}}_h \times \mathcal{Q}_h \times \mathcal{M}_h$ such that:

$$\int_T \Big(\mathrm{i}\omega\mu_0 \int_\Omega G(\boldsymbol{x},\boldsymbol{y})\,\boldsymbol{J}_h(\boldsymbol{y})\,d\boldsymbol{y} + \sigma^{-1}\boldsymbol{J}_h(\boldsymbol{x})\Big)\cdot\overline{\boldsymbol{v}}(\boldsymbol{x})\,d\boldsymbol{x} - \int_T p_h \operatorname{div}\overline{\boldsymbol{v}}\,d\boldsymbol{x} + \sum_{a\in\mathscr{F}_T}\int_a \lambda_h\overline{\boldsymbol{v}}\cdot\boldsymbol{n}\,ds = V\int_S \overline{\boldsymbol{v}}\cdot\boldsymbol{n}\,ds \qquad \forall\, \boldsymbol{v} \in \boldsymbol{\mathcal{X}}_h^T,\ \forall\, T \in \mathscr{T}_h, \tag{7.42}$$

$$\int_T \operatorname{div}\boldsymbol{J}_h\,d\boldsymbol{x} = 0 \qquad \forall\, T \in \mathscr{T}_h, \tag{7.43}$$

$$\int_a [\boldsymbol{J}_h\cdot\boldsymbol{n}]\,ds = 0 \qquad \forall\, a \in \mathscr{F}_h, \tag{7.44}$$

$$\int_a \boldsymbol{J}_h\cdot\boldsymbol{n}\,ds = 0 \qquad \forall\, a \in \mathscr{F}_h^\Gamma. \tag{7.45}$$

We recall here that $\mathscr{F}_T$ is the set of faces of the tetrahedron T and that $\mathscr{T}_a$ is the set of (2) tetrahedra that share the face a.

In the system (7.42)–(7.45) we have two Lagrange multipliers, the first one p_h takes into account the divergence–free constraint, while the second one λ_h is useful to enforce the continuity of the normal components of $\boldsymbol{J}_h$ across internal edges and the homogeneous boundary condition.

In order to obtain a linear system, let us write the unknowns in the form:

$$\boldsymbol{J}_h(\boldsymbol{x}) = \boldsymbol{J}_T + \alpha_T\boldsymbol{x} \qquad \boldsymbol{x} \in T,\ T \in \mathscr{T}_h,$$

$$p_h(\boldsymbol{x}) = p_T \qquad \boldsymbol{x} \in T,\ T \in \mathscr{T}_h,$$

$$\lambda_h(\boldsymbol{x}) = \lambda_a \qquad \boldsymbol{x} \in a,\ a \in \mathscr{F}_h.$$

We first note that (7.43) implies readily $\alpha_T = 0$. We then select test functions in (7.42)–(7.45) in the following way: we choose successively $\boldsymbol{v} = \boldsymbol{b}_T \in \mathbb{C}^3$ and $\boldsymbol{v} = \boldsymbol{x}$. We obtain the linear system of equations:

$$\sum_{T' \in \mathscr{T}_h} A_{TT'} \boldsymbol{J}_{T'} + \sum_{a \in \mathscr{F}_T} \boldsymbol{B}_{aT} \lambda_a = \boldsymbol{c}_T \qquad \forall\, T \in \mathscr{T}_h, \tag{7.46}$$

$$\sum_{T' \in \mathscr{T}_h} \boldsymbol{D}_{TT'} \cdot \boldsymbol{J}_{T'} + N_T\, p_T + \sum_{a \in \mathscr{F}_T} C_{Ta} \lambda_a = d_T \qquad \forall\, T \in \mathscr{T}_h, \tag{7.47}$$

$$\sum_{T \in \mathscr{F}_a} \boldsymbol{B}_{aT} \cdot \boldsymbol{J}_T = 0 \qquad \forall\, a \in \mathscr{F}_h \setminus \mathscr{T}_h^{\Gamma}, \tag{7.48}$$

where $\boldsymbol{n}_T^a$ is the normal to the face a pointing outside the tetrahedron T, and

$$A_{TT'} = \mathrm{i}\omega\mu_0 \int_T \Big(\int_{T'} G(\boldsymbol{x}, \boldsymbol{y})\, d\boldsymbol{y} \Big)\, d\boldsymbol{x} + \delta_{TT'} \int_T \sigma^{-1}\, d\boldsymbol{x},$$

$$\boldsymbol{D}_{TT'} = \mathrm{i}\omega\mu_0 \int_T \Big(\int_{T'} G(\boldsymbol{x}, \boldsymbol{y})\, d\boldsymbol{y} \Big) \boldsymbol{x}\, d\boldsymbol{x} + \delta_{TT'} \int_T \sigma^{-1} \boldsymbol{x}\, d\boldsymbol{x},$$

$$N_T = -3\,|T|,$$

$$\boldsymbol{B}_{aT} = \int_a \boldsymbol{n}_a^T\, ds = |a|\, \boldsymbol{n}_a^T,$$

$$C_{Ta} = \int_a \boldsymbol{x} \cdot \boldsymbol{n}_a^T\, ds,$$

$$\boldsymbol{c}_T = V \int_{S \cap T} \boldsymbol{n}\, ds,$$

$$d_T = V \int_{S \cap T} \boldsymbol{n} \cdot \boldsymbol{x}\, ds.$$

Let us put the system (7.46)–(7.48) in the matrix form:

$$\begin{pmatrix} \mathbf{A} & \mathbf{0} & \mathbf{B}^{\mathsf{T}} \\ \mathbf{D} & \mathbf{N} & \mathbf{C} \\ \mathbf{B} & \mathbf{0} & \mathbf{0} \end{pmatrix} \begin{pmatrix} \mathbf{u} \\ \mathbf{p} \\ \boldsymbol{\lambda} \end{pmatrix} = \begin{pmatrix} \mathbf{c} \\ \mathbf{d} \\ \mathbf{0} \end{pmatrix}$$

Numerical solution of this system can be tackled by a block Gauss elimination procedure, i.e. by solving successively the linear systems:

$$\mathbf{B}\mathbf{A}^{-1}\mathbf{B}^{\mathsf{T}} \boldsymbol{\lambda} = \mathbf{B}\mathbf{A}^{-1}\mathbf{c},$$

$$\mathbf{N}\mathbf{p} = \mathbf{d} - \mathbf{D}\mathbf{A}^{-1}\mathbf{C} - (\mathbf{C} - \mathbf{D}\mathbf{A}^{-1}\mathbf{B}^{\mathsf{T}})\, \boldsymbol{\lambda},$$

$$\mathbf{A}\mathbf{u} = \mathbf{c} - \mathbf{B}^{\mathsf{T}} \boldsymbol{\lambda}.$$

7.7 Coupled Finite Element/Boundary Element Methods

Once we have seen how (interior) boundary value problems and (exterior) boundary integral problems can be numerically solved, we turn now to the most encountered case, i.e. where these formulations are coupled. We shall present then the numerical discretization of the involved coupled problems and the structure of the linear systems this discretization induces.

7.7.1 The 2-D Transversal Model

We consider in this section the numerical solution of the two-dimensional model presented in Sect. 3.3, i.e. we deal with Eqs. (3.38)–(3.41). As we have seen, we are typically in presence of three domains Λ_1, Λ_2, Λ_3. Let us then consider a triangulation $\mathscr{T}_h^k$ of each domain Λ_k. We assume that these triangulations fulfill the property (7.6) and define

$$\mathscr{T}_h = \mathscr{T}_h^1 \cup \mathscr{T}_h^2 \cup \mathscr{T}_h^3.$$

Since the boundary is assumed polygonal, we denote by $\mathscr{E}_h^k$ the set of edges on the boundary of Λ_k and set similarly

$$\mathscr{E}_h = \mathscr{E}_h^1 \cup \mathscr{E}_h^2 \cup \mathscr{E}_h^3.$$

Let us first note that in (3.38)–(3.41) and then in its variational formulation (3.45), the presence of the mean values M_k makes the problem nonlocal. To avoid this difficulty, we reformulate the problem in the following way: The variational formulation (3.49) can be interpreted as (see the proof of Theorem 3.3.1):

$$\begin{cases} -\operatorname{div}(\mu^{-1}\nabla A) + \mathrm{i}\omega\sigma A = \sigma \displaystyle\sum_{k=1}^{3} C_k \chi_k & \text{in } \mathbb{R}^2, \\ A(\boldsymbol{x}) = \beta + \mathcal{O}(|\boldsymbol{x}|^{-1}) & |\boldsymbol{x}| \to \infty, \end{cases} \tag{7.49}$$

for $\beta \in \mathbb{C}$, where χ_k is the characteristic function of the domain Λ_k (i.e. $\chi_k = 1$ in Λ_k, 0 elsewhere), and

$$C_k = \frac{I_k + \mathrm{i}\omega|\Lambda_k|M_k(\sigma A)}{|\Lambda_k|M_k(\sigma)}, \quad k = 1, 2, 3, \tag{7.50}$$

and where we recall that

$$M_k(\phi) := \frac{1}{|\Lambda_k|}\int_{\Lambda_k} \phi \, d\boldsymbol{x}, \quad k = 1, 2, 3.$$

Let A_k stand for the solution of the problem:

$$\begin{cases} -\operatorname{div}(\mu^{-1}\nabla A_k) + \mathrm{i}\omega\sigma A_k = \sigma\chi_k & \text{in } \mathbb{R}^2, \\ A_k(\boldsymbol{x}) = \beta + \mathcal{O}(|\boldsymbol{x}|^{-1}) & |\boldsymbol{x}| \to \infty, \end{cases} \tag{7.51}$$

for $\beta \in \mathbb{C}$. It is clear that the function $A = C_1 A_1 + C_2 A_2 + C_3 A_3$ is a solution of (7.49). It remains to give an expression of C_k in function of A_1, A_2, A_3. We have from (7.50):

$$|\Lambda_k| M_k(\sigma)\, C_k = I_k + \mathrm{i}\omega|\Lambda_k| \big(C_1 M_k(\sigma A_1) + C_2 M_k(\sigma A_2) + C_3 M_k(\sigma A_3)\big),$$

for $k = 1, 2, 3$. This yields the linear system of equations:

$$\begin{aligned} \big(\mathrm{i}\omega M_1(\sigma A_1) - M_1(\sigma)\big)\, C_1 + \mathrm{i}\omega M_1(\sigma A_2)\, C_2 + \mathrm{i}\omega M_1(\sigma A_3)\, C_3 &= -\frac{I_1}{|\Lambda_1|}, \\ \mathrm{i}\omega M_2(\sigma A_1)\, C_1 + \big(\mathrm{i}\omega M_2(\sigma A_2) - M_2(\sigma)\big)\, C_2 + \mathrm{i}\omega M_2(\sigma A_3)\, C_3 &= -\frac{I_2}{|\Lambda_2|}, \\ \mathrm{i}\omega M_3(\sigma A_1)\, C_1 + \mathrm{i}\omega M_3(\sigma A_2)\, C_2 + \big(\mathrm{i}\omega M_3(\sigma A_3) - M_3(\sigma)\big)\, C_3 &= -\frac{I_3}{|\Lambda_3|}. \end{aligned}$$

Consequently, solving (3.49) can be performed by solving (7.51) for each $k = 1, 2, 3$. Let us address the numerical solution of this problem and set for shortness $u = A_k$ for a fixed k. Defining the space

$$\mathcal{V} := \{v \in \mathcal{W}^1(\mathbb{R}^2);\ M_1(v) = 0\},$$

we have for $k = 1, 2, 3$ the variational formulation:

$$\begin{cases} \text{Find } u \in \mathcal{V} \text{ such that} \\ \displaystyle\int_{\mathbb{R}^2} \mu^{-1}\nabla u \cdot \nabla\overline{v}\, d\boldsymbol{x} + \mathrm{i}\omega \int_{\Lambda} \sigma\, u\overline{v}\, d\boldsymbol{x} = \int_{\Lambda_k} \sigma\overline{v}\, d\boldsymbol{x} \qquad \forall\, v \in \mathcal{V}. \end{cases} \tag{7.52}$$

7.7.1.1 A Formulation Using the Steklov–Poincaré Operator

Let us investigate the formulation defined in (3.3.1) applied to (7.52). We have the variational formulation:

$$\begin{cases} \text{Find } u \in \mathcal{W} \text{ such that} \\ \mathscr{B}_\Lambda(u, v) + \mathscr{P}(u, v) = \mathscr{L}(v) \qquad \forall\, v \in \mathcal{W}, \end{cases} \tag{7.53}$$

where

$$\mathcal{W} := \{v \in \mathcal{H}^1(\Lambda);\; M_1(v) = 0\},$$

$$\mathscr{B}_\Lambda(v,w) := \int_\Lambda \mu^{-1}\nabla v \cdot \nabla \overline{w}\, d\boldsymbol{x} + \mathrm{i}\omega \int_{\Lambda_k} \sigma v \overline{w}\, d\boldsymbol{x},$$

$$\mathscr{P}(v,w) := \mu_0^{-1} \int_\gamma (Pv)\,\overline{w}\, ds,$$

$$\mathscr{L}(v) := \int_{\Lambda_k} \sigma \overline{v}\, d\boldsymbol{x}.$$

Here P stands for the Steklov–Poincaré operator defined in Sect. 1.4.4. We define the finite–dimensional space

$$\mathcal{W}_h := \{v \in C^0(\overline{\Lambda});\; v_{|T} \in \mathbb{P}_1(T)\ \forall\, T \in \mathscr{T}_h\},$$

and define P_h as the approximation of the operator P defined by (7.35). In order to account for the condition $M_1(v) = 0$ we introduce the Lagrange multiplier λ_h and define the discrete problem:

$$\begin{cases} \text{Find } (u_h, \lambda_h) \in \mathcal{W}_h \times \mathbb{C} \text{ such that:} \\ \mathscr{B}_\Lambda(u_h, v) + \mathscr{P}_h(u_h, v) - \lambda_h M_1(\overline{v}) = \mathscr{L}(v) \qquad \forall\, v \in \mathcal{W}_h, \\ M_1(u_h) = 0. \end{cases} \tag{7.54}$$

where

$$\mathscr{P}_h(v,w) := \mu_0^{-1} \int_\gamma (P_h v)\,\overline{w}\, ds.$$

Let us derive the resulting linear system. We assume for this, that the mesh nodes are ordered such that internal nodes are labelled first and then boundary nodes.[1] Denoting as before by (ϕ_i) the canonical Lagrange basis of $\mathcal{W}_h$, we obtain the linear system

$$\mathbf{B}\mathbf{u} + \mu_0^{-1}\mathbf{P}\mathbf{u} - \mathbf{d}\lambda_h = \mathbf{c},$$

$$\mathbf{d}^{\mathsf{T}}\mathbf{u} = 0.$$

[1] This assumption is not mandatory but simplifies the presentation.

where **P** is the Steklov–Poincaré matrix defined by (7.36) and **B**, **c** and **d** are the matrix and vectors defined by:

$$
\begin{aligned}
B_{ij} &= \int_\Lambda \mu^{-1} \nabla\phi_j \cdot \nabla\phi_i \, d\boldsymbol{x} + \mathrm{i}\omega \int_{\Lambda_k} \sigma\phi_j\phi_i \, d\boldsymbol{x}, \\
c_i &= \int_\Lambda \sigma\phi_i \, d\boldsymbol{x}, \\
d_i &= M_1(\phi_i),
\end{aligned}
$$

for $i, j \in \mathscr{I}_h$, and where **u** is the vector of node values of u_h.

7.7.1.2 A Formulation with Simple and Double Layer Potentials

Let us consider the coupled problem defined in Sect. 3.3.3, i.e. the variational formulation (3.62). Let us consider again a triangulation $\mathscr{T}_h$ of the domain Λ that consists in the union of triangulations $\mathscr{T}_h^k$ of the domains Λ_k for $k = 0, 1, 2$. We start by defining finite–dimensional spaces:

$$
\begin{aligned}
\mathcal{W}_h &:= \big\{\phi \in C^0(\overline{\Lambda});\ \phi_{|T} \in \mathbb{P}_1(T)\ \forall\, T \in \mathscr{T}_h\big\}, \\
\mathcal{H}_h &:= \big\{q \in \mathcal{L}^2(\gamma);\ q_{|e} \in \mathbb{P}_0(e)\ \forall\, e \in \mathscr{E}_h\big\}.
\end{aligned}
$$

We define the discrete problem:

$$
\begin{cases}
\text{Find } (A_h, p_h, \lambda_h, \xi_h) \in \mathcal{W}_h \times \mathcal{H}_h \times \mathbb{C} \times \mathbb{C} \text{ such that:} \\
\mathscr{B}_\Lambda(A_h, \phi) - \overline{\mathscr{E}(\phi, p_h)} - \lambda_h \displaystyle\int_{\Lambda_0} \overline{v}\, d\boldsymbol{x} = \mathscr{L}(\phi) \qquad \forall\, \phi \in \mathcal{W}_h, \\
\mathscr{D}(p_h, q) + \mathscr{E}(A_h, q) - \mathscr{R}(A_h, q) - \theta_h \displaystyle\int_\gamma \overline{q}\, ds = 0 \quad \forall\, q \in \mathcal{H}_h, \\
\displaystyle\int_{\Lambda_0} A_h \, d\boldsymbol{x} = 0, \\
\displaystyle\int_\gamma p_h \, ds = 0.
\end{cases}
\tag{7.55}
$$

where we recall:

$$
\begin{aligned}
&\mathscr{E}(\phi, q) := \int_\gamma \phi\, \overline{q}\, ds, \\
&\mathscr{R}(\phi, q) := 2\int_\gamma (R\phi)\overline{q}\, ds := 2\int_\gamma\!\int_\gamma \frac{\partial G}{\partial n_y}(\boldsymbol{x}, \boldsymbol{y})\, \phi(\boldsymbol{y})\, \overline{q}(\boldsymbol{x})\, ds(\boldsymbol{y})\, ds(\boldsymbol{x}),
\end{aligned}
$$

$$\mathscr{D}(p,q) := 2\mu_0 \int_\gamma (Kp)\overline{q}\,ds := 2\mu_0 \int_\gamma\int_\gamma G(\boldsymbol{x},\boldsymbol{y})\,p(\boldsymbol{y})\,\overline{q}(\boldsymbol{x})\,ds(\boldsymbol{y})\,ds(\boldsymbol{x}).$$

The matrix formulation of (7.55) is obtained, as usual, by choosing a basis of the spaces $\mathcal{W}_h$ and $\mathcal{H}_h$. For $\mathcal{W}_h$ we choose the standard Lagrange canonical basis (ϕ_i) and for $\mathcal{H}_h$ the space of piecewise constant functions with the basis q_k. We obtain the linear system:

$$\begin{aligned}
&\mathbf{Bu} - \overline{\mathbf{E}}\mathbf{p} - \mathbf{d}\lambda_h = \mathbf{c},\\
&(\mathbf{E} - \mathbf{R})\mathbf{u} + \mathbf{D} - \mathbf{f}\theta_h = \mathbf{0},\\
&\mathbf{d}^{\mathsf{T}}\mathbf{u} = 0,\\
&\mathbf{f}^{\mathsf{T}}\mathbf{p} = 0,
\end{aligned}$$

where $\mathbf{u}$ is the vector having as components the values of A_h at nodes and

$$\begin{aligned}
B_{ij} &= \mathscr{B}(v_j, v_i),\\
E_{i\ell} &= \mathscr{E}(v_i, q_\ell),\\
D_{k\ell} &= \mathscr{D}(q_k, q_\ell),\\
R_{kj} &= \mathscr{R}(v_j, q_k),\\
d_i &= \int_{\Lambda_0} \phi_i\,d\boldsymbol{x},\\
f_k &= \int_\gamma q_k\,ds.
\end{aligned}$$

7.7.2 The 3-D *H*–Model

Chapter 4 gives some models where a coupling between partial differential equations in the conductors and the harmonic equation in the outer space is used. This again suggests the use of coupled finite element/boundary element techniques. We shall focus the presentation on the magnetic field model described in Sect. 4.2. This one is in fact the most popular and is at the origin of the *Trifou* code (see [36]).

Let us recall the magnetic field model (4.34): Denoting by $\mathcal{K}_\Gamma$ the space

$$\mathcal{K}_\Gamma := \big\{(\boldsymbol{w},\psi,\beta) \in \boldsymbol{\mathcal{H}}(\mathbf{curl},\Omega) \times \mathcal{H}^{\frac{1}{2}}(\Gamma) \times \mathbb{C};\\
\boldsymbol{w}\times\boldsymbol{n} + \mathbf{curl}_\Gamma\,\psi + \beta\,\mathbf{curl}_\Gamma\,p = 0 \text{ on } \Gamma\big\},$$

where p is the solution of (1.20), we have the variational formulation:

$$\begin{cases} \text{Find } (\boldsymbol{H}, \phi, \alpha) \in \mathcal{K}_\Gamma \text{ such that} \\ \mathrm{i}\omega \displaystyle\int_\Omega \mu \boldsymbol{H} \cdot \overline{\boldsymbol{w}}\, d\boldsymbol{x} + \mathrm{i}\omega\mu_0 \int_\Gamma (P\phi)\overline{\psi}\, ds + \mathrm{i}\omega L\alpha\overline{\beta} \\ \qquad + \displaystyle\int_\Omega \sigma^{-1}\, \mathbf{curl}\, \boldsymbol{H} \cdot \mathbf{curl}\, \overline{\boldsymbol{w}}\, d\boldsymbol{x} = \tilde{\mathscr{L}}(\boldsymbol{w}) \quad \forall\, (\boldsymbol{w}, \psi, \beta) \in \mathcal{K}_\Gamma. \end{cases} \tag{7.56}$$

Let us also recall that $\mathscr{T}_h$ is the set of tetrahedra constituting the finite element mesh. The set of faces of theses tetrahedra is denoted by $\mathscr{F}_h$ while $\mathscr{E}_h$ is the set of all mesh edges. In addition, $\mathscr{F}_h^\Gamma$ (*resp.* $\mathscr{E}_h^\Gamma$) stands for the set of faces (*resp.* edges) that lie on the boundary Γ.

We consider the construction of a finite–dimensional subspace of $\mathcal{K}_\Gamma$ in order to define a variational finite element/boundary element method.

Let $e_{ij} \in \mathscr{E}_h$ stand for an edge whose ends are the nodes $\boldsymbol{a}_i, \boldsymbol{a}_j \in \mathscr{N}_h$, where $\mathscr{N}_h$ is the set of nodes. As we have already defined, to e_{ij} we can associate the basis function

$$\boldsymbol{\xi}_{ij} := \boldsymbol{\xi}_e = \phi_i \nabla\phi_j - \phi_i \nabla\phi_i,$$

where ϕ_i is the Lagrange basis function associated to $\boldsymbol{a}_i$ in the space of continuous piecewise affine functions. Hence we can define

$$\begin{aligned} \mathcal{W}_h &= \{\boldsymbol{w} \in \mathcal{L}^2(\Omega);\ \boldsymbol{\xi}_{|T} = \boldsymbol{b}_T + \boldsymbol{c}_T \times \boldsymbol{x},\ \boldsymbol{b}_T, \boldsymbol{c}_T \in \mathbb{C}^3,\ \forall\, T \in \mathscr{T}_h, \\ &\qquad [\boldsymbol{w} \times \boldsymbol{n}]_f = 0\ \forall\, f \in \mathscr{F}_h\} \\ &= \mathrm{Span}\{\boldsymbol{\xi}_{ij};\ e_{ij} \in \mathscr{E}_h\}. \end{aligned}$$

Let us also define a finite element subspace of $\mathcal{H}^{\frac{1}{2}}(\Gamma)$ by

$$\begin{aligned} \mathcal{V}_h &= \{\psi \in \mathcal{C}^0(\Gamma);\ \psi_{|f} \in \mathbb{P}_1(f)\ \forall\, f \in \mathscr{F}_h^\Gamma\}, \\ &= \mathrm{Span}\{\phi_j;\ \boldsymbol{a}_j \in \mathscr{N}_\Gamma\}, \end{aligned}$$

where $\mathscr{N}_\Gamma$ is the set of nodes lying on Γ. Thus $\mathcal{K}_\Gamma$ can be approximated by the finite–dimensional space

$$\mathcal{K}_{\Gamma,h} := \{(\boldsymbol{w}, \psi, \beta) \in \mathcal{W}_h \times \mathcal{V}_h \times \mathbb{C};\ \boldsymbol{w} \times \boldsymbol{n} + \mathbf{curl}_\Gamma\, \psi + \beta\, \mathbf{curl}_\Gamma\, p_h = 0 \text{ on } \Gamma\},$$

where p_h is an approximation of p, solution of (1.20), that will be defined later.

Remark 7.7.1. It is easy to see that if we take $p_h = p$ then $\mathcal{K}_{\Gamma,h}$ is a subspace of $\mathcal{K}_\Gamma$. In practice this is impossible since Problem (1.20) cannot be in general solved exactly since p_h is an approximation of p.

We define the discrete variational problem:

$$\left\{\begin{array}{l}\text{Find } (\boldsymbol{H}_h, \varphi_h, \alpha_h) \in \mathcal{K}_{\Gamma,h} \text{ such that}\\ \mathrm{i}\omega \displaystyle\int_\Omega \mu \boldsymbol{H}_h \cdot \overline{\boldsymbol{w}}\, d\boldsymbol{x} + \mathrm{i}\omega\mu_0 \int_\Gamma (P_h\varphi_h)\overline{\psi}\, ds + \mathrm{i}\omega L_h \alpha_h \overline{\beta}\\ \qquad + \displaystyle\int_\Omega \sigma^{-1}\, \mathbf{curl}\, \boldsymbol{H}_h \cdot \mathbf{curl}\, \overline{\boldsymbol{w}}\, d\boldsymbol{x} = V \int_S \mathbf{curl}\, \overline{\boldsymbol{w}} \cdot \boldsymbol{n}\, ds\\ \qquad\qquad \forall\, (\boldsymbol{w}, \psi, \beta) \in \mathcal{K}_{\Gamma,h}.\end{array}\right. \tag{7.57}$$

Our aim is to exhibit the expression of the resulting linear system. To this end, several issues are to be tackled:

(i) An approximation p_h of the potential p.
(ii) A definition of an approximation L_h of the inductance L.
(iii) A definition of the approximation P_h of the Steklov-Poincaré operator following Sect. 7.5.1.3 is to be given.
(iv) An implementation of the interface condition on Γ given in the definition of $\mathcal{K}_{\Gamma,h}$.

7.7.2.1 Approximation of the Potential p

The potential p defined by (1.20) is needed for the computation of the inductance and for the interface condition enforced in the space $\mathcal{K}_{\Gamma,h}$.

Let us recall that by using Theorem 1.3.7 with $p = 0$ in Ω, we have the integral equation:

$$\frac{1}{2}\, p(\boldsymbol{x}) - \int_\Gamma p(\boldsymbol{y}) \frac{\partial G}{\partial n_y}(\boldsymbol{x}, \boldsymbol{y})\, ds(\boldsymbol{y}) = \int_\Sigma \frac{\partial G}{\partial n}(\boldsymbol{x}, \boldsymbol{y})\, ds(\boldsymbol{y}) \qquad \boldsymbol{x} \in \Gamma.$$

In view of a finite element approximation, a variational formulation of this equation is given by the problem:

$$\left\{\begin{array}{l}\text{Find } p \in \mathcal{H}^{\frac{1}{2}}(\Gamma) \text{ such that}\\ \displaystyle\frac{1}{2}\int_\Gamma pq\, ds - \int_\Gamma \Big(\int_\Gamma \frac{\partial G}{\partial n_y}(\boldsymbol{x}, \boldsymbol{y})\, p(\boldsymbol{y})\, q(\boldsymbol{x})\, ds(\boldsymbol{y}) \Big)\, ds(\boldsymbol{x})\\ \qquad = \displaystyle\int_\Gamma \Big(\int_\Sigma \frac{\partial G}{\partial n_y}(\boldsymbol{x}, \boldsymbol{y})\, q(\boldsymbol{x})\, ds(\boldsymbol{y}) \Big)\, ds(\boldsymbol{x}) \qquad \forall\, q \in \mathcal{H}^{\frac{1}{2}}(\Gamma).\end{array}\right. \tag{7.58}$$

We use the space $\mathcal{V}_h$ to approximate functions of $\mathcal{H}^{\frac{1}{2}}(\Gamma)$. Problem (7.58) is hence approximated by the following one:

$$\begin{cases} \text{Find } p_h \in \mathcal{V}_h \text{ such that} \\ \displaystyle \frac{1}{2}\int_\Gamma p_h q\, ds - \int_\Gamma \left(\int_\Gamma \frac{\partial G}{\partial n_y}(x, y)\, p_h(y)\, q(x)\, ds(y) \right) ds(x) \\ \displaystyle \qquad = \int_\Gamma \left(\int_\Sigma \frac{\partial G}{\partial n_y}(x, y)\, q(x)\, ds(y) \right) ds(x) \qquad \forall\, q \in \mathcal{V}_h. \end{cases} \tag{7.59}$$

In other terms, the potential p is approximated by a continuous piecewise linear function p_h. Using the canonical Lagrange basis (ϕ_i) of $\mathcal{V}_h$, we obtain the matrix formulation

$$\mathbf{A}\mathbf{p} = \mathbf{b},$$

where the matrix $\mathbf{A} = (A_{ij})$ and the vectors $\boldsymbol{b} = (b_i)$, $\boldsymbol{p} = (p_i)$ are respectively defined by:

$$\begin{aligned} A_{ij} &= \frac{1}{2}\int_\Gamma \phi_j \phi_i\, ds - \int_\Gamma \left(\int_\Gamma \frac{\partial G}{\partial n_y}(x, y)\, \phi_j(y)\, \phi_i(x)\, ds(x) \right) ds(y), \\ b_i &= \int_\Gamma \left(\int_\Sigma \frac{\partial G}{\partial n_y}(x, y)\, \phi_i(x)\, ds(x) \right) ds(y), \\ p_i &= p_h(\boldsymbol{a}_i), \end{aligned}$$

for all boundary nodes $i, j \in \mathscr{I}_h^\Gamma$. Clearly the main difficulty relies on numerical evaluation of the above integrals, the integrands being singular.

7.7.2.2 Approximation of the Inductance

Applying Theorem 4.2.3, the inductance coefficient is defined by

$$L = \frac{\mu_0}{4\pi}\int_\Gamma \left(\int_\Gamma \frac{\boldsymbol{J}_\Gamma(\boldsymbol{x}) \cdot \boldsymbol{J}_\Gamma(\boldsymbol{y})}{|\boldsymbol{x} - \boldsymbol{y}|}\, ds(\boldsymbol{x}) \right) ds(\boldsymbol{y}),$$

where $\boldsymbol{J}_\Gamma$ is the surface vector-field given by $\mathbf{curl}_\Gamma\, p$, the function p standing for the unique solution of (1.20).

Clearly, the numerical solution of (7.59) provides the vector $\boldsymbol{p}$ having as entries the values of p_h at the nodes $\boldsymbol{a}_i$ of Γ. Since the mesh $\mathscr{T}_h$ is made of tetrahedra T, the faces of $\mathscr{F}_h^\Gamma$ on Γ are triangles $f \in \mathscr{F}_h^\Gamma$. For a face $f \in \mathscr{F}_h^\Gamma$ with vertices $\boldsymbol{a}_1^f$, $\boldsymbol{a}_2^f$, $\boldsymbol{a}_3^f$, the outer normal to f is given by

$$\boldsymbol{n}_f = \frac{(\boldsymbol{a}_2 - \boldsymbol{a}_1) \times (\boldsymbol{a}_3 - \boldsymbol{a}_1)}{|\boldsymbol{a}_2 - \boldsymbol{a}_1|\, |\boldsymbol{a}_3 - \boldsymbol{a}_1|}.$$

Therefore, an approximation of $\mathbf{curl}_\Gamma\, p$ is obtained by

$$\boldsymbol{J}^h_{\Gamma|f} := \boldsymbol{n}_f \times \nabla p_h \qquad \forall\, f \in \mathscr{F}^\Gamma_h .$$

In other words, we have obtained an approximation of $\boldsymbol{J}_\Gamma$ as a piecewise constant vector. An approximation of the inductance is then given by

$$\begin{aligned} L_h &:= \frac{\mu_0}{4\pi} \int_\Gamma \Big(\int_\Gamma \frac{\boldsymbol{J}^h_\Gamma(\boldsymbol{x}) \cdot \boldsymbol{J}^h_\Gamma(\boldsymbol{y})}{|\boldsymbol{x}-\boldsymbol{y}|}\, ds(\boldsymbol{y}) \Big)\, ds(\boldsymbol{x}) \\ &= \frac{\mu_0}{4\pi} \sum_{f,f' \in \mathscr{F}^\Gamma_h} \boldsymbol{J}_f \cdot \boldsymbol{J}_{f'} \int_f \Big(\int_{f'} \frac{1}{|\boldsymbol{x}-\boldsymbol{y}|}\, ds(\boldsymbol{y}) \Big)\, ds(\boldsymbol{x}), \end{aligned}$$

where $\boldsymbol{J}_f = \boldsymbol{J}^h_{\Gamma|f}$.

7.7.2.3 Approximation of the Steklov–Poincaré Operator

We have already presented in Sect. 7.5.1.3 an approximation method for the Steklov–Poincaré operator, where it turns out that this one is given by a dense matrix, as expressed in (7.36).

7.7.2.4 Interface Conditions

The interface conditions contained in the space $\mathcal{K}_{\Gamma,h}$ can be discretized as follows: Let e denote a face on the boundary Γ, that belongs to a tetrahedon $T \in \mathscr{T}_h$.

It can be proven (see [36, 88, 152]) that we have the characterization:

$$\begin{aligned} \mathcal{K}_{\Gamma,h} = \Big\{ &(\boldsymbol{w}, \psi, \beta) \in \boldsymbol{\mathcal{W}}_h \times \mathcal{V}_h \times \mathbb{C}; \\ &\qquad \boldsymbol{w}_{|\Gamma} = \nabla_\Gamma(\psi + \beta p_h) \text{ on } \Gamma \setminus S,\ \int_{\partial S} \boldsymbol{w} \cdot \boldsymbol{t}\, d\ell = \beta \Big\}, \end{aligned}$$

where $\boldsymbol{t}$ is the unit tangent vector to the boundary ∂S of the cut S, and ∇_Γ is the tangential gradient vector on Γ, i.e.

$$\nabla_\Gamma \psi := \nabla \psi - (\nabla \psi \cdot \boldsymbol{n})\, \boldsymbol{n} \qquad \text{on } \Gamma.$$

Let us now denote by $(\boldsymbol{\xi}_e)_{e \in \mathscr{E}_h}$ and $(\phi_i)_i$ the respective bases of the spaces $\boldsymbol{\mathcal{W}}_h$ and $\mathcal{V}_h$. More precisely, we recall that the basis $(\boldsymbol{\xi}_e)$ is defined by (7.19). We also recall the identity

$$\boldsymbol{\xi}_e(\boldsymbol{x}) = \phi_i(\boldsymbol{x}) \nabla \phi_j(\boldsymbol{x}) - \phi_j(\boldsymbol{x}) \nabla \phi_i(\boldsymbol{x}) = \boldsymbol{\xi}_{ij}(\boldsymbol{x}),$$

where $\boldsymbol{a}_i$ and $\boldsymbol{a}_j$ are the ends of the edge e. Let $e_{ij} \in \mathscr{E}_h$ denote an edge with ends $\boldsymbol{a}_i$ and $\boldsymbol{a}_j$. For any node $\boldsymbol{a}_k \in \mathscr{N}_h^\Gamma$ we define the index

$$\delta_{ij}^k := \delta_{jk} - \delta_{ik}.$$

Then we associate to each node $\boldsymbol{a}_k$ the vector function

$$\boldsymbol{\zeta}_k(\boldsymbol{x}) = \sum_{i,j} \delta_{ij}^k \, \boldsymbol{\xi}_{ij}(\boldsymbol{x}).$$

In order to express the resulting linear system of equations, it can be shown (see [88, 152]) that the approximate magnetic field $\boldsymbol{H}_h \in \boldsymbol{\mathcal{W}}_h$ in the conductor can be expanded as

$$\boldsymbol{H}_h(\boldsymbol{x}) = \sum_{e\in\mathscr{E}_h^{\text{in}}} H_e \boldsymbol{\xi}_e(\boldsymbol{x}) - \sum_{e\in\mathscr{E}_h^\Gamma} \phi_j \boldsymbol{\zeta}_j(\boldsymbol{x}) + \alpha_h \boldsymbol{\theta}_h(\boldsymbol{x}) \qquad \boldsymbol{x} \in \overline{\Omega},$$

where $\boldsymbol{\theta}_h = \nabla_\Gamma p_h$.

We are now ready to write down the obtained linear system of equations. Let us define the following vectors:

$$\tilde{\boldsymbol{h}} = (H_e)_{e\in\mathscr{E}_h^{\text{int}}}, \quad \boldsymbol{\varphi} = (\varphi_i)_{i\in\mathscr{I}_h^\Gamma}.$$

Taking in (7.57) successively $\boldsymbol{w} = \boldsymbol{\xi}_e$ for $e \in \mathscr{E}_h^{\text{int}}$, $\psi = \phi_i$ for $i \in \mathscr{I}_h^\Gamma$ and $\beta = 1$, we obtain the linear block system of equations:

$$\begin{pmatrix} \mathsf{A}_{11} & \mathsf{A}_{12} & c_1 \\ \mathsf{A}_{12}^\mathsf{T} & \mathsf{A}_{22} & c_2 \\ c_1^\mathsf{T} & c_2^\mathsf{T} & d \end{pmatrix} \begin{pmatrix} \tilde{\boldsymbol{h}} \\ \boldsymbol{\varphi} \\ \alpha_h \end{pmatrix} = \begin{pmatrix} \mathbf{0} \\ \mathbf{0} \\ V \end{pmatrix} \tag{7.60}$$

Here above the involved matrices are defined as follows:

$$(\mathsf{A}_{11})_{e,e'} = \mathrm{i}\omega \int_\Omega \mu \, \boldsymbol{\xi}_{e'} \cdot \boldsymbol{\xi}_e \, d\boldsymbol{x} + \int_\Omega \sigma^{-1} \, \mathbf{curl}\, \boldsymbol{\xi}_{e'} \cdot \mathbf{curl}\, \boldsymbol{\xi}_e \, d\boldsymbol{x},$$

$$(\mathsf{A}_{12})_{e,j} = -\mathrm{i}\omega \int_\Omega \mu \, \boldsymbol{\zeta}_j \cdot \boldsymbol{\xi}_e \, d\boldsymbol{x} + \int_\Omega \sigma^{-1} \, \mathbf{curl}\, \boldsymbol{\zeta}_j \cdot \mathbf{curl}\, \boldsymbol{\xi}_e \, d\boldsymbol{x},$$

$$(\mathsf{A}_{22})_{i,j} = \mathrm{i}\omega \int_\Omega \mu \, \boldsymbol{\zeta}_j \cdot \boldsymbol{\zeta}_i \, d\boldsymbol{x} + \int_\Omega \sigma^{-1} \, \mathbf{curl}\, \boldsymbol{\zeta}_j \cdot \mathbf{curl}\, \boldsymbol{\zeta}_i \, d\boldsymbol{x} - \mathrm{i}\omega\mu_0 P_{ij},$$

$$(c_1)_e = \mathrm{i}\omega \int_\Omega \mu \, \boldsymbol{\theta}_h \cdot \mathbf{curl}\, \boldsymbol{w}_e \, d\boldsymbol{x} + \int_\Omega \sigma^{-1} \, \mathbf{curl}\, \boldsymbol{\theta}_h \cdot \mathbf{curl}\, \boldsymbol{\xi}_e \, d\boldsymbol{x},$$

$$(c_2)_i = -\mathrm{i}\omega \int_\Omega \mu\, \boldsymbol{\theta}_h \cdot \boldsymbol{\zeta}_i \, d\boldsymbol{x} - \int_\Omega \sigma^{-1} \,\mathbf{curl}\, \boldsymbol{\theta}_h \cdot \mathbf{curl}\, \boldsymbol{\zeta}_i \, d\boldsymbol{x},$$

$$d = \mathrm{i}\omega \int_\Omega \mu\, |\boldsymbol{\theta}_h|^2 \, d\boldsymbol{x} + \int_\Omega \sigma^{-1} |\,\mathbf{curl}\, \boldsymbol{\theta}_h|^2 \, d\boldsymbol{x},$$

for $e, e' \in \mathscr{E}_h^{\mathrm{int}}$ and $i, j \in \mathscr{I}_h^{\Gamma}$. Here $\mathsf{P} = (P_{ij})$ is the discretization of the Steklov–Poincaté operator defined in (7.36).

Numerical implementation of the linear system (7.60) is not a simple task for it involves a collection of block matrices with different storage schemes. For numerical solution of this system, one can resort either to a Schur complement method or to an iterative procedure. It is noteworthy, in addition, that the matrix of the linear system (7.60) is not hermitian but actually symmetric which reduces the set of available iterative methods for numerical solution. In [152], numerical tests show that the QMR (Quasi-Minimal Residual) iterative method gives the best convergence behavior for such a matrix.

Remark 7.7.2. The previous developments can be generalized to the case of many inductors Ω_k, $k = 1, \ldots, N$ with prescribed voltages $V_1, \ldots, V_N$.

Numerical investigations on this problem have been carried out in some scientific computing groups. We mention for this among several studies:

- The code TRIFOU developed at EDF (Électricité de France) by A. Bossavit, J.-C. Vérité et al. and related works (see [32–34, 36, 171, 172]) are considered as basic contributions for numerical solution of 3-D eddy current problems by finite element methods.
- In (Rappaz et al. [152]) this model is developed for an induction heating application. In particular, the authors consider a thin inductor approximation. A software is developed for this purpose and numerical results are compared to experimental measurements.
- In her PhD thesis, S. Gauthier [88] has studied the implementation of the presented 3-D model. Practical aspects are considered in this work.
- In his PhD thesis, Henneron [96] considers development and numerical approximation of various three-dimensional eddy current models.

This list shows the main works that have inspired the authors' contribution but this one remains far from being exhaustive.

Part II
Selected Applications

Chapter 8
Induction Heating Processes

The rest of this book is devoted to a presentation of mathematical modelling and numerical solution of some well known applications of low frequency electromagnetics (eddy current and magnetostatics) in industry. For each application, we present the set of equations that govern the process, give some mathematical results when they are known and present the numerical methods to solve the set of equations as well as a series of numerical experiments.

The present chapter is concerned with induction heating processes. Induction heating is a non-contact heating method that consists in inducing a low frequency alternating current and magnetic field in the workpiece to treat. The induction results in a heating of the treated piece. This energy dissipation has two origins:

- Electric energy due to the *Joule effect*. Its density is expressed in terms of the function $\boldsymbol{J} \cdot \boldsymbol{E}$ which acts as a heat source for the heat transfer equation.
- In ferromagnetic materials, hysteresis cycle represented by the mapping $\boldsymbol{H} \mapsto \boldsymbol{B}$ results in energy losses that contribute to conductor heating. Induction heating is therefore better suited for ferromagnetic media.

We shall in the following focus on the first issue. Applications that use induction heating include:

- *Metal melting*: Induction furnaces are industrial devices that heat metals to their melting point. This melting can be combined with the use of a moderately high frequency current to stir the liquid metal in order to homogenize its metallurgical properties. Other applications that involve metal melting include forging, welding, thixoforming and other similar processes.
- *Metal hardening*: The flexibility and the power provided by induction heating allow its use in metal hardening. Fast rates of cooling enable producing harder or softer metals in the piece depending on the nature of the alloy we are working with.

R. Touzani and J. Rappaz, *Mathematical Models for Eddy Currents and Magnetostatics: With Selected Applications*, Scientific Computation, DOI 10.1007/978-94-007-0202-8_8,

In the following we address two topics in induction heating modelling:

1. We present and describe some mathematical models, quote mathematical results about of these problems. The proofs of the results are not given. This is because mathematical analysis of these coupled nonlinear problems requires mathematical tools that are beyond the scope of this monograph. We however mention references where the interested reader can find more details.
2. In view of practical applications, we consider some optimization issues of the process. More precisely, we formulate and solve the problem of finding process parameters, like frequency, voltage, ..., that achieve a desired goal. This approach is generally referred to as optimal control. We present then some numerical experiments related to *thixoforming* of metals.

8.1 A Mathematical Model

Induction heating devices can be schematically represented by a workpiece to treat surrounded by an inductor, which is generally a thin metal piece made of highly electrically conducting material (e.g. a Copper coil). Such devices can be *cold crucibles* or *Induction ovens* which are containers that hold a quantity of metal to be melted. In the case of metal hardening, the device can consist in a solid workpiece which is locally heated then suddenly cooled to obtain the desired phase transition phenomenon.

From a mathematical modelling point of view, induction heating involves a heat equation that handles phase change either in the case where a transition from solid to liquid (or even gas if an ablation process is present) occurs or if a metallurgical phase transition is to be considered. From a numerical simulation point of view, it is clear that the first class of applications generally involves a temperature field, or equivalently an enthalpy, distributed in the whole workpiece whereas heat treatment mostly deals with surface treatment which implies that the temperature and all related quantities exhibit large gradient at the surface of the considered workpiece. This can generally require a specific treatment due to the resulting boundary layer.

The source density produced by the Joule power is given by

$$f_{\text{Joule}}(\boldsymbol{x},t) = \boldsymbol{J}(\boldsymbol{x},t)\cdot\boldsymbol{E}(\boldsymbol{x},t).$$

To consider heat transfer with phase change, we resort to a formulation in terms of enthalpy. This formulation has indeed the advantage of dealing with one fixed domain rather than using a free boundary formulation. We have the equation (see Rappaz et al. [151], p. 236),

$$\frac{\partial h}{\partial t} - \operatorname{div}(k\nabla\theta) = f^*_{\text{Joule}},$$
$$\theta = \beta(h),$$

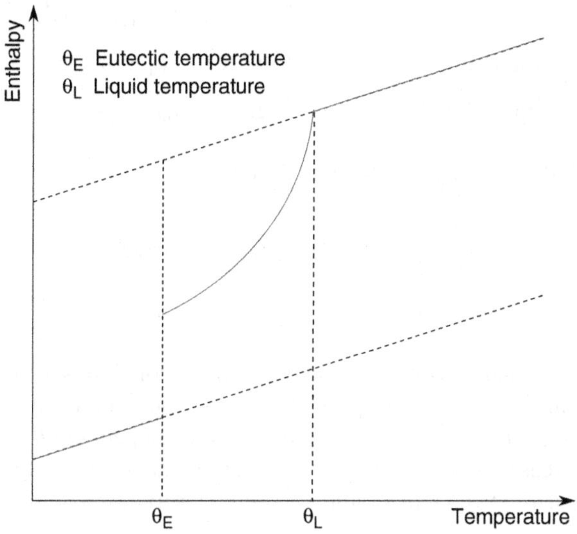

Fig. 8.1 A typical temperature–enthalpy diagram

where t is the time variable $t \in [0, t_{\max}]$, θ is the temperature, h is the enthalpy, and β is a given mapping that describes the relation between temperature and enthalpy. Now since we deal, as we did it so far, with time-harmonic magnetic fields, the Joule power density is approximated by the term f^*_{Joule} defined as the average source density on one time period. This means that we admit as an approximation that thermal quantities (h and θ) are influenced by time averaged Joule losses rather than their instantaneous values. The averaged source density is defined by

$$f^*_{\text{Joule}} := \frac{\omega}{2\pi} \int_0^{\frac{2\pi}{\omega}} \text{Re}(e^{i\omega t} \boldsymbol{J}(\cdot, t)) \cdot \text{Re}(e^{i\omega t} \boldsymbol{E}(\cdot, t))\, dt = \frac{1}{2}\, \text{Re}\,(\boldsymbol{J} \cdot \overline{\boldsymbol{E}}).$$

The mapping β depends on the material and often experiences some critical behaviour that requires a suitable numerical treatment. As an example, we give here a typical relation between θ and h for an alloy of Aluminium and Silicium (see [151], p. 238) (Fig. 8.1).

An alternative to the enthalpy formulation would consist in solving the heat equation in each domain representing a phase. But this approach has the drawback to transform the problem into a free boundary one which requires heavy front tracking techniques for numerical solution.

Let us now present a complete three-dimensional induction heating model based on the problem (4.32). This one consists in looking for

A magnetic field	$\boldsymbol{H} : \mathbb{R}^3 \times (0, t_{\max}) \to \mathbb{C}^3$,
A temperature	$\theta : \Omega \times (0, t_{\max}) \to \mathbb{R}$,
and an enthalpy	$h : \Omega \times (0, t_{\max}) \to \mathbb{R}$

such that for all $0 < t \leq t_{\max}$, (4.24) holds and

$$\frac{\partial h}{\partial t} - \operatorname{div}(k \nabla \theta) = f^*_{\text{Joule}} \quad \text{in } \Omega, \tag{8.1}$$

$$\theta = \beta(h) \quad \text{in } \Omega, \tag{8.2}$$

with appropriate boundary conditions on θ and initial conditions on h. For a general setting of the model, it is to be pointed out that the coefficients of Eqs. (4.24) and (8.1)–(8.2), σ, μ, ϱ, k generally depend on the temperature θ or the enthalpy h and, for the case of a ferromagnetic material, the magnetic permeability μ depends on $\boldsymbol{H}$ and on the temperature. More precisely, we assume a dependency $\mu = \mu(|\boldsymbol{H}|, \theta)$.

Problem (4.24), (8.1)–(8.2) is therefore a nonlinear system of equations where the nonlinearities occur in the dependencies:

$$\sigma = \sigma(\theta), \ \varrho = \varrho(\theta), \ k = k(\theta), \ \mu = \mu(|\boldsymbol{H}|, \theta),$$

$$\theta = \beta(h),$$

$$f^*_{\text{Joule}} = \frac{1}{2} \sigma^{-1}(\theta) \ \mathbf{curl}\, \boldsymbol{H} \cdot \mathbf{curl}\, \overline{\boldsymbol{H}}.$$

Note that we have expressed the Joule source term in function of the magnetic field $\boldsymbol{H}$ by using (4.1).

Clearly, Problem (4.24), (8.1)–(8.2) exhibits a time evolution in (8.1) while the electromagnetic equations (4.24) use the time variable as a parameter. However, in the simple case where the coefficients σ and μ do not depend on the temperature or enthalpy, Problem (4.24) is decoupled from the system (8.1)–(8.2).

One important issue in the modelling of coupled equations when eddy current problems are involved is that, in the general case, one cannot assume that the electromagnetic fields are time periodic. This is due to the nonlinearity of the system. In addition, it seems reasonable to assume that the time scale for temperature evolution is much larger than the period $2\pi/\omega$. To handle this difficulty a possible remedy procedure is described in [55] where the eddy current equations do not use time harmonic fields, i.e. a time dependent eddy current problem is considered. To take into account the presence of two time scales, a double time stepping technique is adopted in the time discretization scheme. More precisely, each time step of the heat equation is divided into a given (large) number of time steps to solve the eddy current equations. For these time steps, the material properties μ and σ are approximated by a constant in time.

Let us give some remarks about this model:

1. Although mathematical analysis of induction heating problems experiences some difficulties due to the lack of regularity of the Joule heating term (see for this [53, 54] for instance), this issue has no influence on the discretization schemes. Classical finite element formulations are well defined in standard finite element spaces. Let us mention here the work in [45] where a finite element approximation is defined for this class of problems and its convergence is proved.
2. The enthalpy formulation has proven to be efficient in the sense that the resulting model handles accurately the Curie point passage. In [47, 48, 55], numerical simulations show good agreement with experimental data.

8.2 Bibliographical Comments

Let us mention some references in the mathematical modelling and numerical simulation of Induction Heating processes. This list is rather a selection of papers and is by no means exhaustive.

For a mathematical analysis of the equations, we refer to [35] where an induction heating problem in a bounded 3-D domain is formulated and existence of solutions is established. In [52–54] the authors consider a two-dimensional geometry like in Sect. 3.2. For this model, existence results are proved either in the stationary [53, 54] or the evolution cases [52]. In [140, 141] a model for induction heating for 2-D geometries as in Sect. 3.3 is analyzed. Numerical solution techniques and industrial applications are considered in [17, 47, 48, 55, 72, 115, 125, 146, 174]. More specifically, [47,48,55,146] consider a 2-D model for induction heating with numerical solutions compared to experimental data. In [17] the authors treat an axisymmetric geometry including thermomechanical deformations, i.e. the system is coupled with continuum mechanics (elasticity equation). In [174] a 3-D model for induction heating is numerically solved. Earlier papers have developed numerical methods for the axisymmetric case [72] or in 1–D for a ferromagnetic material [115]. Let us also mention other references that use Optimal Control to adjust process parameters for induction heating like in [30, 73, 126–128]. This issue will be addressed at the end of this chapter.

Let us now investigate some mathematical models used in induction heating. Our goal is rather to point out some of the difficulties related to mathematical analysis of these problems rather than making detailed study of these types of equations. We shall mention however the main references to these studies. For this end the choice of two-dimensional models seems to be judicious since it mentions the main difficulties without going into technical details mostly related to three-dimensional modelling of eddy currents.

8.3 A 2-D Stationary Problem

The simplest model to consider in this section is related to induction heating of long workpieces using two-dimensional solenoidal eddy current modelling (see Sect. 3.2, Fig. 3.1). More specifically, using the eddy current model (3.22) and the steady state version of (8.1)–(8.2) where the source Joule heating term is obtained by

$$f^*_{\text{Joule}} = \frac{1}{2\sigma}\,\mathbf{curl}\,H \cdot \mathbf{curl}\,\overline{H} = \frac{1}{2\sigma}\nabla H \cdot \nabla \overline{H},$$

we obtain the system of equations:

$$-\operatorname{div}(\sigma^{-1}(\theta)\nabla H) + \mathrm{i}\omega\mu(\theta)H = 0 \qquad \text{in } \Lambda, \tag{8.3}$$

$$H = H_0 \qquad \text{on } \gamma, \tag{8.4}$$

$$-\operatorname{div}(k(\theta)\nabla\theta) = \frac{1}{2\sigma(\theta)}\nabla H \cdot \nabla\overline{H} \qquad \text{in } \Lambda, \tag{8.5}$$

$$\theta = 0 \qquad \text{on } \gamma. \tag{8.6}$$

Some comments about the above system are in order:

1. We have restricted ourselves, for the electromagnetic problem, to the case of only one conductor Λ with boundary γ. It was shown indeed in Sect. 3.2 that it is always possible to reduce the problem to this case.
2. For this stationary case, we have chosen for simplicity to impose a Dirichlet boundary condition. In the case of time dependent problems Neumann boundary conditions, corresponding for instance to thermal insulation can be prescribed.
3. Since the prescribed magnetic field H_0 is constant (see Sect. 3.2), we can replace the magnetic field H by $\tilde{H} := H - H_0$. Then (8.3) can be replaced by an equation with a constant right-hand side $f := -\mathrm{i}\omega\mu H_0$.
4. Equations (8.3) and (8.5) are coupled via the source term in (8.5) and the fact that, in general, the material properties k, μ and σ depend on the temperature.

Let us complete this system with appropriate assumptions on the coefficients. The electric conductivity σ, the magnetic permeability μ and the thermal conductivity k are assumed such that:

$$0 < \sigma_m < \sigma(\theta(\boldsymbol{x})) \le \sigma_M, \tag{8.7}$$

$$0 < \mu_0 \le \mu(\theta(\boldsymbol{x})) \le \mu_M, \tag{8.8}$$

$$0 < k_m \le k(\theta(\boldsymbol{x})) \le k_M, \tag{8.9}$$

for $\boldsymbol{x} \in \Lambda$.

Mathematical analysis of the nonlinear system of partial differential equations (8.7)–(8.9) exhibits difficulties that are mainly related to the structure of the Joule power density nonlinear term. More precisely, if one seeks solutions such that the function H is in the space $\mathcal{H}^1$, which is the space of finite energy for the magnetic field, then the function $\nabla H \cdot \nabla \overline{H}$ lies in $\mathcal{L}^1$ and the temperature θ cannot be sought in $\mathcal{H}^1$, i.e. the Lax–Milgram theorem (Theorem 1.2.1) cannot be applied. To overcome this difficulty, earlier works [52–54] have considered the formulation of the problem (8.3)–(8.6) in Sobolev spaces $\mathcal{W}^{1,p}$ for $p > 2$ and the use of the Schauder fixed point theorem to prove existence of a solution. Let us quote the result obtained in [54]. This one adds to (8.7)–(8.9) the choice $\mu = \mu_0$. Note that these additional hypotheses are not essential in the proof. We have the following result [54]:

Theorem 8.3.1. *There exists* $p > 2$ *such that* (8.3)–(8.6) *has at least one solution* (H, θ) *satisfying*

$$(H, \theta) \in \mathcal{W}^{1,p}(\Lambda) \times \mathcal{W}_0^{1,q}(\Lambda),$$

where q *is arbitrary (in* $[1, \infty]$*) if* $p \geq 4$ *and satisfies* $2 < q < 2p/(4-p)$ *if* $2 < p < 4$. *Moreover, there exists a constant* C *such that*

$$\|H\|_{\mathcal{W}^{\infty,\checkmark}(\Lambda)} + \|\theta\|_{\mathcal{W}^{\infty,\amalg}(\Lambda)} \leq C\, |H_0|.$$

More recent works on nonlinear elliptic equations with $\mathcal{L}^1$ right-hand side using the so-called renormalized solutions (cf. [19, 29, 86]) have been considered as a natural framework for the formulation of problems like (8.3)–(8.6). In view of applying this theory to our problem, the work of Gallouët–Herbin [86] can be adapted to prove that under less regularity assumptions on the domain and with extending the assumptions on k and μ, by assuming that they depend also on the position $\boldsymbol{x}$, i.e. by allowing heterogeneous material, the solution (H, θ) satisfies

$$H \in \mathcal{H}^1(\Lambda), \quad \theta \in \mathcal{W}^{1,p}(\Lambda) \qquad \forall\, p \in [1, 2).$$

We also mention here the paper of Parietti–Rappaz [141] and the PhD thesis of Parietti [140] where the two-dimensional induction heating problem using the transversal model for electromagnetics (see Sect. 3.3) is analyzed. Similar results for existence of solutions are obtained. This work is completed by considering numerical analysis of the problem by using a finite element method (see [142]).

8.4 A 2-D Time Dependent Problem

We consider here the time dependent version of the two-dimensional model presented in the previous section. The system of equations reads:

$$
\begin{cases}
-\operatorname{div}(\sigma^{-1}(\theta)\nabla H) + \mathrm{i}\omega\mu(\theta)H = 0 & \text{in } \Lambda, \\
\dfrac{\partial h}{\partial t} - \operatorname{div}(k(\theta)\nabla\theta) = \dfrac{1}{2}\sigma^{-1}(\theta)\nabla H \cdot \nabla\overline{H} & \text{in } \Lambda, \\
\theta = \beta(h) & \text{in } \Lambda, \\
h(t=0) = h_0 & \text{in } \Lambda, \\
H = H_0 & \text{on } \gamma, \\
k(\theta)\,\nabla\theta \cdot \boldsymbol{n} = 0 & \text{on } \gamma.
\end{cases}
\tag{8.10}
$$

We may notice that the use of a Neumann boundary condition for the temperature (i.e., assumption of insulated boundary) is allowed since we are dealing with a time dependent problem and, in addition, this is more justified from a physical point of view. In this form, there is no known result of existence of a solution to this system, up to our knowledge. We shall however mention connected works where an evolution equation is considered for the electromagnetic problem. In other words, we do not assume a time harmonic behaviour evolution for the electromagnetic process. We mention for this two major works:

- In the paper by Bossavit and Rodrigues [35] the electromagnetic problem is stated in a bounded domain assuming either a perfectly conductive or a perfectly permeable boundary. The authors prove existence of weak solution of the system of partial differential equations.
- In the PhD thesis of S. Clain [52], the case of the electromagnetic problem in the whole space is considered. Using a regularization technique, existence of a less weak solution is proved for the coupled system of parabolic equations.

One can reasonably conjecture that existence of a solution in the same type of spaces can be obtained for (8.10) using either the technique of [52] or the one developed in [19, 86].

8.5 Numerical Experiments

Let us present some numerical experiments illustrating an induction heating process. We report here numerical tests as described in [152] where a three-dimensional model is compared to an axisymmetric one to test the code effectiveness. The simulation uses an axisymmetric toroidal workpiece as depicted in Fig. 8.2, where the multivalued external potential p, solution of (1.20), is presented. Figure 8.3 shows temperature contours and Fig. 8.4 the current density vector.

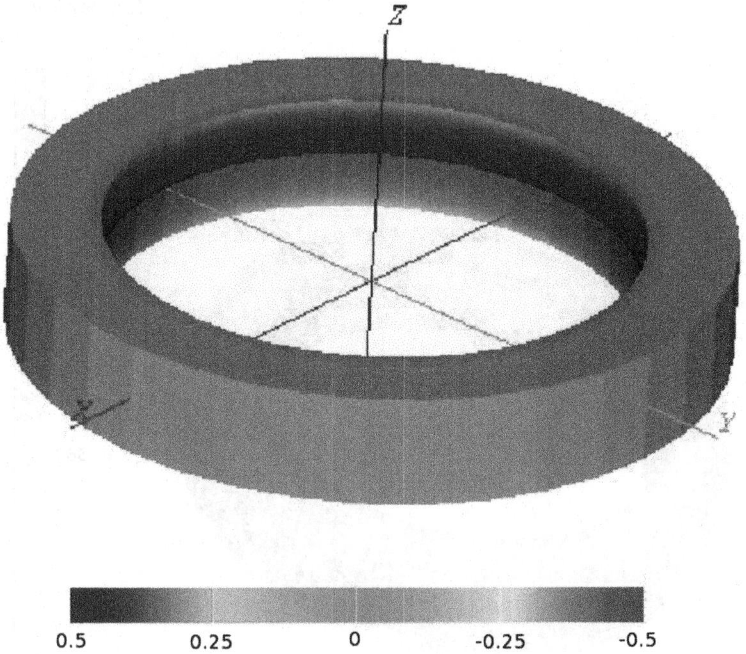

Fig. 8.2 The multivalued potential

8.6 An Optimal Control Problem

Eddy current processes in general and induction heating in particular involve a wide variety of optimization problems (shape optimization, optimal control, inverse problems, ...). We have chosen to illustrate here an optimal control problem of an induction heating process. This problem is mainly associated to a particular industrial application. The process can be described in its general setting in the following form: Assume we want to heat a workpiece by means of electromagnetic induction but we want to control the process in such a way that:

(i) The temperature, at a given final time, is as close as possible to a given value.
(ii) The temperature does not exceed, at any time, a given prescribed value.

To realize such an objective we assume that we have control on the frequency and the voltage. As a typical practical application we can think of the optimization of the thixoforming process (see [126–128]) where the metal has to be heated without reaching the melting point. To give further flexibility to the process, we assume that we have time intervals in which the voltage can be chosen constant. In other words, we consider a partitioning of the time interval $[0, T]$:

$$0 = t_0 < t_1 < t_2 < \ldots < t_M = t_{\max},$$

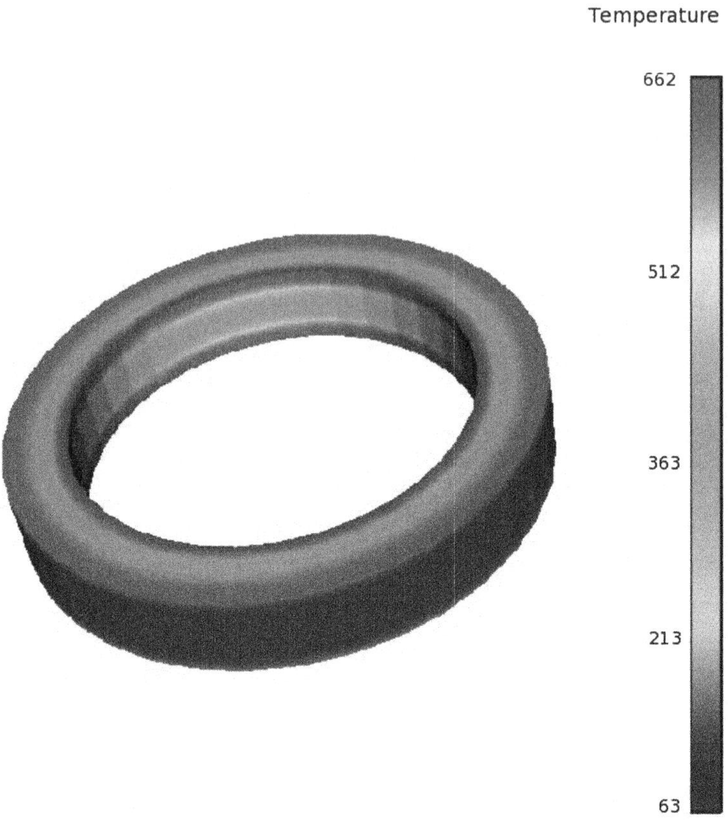

Fig. 8.3 Temperature contours

where the voltage is assumed of the form

$$V(t) = \mathrm{Re}\left(e^{\mathrm{i}\omega t} \sum_{j=1}^{M} V_j \chi_j(t)\right), \qquad 0 \le t \le t_{\max}. \tag{8.11}$$

Here $\chi_j(t)$ is the characteristic function of the interval $[t_{j-1}, t_j)$, i.e. $\chi_j(t) = 1$ on $[t_{j-1}, t_j)$ and 0 on the other intervals, and V_j are complex values. In this presentation, we consider the two-dimensional induction heating model defined in (8.10) as a typical problem for which some industrial investigations have been made (see [126–128]). Extension to other eddy current models as in Chaps. 3, 4 or 5 is formally easy and requires only some technical developments. Since the voltage

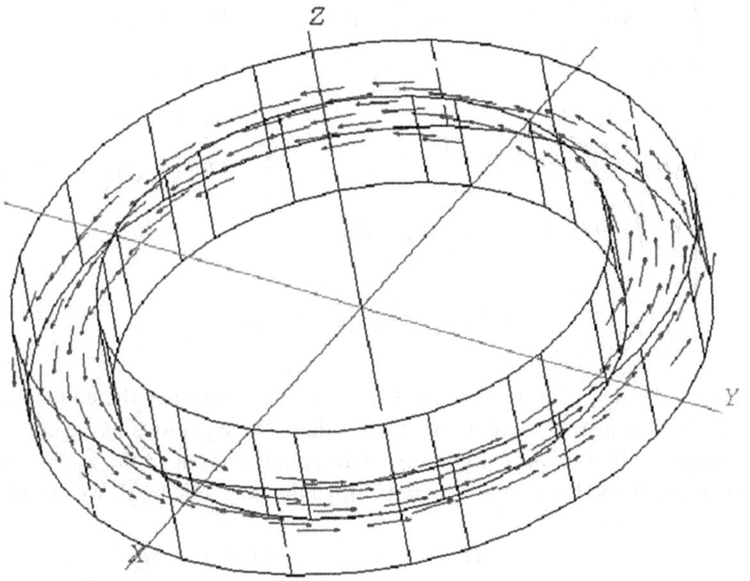

Fig. 8.4 Current density vector

Fig. 8.5 Notation for conductor domains

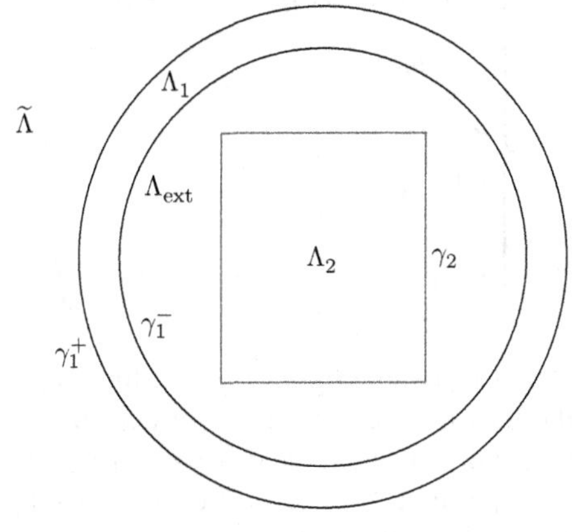

has the form (8.11), we have to start from time dependent eddy current equations, i.e. we consider (2.14)–(2.17) where we look for a magnetic field of the form:

$$\boldsymbol{H}(\boldsymbol{x},t) = \mathrm{Re}(e^{\mathrm{i}\omega t} H_j(x_1,x_2))\,\boldsymbol{e}_3 \qquad \text{for } t_{j-1} \le t < t_j .$$

Recalling the definition of the conductors stated in Chap. 3 as presented in Fig. 8.5, we obtain for each $j = 1,\ldots,M$ the problem arising for $t \in (t_{j-1},t_j)$ (see (3.22), (3.23), (3.32)):

$$
\begin{cases}
-\operatorname{div}(\sigma^{-1}\nabla H_j) + \mathrm{i}\omega\mu H_j = 0 & \text{in } \Lambda_1 \cup \Lambda_2, \\
H_j = H_{0j} & \text{in } \Lambda_{\text{ext}}, \\
H_j = 0 & \text{on } \gamma_1^+, \\
[H_j] = 0 & \text{on } \gamma_1^- \cup \gamma_2, \\
\mathrm{i}\omega \displaystyle\int_{\Lambda_2} \mu H_j \, d\boldsymbol{x} + \mathrm{i}\omega\mu_0|\Lambda_{\text{ext}}|H_{0j} & \\
\qquad\qquad + \displaystyle\int_{\gamma_1^-} \sigma^{-1}\frac{\partial H_j}{\partial n}\, ds = V_j. &
\end{cases}
\tag{8.12}
$$

Here H_{0j} is the magnetic field in the free space that separates the conductors Λ_1 and Λ_2, which is also a constant unknown value. The equations in the conductors can be arranged by a scaling argument. Here unlike in Chap. 3 the scaling will be made by using the voltage values V_j rather than the magnetic field. We clearly have

$$
H_j = uV_j, \; H_{0j} = u_0 V_j \qquad \text{for } 1 \le j \le M, \tag{8.13}
$$

where (u, u_0) is the unique solution of the following problem:

$$
\begin{cases}
-\operatorname{div}(\sigma^{-1}\nabla u) + \mathrm{i}\omega\mu u = 0 & \text{in } \Lambda_1 \cup \Lambda_2, \\
u = u_0 & \text{in } \Lambda', \\
u = 0 & \text{on } \gamma_1^+, \\
[u] = 0 & \text{on } \gamma_1^- \cup \gamma_2, \\
\mathrm{i}\omega \displaystyle\int_{\Lambda_2} \mu u \, d\boldsymbol{x} + \mathrm{i}\omega\mu_0|\Lambda'|u_0 + \displaystyle\int_{\gamma_1^-} \sigma^{-1}\frac{\partial u}{\partial n}\, ds = 1. &
\end{cases}
\tag{8.14}
$$

Let us denote by $u = u(\omega)$, the unique solution of (8.14). Let us now turn to the heat transfer problem. We consider the enthalpy formulation given by (8.10) where the magnetic field is written in the form (8.13) and where the heat problem takes place in Λ_2. The source term in the heat equation is given by averaging on one period for each time interval $[t_{j-1}, t_j)$. We obtain

$$
\begin{cases}
\dfrac{\partial h}{\partial t} - \operatorname{div}(k\nabla\theta) = \dfrac{1}{2\sigma}\nabla u \cdot \nabla\bar{u}\, V_j^2 & \text{in } \Lambda_2, \\
\theta = \beta(h) & \text{in } \Lambda_2, \\
k\,\nabla\theta \cdot \boldsymbol{n} = 0 & \text{on } \gamma_2,
\end{cases}
\tag{8.15}
$$

for $t \in [t_{j-1}, t_j)$ with $j = 1, \ldots, M$. The initial condition is given by

$$
h(\boldsymbol{x}, 0) = h_0 \qquad \text{in } \Lambda_2.
$$

Here we have assumed, for simplicity, that physical properties of the media (electric conductivity, thermal conductivity, magnetic permeability) do not depend on the temperature. This assumption simplifies the problem setting in the sense that the variation of these properties with respect to control parameters will not be involved in the optimality system.

Let us turn now to the optimal control problem. We define the cost functional

$$J(\omega,(V_j)) := \int_{\Lambda_2} (h(\boldsymbol{x},T)-\tilde{h})^2\,d\boldsymbol{x} \\ + \lambda \sum_{j=1}^{M} \int_{t_{j-1}}^{t_j} \Big(\int_{\Lambda_2} \big((h(\boldsymbol{x},t)-h_{\max})^+\big)^2\,d\boldsymbol{x} \Big)\,dt, \qquad (8.16)$$

where the superscript '+' stands for the positive part of a real number ($z^+ := \max(z,0)$). Here $\tilde{h}$ is the constant target enthalpy that we want to reach, $h_{\max}$ is a critical value of the enthalpy. By minimizing the cost functional, the first integral in the functional imposes that the final enthalpy must be as close as possible to $\tilde{h}$, in the L^2–norm, while the second integral imposes, by penalty, that at all times the enthalpy must be smaller that a critical enthalpy, which is typically the fusion enthalpy. The real positive number λ is large enough to enforce this condition. Note in addition that we have omitted to mention the dependency of h on ω and the voltages (V_j) in order to simplify the notation.

The optimal control problem can now be stated as the following:

$$\begin{cases} \text{Find } (\omega,(V_j)) \in \mathcal{U}_{\text{ad}} \text{ such that} \\ J(\omega,(V_j)) = \inf\limits_{(\omega^*,(V_j)^*)\in\mathcal{U}_{\text{ad}}} J(\omega^*,(V_j)^*) \end{cases} \qquad (8.17)$$

where $\mathcal{U}_{\text{ad}}$ is the set of admissible controls defined by

$$\mathcal{U}_{\text{ad}} := \{(\omega,(V_j)_{j=1}^M) \in [\omega_m,\omega_M]\times\mathbb{R}^M;\ 0 \le V_j \le V_M,\ 1 \le j \le M\}.$$

This set is simply defined by prescribing bound constraints on the control variables. This enables using standard optimization libraries for the numerical solution.

From a theoretical viewpoint, Problem (8.17) is not easy to solve. This one involves indeed a penalty formulation of a problem with constraints on the state variable h. More clearly, the term using the parameter λ should be removed and replaced by the additional constraint $h \le h_{\max}$ in the set $\mathcal{U}_{\text{ad}}$. This implies in general singular adjoint problems (see [46] for a mathematical analysis). The formulation we have adopted here gives, as numerical experiments show, satisfactory results for our application.

Numerical solution of (8.17) is handled by using the well known adjoint problem technique (see [31] for instance). For this, we have chosen to apply the method on the time discretized problem. It well known indeed that it is not always equivalent to first discretize then derive the optimality system and to do the reverse approach.

8.6.1 Time Discretization

Let us define a discretization scheme that avoids dealing with the degeneracy of the mapping β. This scheme was introduced in [7].

For an integer $j = 1, \dots, M$, we denote by N_j and $\delta t_j > 0$ the number of time steps in the time interval $[t_{j-1}, t_j)$ and the time step, so that

$$N_j \delta t_j = t_j - t_{j-1}.$$

We define a time subdivision (t_j^n) by $t_j^n = t_{j-1} + n\delta t_j$ for $n = 1, \dots, N_j$, $1 \le j \le M$. The approximation h_j^n of $h(t_j^n)$ is defined by the following Chernoff scheme:

$$\gamma \frac{\theta_j^{n+1} - \beta(h_j^n)}{\delta t_j} - \operatorname{div}(k \nabla \theta_j^{n+1}) = \frac{1}{2\sigma} \nabla u \cdot \nabla \overline{u} \, V_j^2 \qquad \text{in } \Lambda_2, \tag{8.18}$$

$$k \nabla \theta_j^{n+1} \cdot \boldsymbol{n} = 0 \qquad \text{on } \gamma_2, \tag{8.19}$$

$$h_j^{n+1} = h_j^n + \gamma(\theta_j^{n+1} - \beta(h_j^n)) \qquad \text{in } \Lambda_2, \tag{8.20}$$

$$h_j^0 = h_{j-1}^{N_{j-1}} \qquad \text{in } \Lambda_2, \tag{8.21}$$

where γ is a positive parameter that must be chosen (see [7]) such that

$$0 < \gamma < \frac{1}{\sup_{s \in \mathbb{R}} \beta'(s)}.$$

In (8.18) and (8.20), θ_j^{n+1} is an auxiliary variable at time t_j^{n+1} that has the same dimension as a temperature but is not the actual temperature, which is $\beta(h_j^{n+1})$.

8.6.2 The Optimality System

The optimality system associated to the optimal control problem (8.17) will be derived by using the so-called adjoint state technique (see [46] for instance). Without detailing the theory of adjoint problems we directly consider the development for our problem. We start by defining the associated lagrangian to the functional J where h and θ are considered as independent variables from control variables. Let w (*resp.* ϑ) stand for the adjoint variable to h (*resp.* θ). In other words, w (*resp.* ϑ) can be seen as the Lagrange multiplier associated to the state equation (8.20) (*resp.* (8.18)–(8.19)) viewed as a constraint. The time discrete version of the Lagrangian is defined by

$$
\begin{aligned}
&\mathscr{L}(\omega,(V_j),(h_j^n),(\theta_j^n);(w_j^n),(\vartheta_j^n)) \\
&\quad := \int_{\Lambda_2} (h_M^{N_M}(\boldsymbol{x}) - \tilde{h})^2\,d\boldsymbol{x} \\
&\qquad + \lambda \sum_{j=1}^{M} \delta t_j \sum_{n=0}^{N_j-1} \int_{\Lambda_2} \left((h_j^n(\boldsymbol{x}) - h_{\max})^+\right)^2 d\boldsymbol{x} \\
&\qquad + \sum_{j=1}^{M} \Bigg(\delta t_j \sum_{n=0}^{N_j-1} \Big(\frac{\gamma}{\delta t_j} \int_{\Lambda_2} (\theta_j^{n+1} - \beta(h_j^n))\vartheta_j^{n+1}\,d\boldsymbol{x} \\
&\qquad\qquad + \int_{\Lambda_2} k\nabla\theta_j^{n+1}\cdot\nabla\vartheta_j^{n+1}\,d\boldsymbol{x} \\
&\qquad\qquad - \frac{V_j^2}{2}\int_{\Lambda_2} \sigma^{-1}\nabla H\cdot\nabla\overline{H}\,\vartheta_j^{n+1}\,d\boldsymbol{x} \\
&\qquad\qquad + \frac{1}{\delta t_j}\int_{\Lambda_2} (h_j^{n+1} - h_j^n)\,w_j^{n+1}\,d\boldsymbol{x} \\
&\qquad\qquad - \frac{\gamma}{\delta t_j}\int_{\Lambda_2} (\theta_j^{n+1} - \beta(h_j^n))\,w_j^{n+1}\,d\boldsymbol{x}\Big)\Bigg),
\end{aligned}
\tag{8.22}
$$

where for a function $u(\boldsymbol{x},t)$, $u^n(\boldsymbol{x})$ is an approximation of $u(\boldsymbol{x},t^n)$. Using the identity

$$
\begin{aligned}
\sum_{n=0}^{N_j-1}\int_{\Lambda_2} (h_j^{n+1} - h_j^n)w_j^{n+1}\,d\boldsymbol{x} &= \sum_{n=0}^{N_j-1}\int_{\Lambda_2} h_j^n(w_j^n - w_j^{n+1})\,d\boldsymbol{x} \\
&\quad + \int_{\Lambda_2} h_j^{N_j} w_j^{N_j}\,d\boldsymbol{x} - \int_{\Lambda_2} h_j^0 w_j^0\,d\boldsymbol{x},
\end{aligned}
$$

we obtain

$$
\begin{aligned}
&\mathscr{L}(\omega,(V_j),(h_j^n),(\theta_j^n);(w_j^n),(\vartheta_j^n))\\
&\quad=\int_{\Lambda_2}(h_M^{N_M}(\boldsymbol{x})-\tilde{h})^2\,d\boldsymbol{x}\\
&\qquad+\lambda\sum_{j=1}^{M}\delta t_j\sum_{n=0}^{N_j-1}\int_{\Lambda_2}\big((h_j^n(\boldsymbol{x})-h_{\max})^+\big)^2\,d\boldsymbol{x}\\
&\qquad\quad+\sum_{j=1}^{M}\Big(\delta t_j\sum_{n=0}^{N_j-1}\Big(\frac{\gamma}{\delta t_j}\int_{\Lambda_2}(\theta_j^{n+1}-\beta(h_j^n))\vartheta_j^{n+1}\,d\boldsymbol{x}\\
&\qquad\qquad+\int_{\Lambda_2}k\nabla\theta_j^{n+1}\cdot\nabla\vartheta_j^{n+1}\,d\boldsymbol{x}\\
&\qquad\qquad-\frac{V_j^2}{2}\int_{\Lambda_2}\sigma^{-1}\nabla H\cdot\nabla\overline{H}\,\vartheta_j^{n+1}\,d\boldsymbol{x}\\
&\qquad\qquad+\frac{1}{\delta t_j}\int_{\Lambda_2}h_j^n(w_j^n-w_j^{n+1})\,d\boldsymbol{x}\\
&\qquad\qquad\quad-\frac{\gamma}{\delta t_j}\int_{\Lambda_2}(\theta_j^{n+1}-\beta(h_j^n))\,w_j^{n+1}\,d\boldsymbol{x}\Big)\Big)\\
&\qquad\qquad\qquad+\frac{1}{\delta t_j}\int_{\Lambda_2}h_M^{N_M}w_M^{N_M}\,d\boldsymbol{x}-\frac{1}{\delta t_j}\int_{\Lambda_2}h_1^0w_1^0\,d\boldsymbol{x}.
\end{aligned}
$$

The optimality system is given by:

$$\frac{\partial\mathscr{L}}{\partial(h_j)}\Big(\omega,(V_j),(h_j^n),(\theta_j^n);(w_j^n),(\vartheta_j^n)\Big)=0, \tag{8.23}$$

$$\frac{\partial\mathscr{L}}{\partial(w_j)}\Big(\omega,(V_j),(h_j^n),(\theta_j^n);(w_j^n),(\vartheta_j^n)\Big)=0, \tag{8.24}$$

$$\frac{\partial\mathscr{L}}{\partial(\theta_j)}\Big(\omega,(V_j),(h_j^n),(\theta_j^n);(w_j^n),(\vartheta_j^n)\Big)=0, \tag{8.25}$$

$$\frac{\partial\mathscr{L}}{\partial(\vartheta_j)}\Big(\omega,(V_j),(h_j^n),(\theta_j^n);(w_j^n),(\vartheta_j^n)\Big)=0, \tag{8.26}$$

$$\frac{\partial\mathscr{L}}{\partial\omega}\Big(\omega,(V_j),(h_j^n),(\theta_j^n);(w_j^n),(\vartheta_j^n)\Big)(\omega^*-\omega)\geq 0, \tag{8.27}$$

$$\frac{\partial\mathscr{L}}{\partial(V_j)}\Big(\omega,(V_j),(h_j^n),(\theta_j^n);(w_j^n),(\vartheta_j^n)\Big)(V_i^*-V_i)\geq 0 \qquad 1\leq i\leq M, \tag{8.28}$$

for all $(\omega^*, (V_j)^*) \in \mathcal{U}_{\text{ad}}$, where $\big(\omega, (V_j), (h_j^n), (\theta_j^n); (w_j^n), (\vartheta_j^n)\big)$ is a solution of the optimal control problem.

Let us write down the expression of these directional derivatives to get the final system to solve. We skip here computation details that can be found in [126]:

- From (8.23) we deduce for all $n = 1, \ldots, N_j$ and $j = 1, \ldots, M$ the following problem:

$$\begin{cases} \gamma\, \beta'(h_i^n) \dfrac{w_i^{n+1} - \vartheta_i^n}{\delta t_i} + \dfrac{w_i^n - w_i^{n+1}}{\delta t_i} + \lambda\,(h_i^n - h_M)^+ = 0 & \text{in } \Lambda_2, \\ w_M^{N_M} = \tilde{h} - h_M^{N_M} & \text{in } \Lambda_2, \\ w_i^{N_i} = h_{i+1}^0 & \text{in } \Lambda_2. \end{cases} \tag{8.29}$$

- Equation (8.25) yields the adjoint equation for (ϑ_j):

$$\begin{cases} -\operatorname{div}(k \nabla \vartheta_i^{n+1}) + \dfrac{\alpha}{\delta t_i} \vartheta_i^{n+1} = \dfrac{\alpha}{\delta t_i} w_i^{n+1} & \text{in } \Lambda_2, \\ k \dfrac{\partial \vartheta_i^{n+1}}{\partial n} = 0 & \text{on } \gamma_2. \end{cases} \tag{8.30}$$

- Equation (8.26) results in the electromagnetic problem
- To express (8.27) we obtain after differentiating:

$$\frac{\partial \mathcal{L}}{\partial (V_j)} \big(\omega, (V_j), (h_j^n), (\theta_j^n); (w_j^n), (\vartheta_j^n)\big) = -\delta t_i \sum_{n=0}^{N_i - 1} V_i \int_{\Lambda_2} \sigma^{-1} \vartheta_i^{n+1} \nabla H \cdot \nabla \overline{H} \, d\boldsymbol{x}. \tag{8.31}$$

- We obtain from (8.28) after differentiating:

$$\frac{\partial \mathcal{L}}{\partial \omega} \big(\omega, (V_j), (h_j^n), (\theta_j^n); (w_j^n), (\vartheta_j^n)\big) = -\sum_{j=1}^{M} \delta t_j V_j^2 \sum_{n=0}^{N_j - 1} \int_{\Lambda_2} \sigma^{-1} \operatorname{Re}(\nabla H \cdot \nabla \overline{H}_\omega) \vartheta_j^{n+1} \, d\boldsymbol{x}, \tag{8.32}$$

where H_ω is the solution of the following problem:

$$\begin{cases} -\operatorname{div}(\sigma^{-1}\nabla H_\omega) + \mathrm{i}\omega\mu H_\omega = -\mathrm{i}\mu H & \text{in } \Lambda_1 \cup \Lambda_2, \\ H_\omega = 0 & \text{on } \gamma_2, \\ H_\omega = 0 & \text{on } \gamma_1^+ \cup \gamma_1^-. \end{cases} \tag{8.33}$$

Here (8.33) is obtained by differentiating (8.14) with respect to ω.

8.6.3 An Iterative Procedure

Let us give now an iterative procedure to compute an optimal solution. We define the following flowchart of an iterative algorithm that was extensively tested:

1. Give an initial guess of the control parameters $(\omega^0, (V_j^0)) \in \mathcal{U}_{\text{ad}}$, and set iteration counter r to 0.
2. Compute the pair (u, u_0) by solving (8.14), and deduce $H_j = uV_j$ for $j = 1, \dots, M$.
3. Compute the enthalpy h and the temperature θ by solving (8.18)–(8.21).
4. Compute w and ϑ by solving the adjoint problems (8.29) and (8.30) successively.
5. Evaluate J as given by (8.16).
6. Calculate partial derivatives of $\mathscr{L}$ with respect to (V_j) and ω from (8.31) to (8.32).
7. Update control parameters by calculating ω^{k+1} and (V_j^{k+1}) by a descent algorithm.
8. Project control parameters on the set $\mathcal{U}_{\text{ad}}$.
9. If convergence is achieved then stop; otherwise set $r := r + 1$, go to 2.

From a practical viewpoint, we perform steps 7 and 8 using as a descent algorithm the BFGS method (see [31] for instance).

8.6.4 A Numerical Test

We report in this section a numerical experiment concerning a case with simple geometry. The details of the experiment as well as physical properties of the used materials are presented in [126,128]. The inductor consists in a cylinder with annular section and the heated conductor has a square section (see Fig. 8.6).

The inductor has as internal diameter 175 mm and thickness 10 mm and is made of copper. The inductor is also assumed to be at a constant temperature, which is a reasonable approximation since in a realistic situation this one is cooled by a water circuit. For the test we have chosen the following values for electric conductivity and magnetic permeability:

$$\sigma = 0.59 \times 10^8\,\Omega^{-1}\,\mathrm{m}^{-1}, \quad \mu = \mu_0 = 4\pi \times 10^{-7}\,\mathrm{H\,m}^{-1}.$$

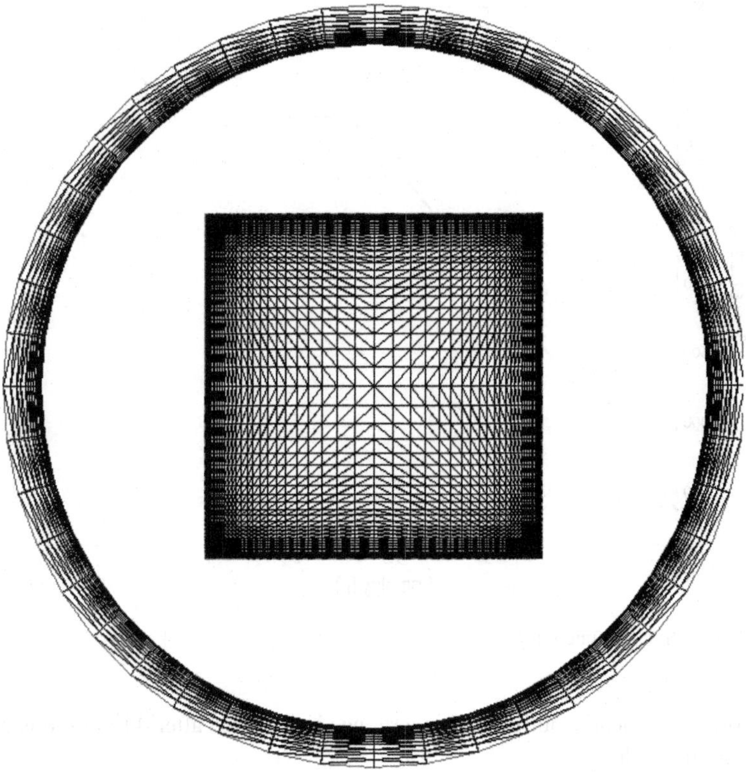

Fig. 8.6 Domain and finite element mesh

The square–section conductor has a side size of 88 mm and is made of aluminium. Its electric conductivity is $\sigma = 0.37 \times 10^8\,\Omega^{-1}\,\mathrm{m}^{-1}$ and the magnetic permeability is $\mu = \mu_0$. The thermal conductivity is $k = 140\,\mathrm{W\,m^{-1}\,K^{-1}}$ and the graph of the function β is given in Fig. 8.7.

For the tests, we have chosen the initial temperature $\theta_0 = 293$ K and the target temperature $\tilde{\theta} = 847$ K, which implies the target enthalpy $\tilde{h} = 2.18 \times 10^9\,\mathrm{J\,m^{-3}}$.

Numerical simulations are carried out with two types of control data:

1. In the first test, we consider two control variables: the frequency and one voltage value. This results in a 2-variable optimization problem. To run this test we use as initial angular frequency $\omega_0 = 2\pi \times 10^4\,\mathrm{rad\,s^{-1}}$. In addition, no upper bound for the voltage is used ($V_M = \infty$). For the frequency, we used the lower and upper bounds $\omega_m = 2\pi \times 10^3\,\mathrm{rad\,s^{-1}}$ and $\omega_M = 2\pi \times 10^4\mathrm{rad\,s^{-1}}$. Finally, the penalty parameter λ was chosen equal to zero.

   Figure 8.8 gives isocontour lines of the cost function for a frequency varying from 1,000 to 10,000 Hz and a voltage varying from 1 to 60 V. This figure shows also the optimization path (white line) where the descent behaviour of the

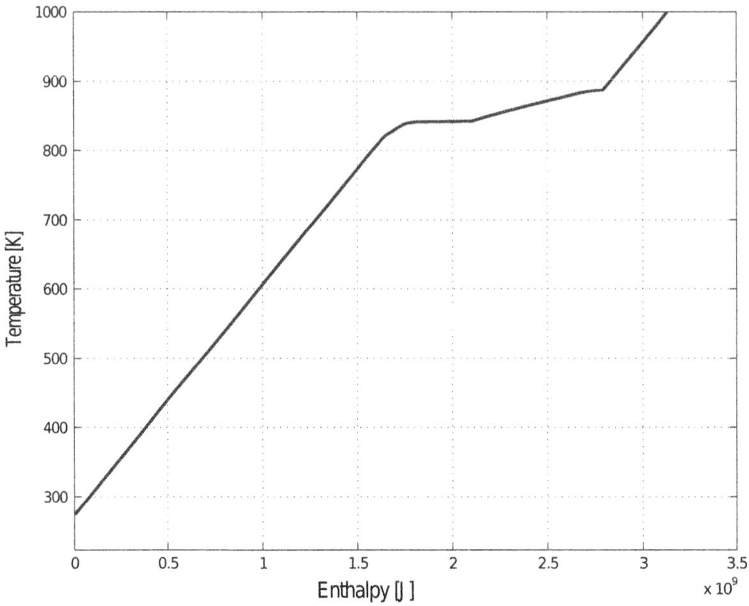

Fig. 8.7 Graph of the function β

algorithm is clearly outlined. Convergence is obtained after 44 iterations giving the control values:

$$\omega = 6{,}280\,\text{rad s}^{-1}, \quad V = 10.7\,\text{V}.$$

Figure 8.9 shows the iterated values of the cost function. Clearly, the cost function decreases rapidly to a given value corresponding to the lower bound of frequency. This shows that the choice of control variables is not relevant. The second choice of control variables confirms this observation.

2. The second test attempts to give more flexibility to the process in order to try to obtain a better heating behaviour. For this, we fix the frequency to a constant value, that is the optimal one obtained in Test 1 ($\omega = 6{,}280\,\text{rad s}^{-1}$), and choose to select a series of 4 voltage values each acting during 100 s. We impose that the voltage values remain lower than a fixed maximal value 17 V. In addition we impose a maximal value for the temperature that is $\theta_M = 868$ K, which results a maximal enthalpy equal to $2.53 \times 10^9\,\text{J m}^{-3}$. We also choose a penalty parameter $\lambda = 100$.

   After 64 iterations, numerical solution of the optimal control problem gives the control values:

$$V_1 = 17\,\text{V}, \; V_2 = 11.6\,\text{V}, \; V_3 = 4.8\,\text{V}, \; V_4 = 0\,\text{V}.$$

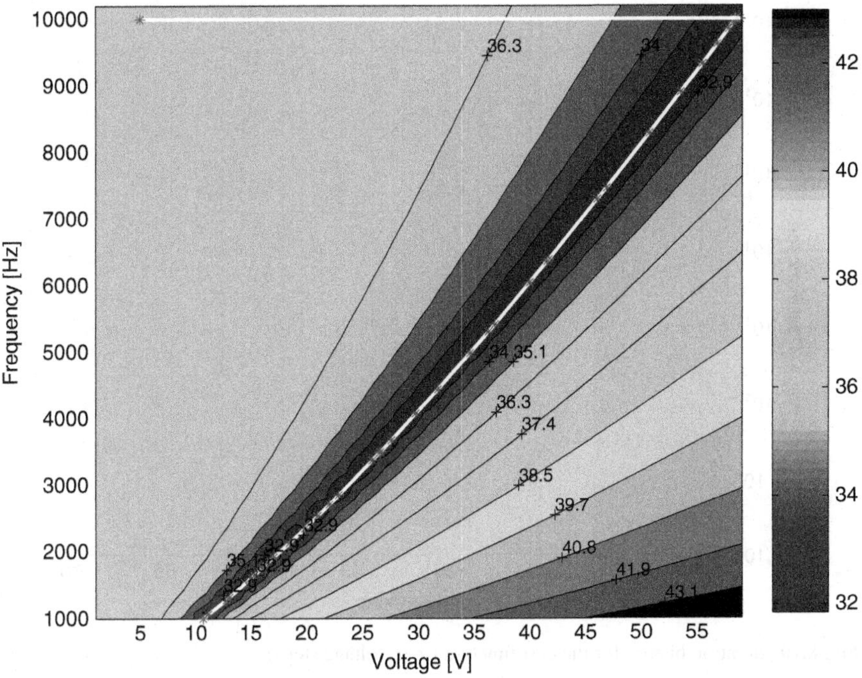

Fig. 8.8 Contours of the cost function (two control variables)

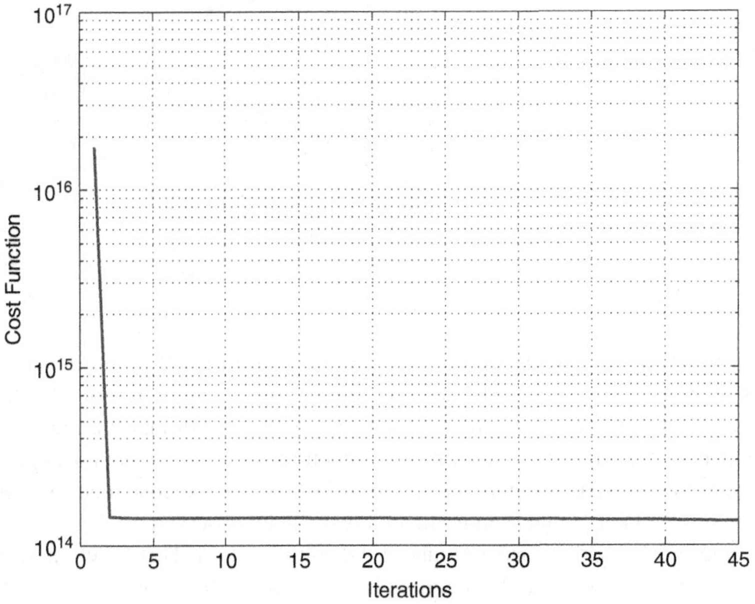

Fig. 8.9 Iteration history for the cost function (two control variables)

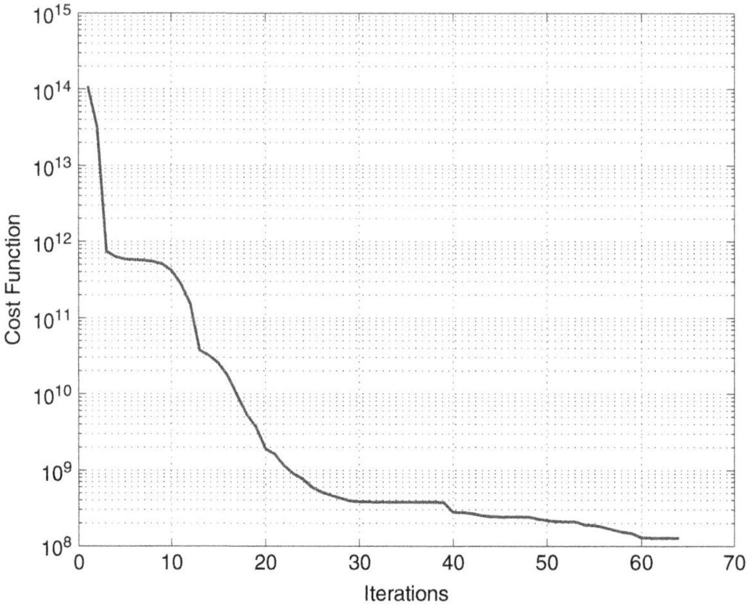

Fig. 8.10 Iteration history for the cost function (four voltage steps)

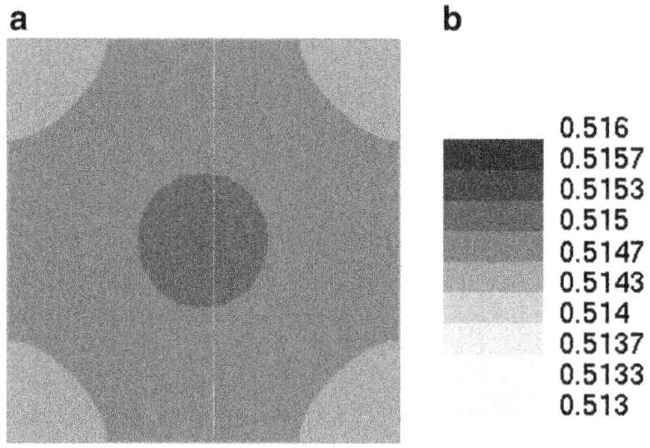

Fig. 8.11 Isocontours of the solid fraction function

It is noteworthy that the optimization process confirms the intuition that the best way to obtain a close to uniform temperature is to switch off the current for a given time period in order to privilege diffusion. In fact the higher is the frequency, the higher the heating is concentrated near the boundary (Screen effect). Figure 8.10 presents the iteration history for the cost function. Here a more relevant behaviour than in the first test can be observed.

Figure 8.11 displays isocontours of the solid fraction. This function, denoted by f_s, is defined through the mapping $\theta \mapsto h$ (inverse of β) by

$$h = \int_0^\theta C_p(s)\,ds + L(1 - f_s(\theta)),$$

where C_p is the specific heat at constant pressure and L is the latent heat. We note that the solid fraction values are all close to a constant (0.5), which means that the temperature, and then the enthalpy, is close to a uniform field, which is the desired goal. We notice here that, since the current density is small in the angles of the conductor, their heating is more difficult.

Chapter 9
Magnetohydrodynamics and Magnetic Shaping

The term *Magnetohydrodynamics* (MHD) refers in general to mathematical models that couple electromagnetic phenomena, including wave propagation, with fluid dynamics. These models are generally involved in plasma physics, stellar dynamics, metal liquid flows and many other applications. Here we are concerned, as this is the case throughout this textbook, with the particular case of eddy current problems, i.e. when wave propagation is neglected. We consider again only time harmonic currents, which result in a time–independent (or quasi-static) problem for electromagnetics. Moreover, we restrict ourselves in the present chapter to the incompressible fluid dynamics case. An application with compressible fluid flow is presented in Chap. 10. Let us notice that [91] contains a family of magnetohydrodynamic problems with applications in aluminium electrolysis.

An important feature of magnetohydrodynamics is that when the fluid, or a part of it, is not confined in a container, i.e. in contact with the ambient air, the shape of the fluid domain is a priori unknown and we are faced with a *free boundary problem*. This is at the origin of magnetic shaping problems with interesting industrial applications such as electromagnetic casting.

We present in this chapter a collection of mathematical models borrowed from industrial applications. In a first step, we consider fixed domain magnetohydrodynamics where the coupling between eddy current and incompressible fluid dynamics is described. We report the most known mathematical results, up to our knowledge. In a second step, we consider magnetic shaping applications: We start by considering simple models where no fluid flow is present, the position of the free boundary being described by a pressure balance equation. We see how such models can be formulated in terms of optimization problems enabling the use of optimization and optimal control theory for their analysis. More complete setting of more sophisticated models where coupling with incompressible Navier-Stokes equations is then presented.

When dealing with liquid metal flow induced by the use of Lorentz forces, two major types of applications are encountered:

R. Touzani and J. Rappaz, *Mathematical Models for Eddy Currents and Magnetostatics: With Selected Applications*, Scientific Computation, DOI 10.1007/978-94-007-0202-8_9,

- The Lorentz force helps maintaining the fluid in levitation and in particular giving a particular shape of the fluid domain. Applications of this feature include magnetic shaping. In this case, it is desirable that the gradient part of the Lorentz force is more significant than the remaining part. It is clear indeed that the gradient part of an external force field in the Navier-Stokes equations has no effect on the fluid motion.
- The Lorentz force is at the origin of fluid stirring. This property can be used for instance in melting in order to homogenize the liquid metal and avoid formation of undesired microstructures. This feature can be used especially when the non-gradient part of the Lorentz force is significant.

9.1 Incompressible Magnetohydrodynamics

The instantaneous density of the Lorentz force is given by the vector field $\boldsymbol{f}_L := \boldsymbol{J} \times \boldsymbol{B}$ (see [74] for instance). Analogously to the induction heating application, it seems reasonable to assume that fluid motion is not sensitive to the current frequency and then only the time average on each period of the Lorentz force drives this motion. Following this, the applied force $\boldsymbol{f}_L$ is approximated by the vector field

$$\begin{aligned} \boldsymbol{f}_L^* &:= \frac{\omega}{2\pi} \int_0^{\frac{2\pi}{\omega}} \mathrm{Re}(e^{\mathrm{i}\omega t} \boldsymbol{J}(\cdot,t)) \times \mathrm{Re}(e^{\mathrm{i}\omega t} \boldsymbol{B}(\cdot,t))\, dt \\ &= \frac{1}{2}\, \mathrm{Re}\,(\boldsymbol{J} \times \overline{\boldsymbol{B}}). \end{aligned} \tag{9.1}$$

Let us assume that the liquid metal motion is governed by the incompressible Navier-Stokes equations (see [117] for instance):

$$\varrho \frac{\partial \boldsymbol{v}}{\partial t} + \varrho\, (\boldsymbol{v} \cdot \nabla)\boldsymbol{v} = 2\,\eta\, \mathbf{div}\, \mathsf{D}(\boldsymbol{v}) - \nabla p + \frac{1}{2}\, \mathrm{Re}\,(\boldsymbol{J} \times \overline{\boldsymbol{B}}), \tag{9.2}$$

$$\mathrm{div}\, \boldsymbol{v} = 0, \tag{9.3}$$

where $\boldsymbol{v}$ and p are respectively fluid velocity and pressure, η and ϱ stand for molecular viscosity and density of the fluid, and $\mathsf{D}(\boldsymbol{v})$ is the symmetric deformation tensor

$$\mathsf{D}(\boldsymbol{v}) := \frac{1}{2}\, (\nabla \boldsymbol{v} + (\nabla \boldsymbol{v})^{\mathsf{T}}). \tag{9.4}$$

In (9.2), the divergence of the tensor D is defined by

$$\mathbf{div}\, \mathsf{D} := \sum_{i,j=1}^{3} \frac{\partial D_{ij}}{\partial x_j}\, \boldsymbol{e}_i.$$

We can remark that by using (9.3), Eqs. (9.2)–(9.3) reduce to the more common form:

$$\varrho \frac{\partial \boldsymbol{v}}{\partial t} + \varrho\,(\boldsymbol{v}\cdot\nabla)\,\boldsymbol{v} = \eta\Delta\boldsymbol{v} - \nabla p + \frac{1}{2}\,\mathrm{Re}\,(\boldsymbol{J}\times\overline{\boldsymbol{B}}), \tag{9.5}$$

$$\operatorname{div}\boldsymbol{v} = 0. \tag{9.6}$$

This formulation is however less practical when natural (Neumann) boundary conditions are to be imposed.

The electromagnetic problem can be derived by using Eqs. (2.62)–(2.65), (2.67)–(2.68) and replacing (2.66) by Ohm's law (2.10) for a moving conductor.

Let us summarize a three-dimensional magnetohydrodynamic problem based on the $\boldsymbol{H}$–model (4.24) (or equivalently (4.32)). To this end, let us denote by Ω_F the domain in which lies the fluid, with boundary Γ_F. In fact, in most applications, the eddy current setup consists in a container filled with liquid metal to stir and the remaining conductors are "solid" conductors that actually stand either for inductors or any other devices.

Clearly, the $\boldsymbol{H}$–model (4.24) must be updated in order to take into account the use of Ohm's law (2.10). The system of equations (4.25)–(4.27) can now replaced by:

$$\begin{aligned}
\mathrm{i}\omega\mu\boldsymbol{H} + \mathbf{curl}\,\boldsymbol{E} &= V\,\boldsymbol{\delta}_{\partial S},\\
\mathbf{curl}\,\boldsymbol{H} &= \boldsymbol{J},\\
\boldsymbol{J} = \sigma(\boldsymbol{E} &+ \boldsymbol{v}\times\boldsymbol{B}),
\end{aligned}$$

in $\mathbb{R}^3$. Following the same approach as in Chap. 4 we eventually obtain the variational formulation:

$$\left\{\begin{aligned}
&\text{Find } \boldsymbol{H}\in\boldsymbol{\mathcal{H}} \text{ such that}\\
&\mathrm{i}\omega\int_{\mathbb{R}^3}\mu\boldsymbol{H}\cdot\overline{\boldsymbol{k}}\,d\boldsymbol{x} + \int_{\Omega}\sigma^{-1}\,\mathbf{curl}\,\boldsymbol{H}\cdot\mathbf{curl}\,\overline{\boldsymbol{k}}\,d\boldsymbol{x}\\
&\qquad + \int_{\Omega_F}\mu\,(\boldsymbol{H}\times\boldsymbol{v})\cdot\mathbf{curl}\,\overline{\boldsymbol{k}}\,d\boldsymbol{x}\\
&\qquad\qquad = V\int_S \mathrm{curl}_S\,\overline{\boldsymbol{k}}\,ds \qquad \forall\,\boldsymbol{k}\in\boldsymbol{\mathcal{H}},
\end{aligned}\right. \tag{9.7}$$

where V is the current voltage and $\boldsymbol{\mathcal{H}}$ is the space defined in (4.23), and where a Green formula has been used to obtain the term involving $\boldsymbol{v}$ with minimal regularity requirement on it.

Some remarks can be made about (9.7):

1. As far as numerical solution of this problem is concerned, we can resort like for (4.24) to the formulation (4.32).

2. The coupling term is a nonlinear convection-like term. However, it is common to neglect this term in most melt flows. In this case, problems (9.7) and (9.5)–(9.6) are decoupled.

   The final model consists then in looking for:

A magnetic field	$\boldsymbol{H} : \mathbb{R}^3 \times (0, t_{\max}) \to \mathbb{C}^3$,
A velocity	$\boldsymbol{v} : \Omega_F \times (0, t_{\max}) \to \mathbb{R}^3$,
and a pressure	$p : \Omega_F \times (0, t_{\max}) \to \mathbb{R}$

such that for $t \in (0, t_{\max})$, the pair $(\boldsymbol{H}, \boldsymbol{v})$ satisfies the variational problem (9.7) and

$$\varrho \frac{\partial \boldsymbol{v}}{\partial t} + \varrho\,(\boldsymbol{v} \cdot \nabla)\,\boldsymbol{v} - 2\eta\,\mathbf{div}\,\mathsf{D}(\boldsymbol{v}) + \nabla p = \frac{\mu}{2}\,\mathrm{Re}(\mathbf{curl}\,\boldsymbol{H} \times \overline{\boldsymbol{H}}) \quad \text{in } \Omega_F, \tag{9.8}$$

$$\operatorname{div} \boldsymbol{v} = 0 \quad \text{in } \Omega_F. \tag{9.9}$$

Here $t_{\max} > 0$ is the maximal time value for a simulation.

The system of equations (9.8)–(9.9) must be supplied with the condition at the infinity (4.5) and appropriate boundary conditions for $\boldsymbol{v}$ on the boundary Γ_F as well.

Let us give some additional remarks about the system (9.7)–(9.9) and its properties:

1. As it can be noticed, the hydrodynamic problem is time dependent, whereas the electromagnetic problem is quasistatic in the sense that the electromagnetic fields are assumed periodic in time. This means that we have assumed, for this type of problem, two time scales: The small time scale for electromagnetics in which the fields are assumed to oscillate with a frequency $\omega/2\pi$ and the large time scale that governs evolution of velocity and pressure. Due to the nonlinearity of the problem, a rigorous formulation would not enable time periodic electromagnetic fields but, as a first approximation, it is reasonable to assume that the large time scale does not affect this periodicity.
2. Considering the Lorentz force term, we have the classical vector identity:

$$\mathrm{Re}\,(\mathbf{curl}\,\boldsymbol{H} \times \overline{\boldsymbol{H}}) = \mathrm{Re}\,((\boldsymbol{H} \cdot \nabla)\,\overline{\boldsymbol{H}}) - \frac{1}{2}\nabla(\boldsymbol{H} \cdot \overline{\boldsymbol{H}}).$$

   The second term in the right-hand side can be viewed as a *Magnetic Pressure*. Indeed, if we define a new pressure field $\tilde{p}$ by

$$\tilde{p} := p + \frac{\mu}{4}\,\boldsymbol{H} \cdot \overline{\boldsymbol{H}},$$

Equation (9.8) becomes

$$\frac{\partial \boldsymbol{v}}{\partial t} + (\boldsymbol{v} \cdot \nabla)\, \boldsymbol{v} - 2\eta\, \mathbf{div}\, \mathsf{D}(\boldsymbol{v}) + \nabla \tilde{p} = \frac{\mu}{2}\, \mathrm{Re}\, ((\boldsymbol{H} \cdot \nabla)\, \overline{\boldsymbol{H}}) \qquad \text{in } \Omega_F, \quad (9.10)$$

for all t with $0 < t \le t_{\max}$.

It is well known that, from a numerical point of view (see for instance [90]), using (9.10) is more advantageous than (9.8). This is due to the fact that the gradient part of the body force term can be large and consequently produces a significant numerical error on the velocity. Proceeding in this way does not however remove completely the difficulty. Indeed the right-hand side of (9.10) can still contain a gradient contribution. To handle this, we use the Helmholtz decomposition of this vector

$$\frac{\mu}{2}\, \mathrm{Re}\, ((\boldsymbol{H} \cdot \nabla)\, \overline{\boldsymbol{H}}) = \nabla \varphi + \mathbf{curl}\, \boldsymbol{w},$$

by minimizing the Euclidean norm of $\mathbf{curl}\, \boldsymbol{w}$. It is then well known that if $(\boldsymbol{H} \cdot \nabla)\, \overline{\boldsymbol{H}} \in \mathcal{L}^2(\Omega_F)$ then the vector $\nabla\varphi$ is given by the projection of the force term on all the gradients of $\mathcal{H}^1$–functions (see [26] for instance). This is obtained by solving the elliptic problem:

$$\begin{cases} -\Delta \varphi = -\dfrac{\mu}{2}\, \mathrm{Re}\, \big(\mathrm{div}((\boldsymbol{H} \cdot \nabla)\, \overline{\boldsymbol{H}})\big) & \text{in } \Omega_F, \\[2ex] \dfrac{\partial \varphi}{\partial n} = \dfrac{\mu}{2} \big(\mathrm{Re}(\boldsymbol{H} \cdot \nabla)\boldsymbol{H}\big) \cdot \boldsymbol{n} & \text{on } \Gamma_F. \end{cases}$$

9.1.1 A 3-D MHD Problem Using the Magnetic Field

Let us consider the model given by (9.7)–(9.9). To this system we add the boundary condition

$$\boldsymbol{v} = 0 \qquad \text{on } \Gamma_F,$$

that expresses that the fluid is confined in its container. Of course, other types of boundary conditions can be envisaged.

Let us define the space in which lies the velocity field

$$\mathcal{V} := \{\boldsymbol{v} \in \mathcal{H}_0^1(\Omega_F);\ \mathrm{div}\, \boldsymbol{v} = 0\}.$$

We have the variational problem:

$$
\begin{cases}
\text{Find } (\boldsymbol{H}, \boldsymbol{v}) \in \mathcal{H} \times \mathcal{V} \text{ such that for all } (\boldsymbol{k}, \boldsymbol{w}) \in \mathcal{H} \times \mathcal{V}: \\
\mathrm{i}\omega \displaystyle\int_{\mathbb{R}^3} \mu \boldsymbol{H} \cdot \overline{\boldsymbol{k}}\, d\boldsymbol{x} + \int_{\Omega} \sigma^{-1}\, \mathbf{curl}\, \boldsymbol{H} \cdot \mathbf{curl}\, \overline{\boldsymbol{k}}\, d\boldsymbol{x} \\
\qquad + \displaystyle\int_{\Omega_F} \mu\, (\boldsymbol{H} \times \boldsymbol{v}) \cdot \mathbf{curl}\, \overline{\boldsymbol{k}}\, d\boldsymbol{x} \\
\qquad = V \displaystyle\int_S \mathrm{curl}_S\, \overline{\boldsymbol{k}}\, ds, \\
2\eta \displaystyle\int_{\Omega_F} \mathsf{D}(\boldsymbol{v}) : \mathsf{D}(\boldsymbol{w})\, d\boldsymbol{x} + \int_{\Omega_F} (\boldsymbol{v} \cdot \nabla \boldsymbol{v}) \cdot \boldsymbol{w}\, d\boldsymbol{x} \\
\qquad = \dfrac{1}{2} \displaystyle\int_{\Omega_F} \mu\, \mathrm{Re}(\,\mathbf{curl}\, \boldsymbol{H} \times \overline{\boldsymbol{H}}) \cdot \boldsymbol{w}\, d\boldsymbol{x}.
\end{cases}
\tag{9.11}
$$

Note that, as this is usually the case for Navier-Stokes equations (see [93] for instance), an equivalent formulation is given by:

$$
\begin{cases}
\text{Find } (\boldsymbol{H}, \boldsymbol{v}, p) \in \mathcal{H} \times \mathcal{H}_0^1(\Omega_F) \times \mathcal{L}_0^2(\Omega_F) \text{ such that for all} \\
\qquad (\boldsymbol{k}, \boldsymbol{w}, q) \in \mathcal{H} \times \mathcal{H}_0^1(\Omega_F) \times \mathcal{L}_0^2(\Omega_F): \\
\mathrm{i}\omega \displaystyle\int_{\mathbb{R}^3} \mu \boldsymbol{H} \cdot \overline{\boldsymbol{k}}\, d\boldsymbol{x} + \int_{\Omega} \sigma^{-1}\, \mathbf{curl}\, \boldsymbol{H} \cdot \mathbf{curl}\, \overline{\boldsymbol{k}}\, d\boldsymbol{x} \\
\qquad + \displaystyle\int_{\Omega_F} \mu\, (\boldsymbol{H} \times \boldsymbol{v}) \cdot \mathbf{curl}\, \overline{\boldsymbol{k}}\, d\boldsymbol{x} \\
\qquad = V \displaystyle\int_S \mathrm{curl}_S\, \overline{\boldsymbol{k}}\, ds, \\
2\eta \displaystyle\int_{\Omega_F} \mathsf{D}(\boldsymbol{v}) : \mathsf{D}(\boldsymbol{w})\, d\boldsymbol{x} + \int_{\Omega_F} (\boldsymbol{v} \cdot \nabla \boldsymbol{v}) \cdot \boldsymbol{w}\, d\boldsymbol{x} - \int_{\Omega_F} p \,\mathrm{div}\, \boldsymbol{w}\, d\boldsymbol{x} \\
\qquad = \dfrac{1}{2} \displaystyle\int_{\Omega_F} \mu\, \mathrm{Re}(\mathbf{curl}\, \boldsymbol{H} \times \overline{\boldsymbol{H}}) \cdot \boldsymbol{w}\, d\boldsymbol{x}, \\
\displaystyle\int_{\Omega_F} q \,\mathrm{div}\, \boldsymbol{v}\, d\boldsymbol{x} = 0,
\end{cases}
$$

where

$$
\mathcal{L}_0^2(\Omega_F) = \big\{q \in \mathcal{L}^2(\Omega_F);\ \int_{\Omega_F} q\, d\boldsymbol{x} = 0\big\}.
$$

9.1.2 A 2-D MHD Problem Using the Magnetic Potential

To illustrate a system of equations in incompressible magnetohydrodynamics, we present a two-dimensional problem that is considered in [149] for a mathematical analysis. The electromagnetic model is the one presented in Sect. 3.3. More precisely, if we assume we are given a collection of conductors as schematically presented in Fig. 3.3 where the liquid metal is confined in the domain Λ_1 and where Λ_2 and Λ_3 stand for the cross-section on the Ox_1x_2 plane of a unique inductor parallel to the Ox_3–axis. Recalling that when the current density $\boldsymbol{J}$ is assumed aligned with the x_3–axis, there exists a scalar potential A that satisfies (3.36). We can then derive an analogous to the system (3.38)–(3.41) by using Ohm's law (2.10) in Λ_1. Since in this case we have $\boldsymbol{B} = B_1\,\boldsymbol{e}_1 + B_2\,\boldsymbol{e}_2$ and $\boldsymbol{E} = E\,\boldsymbol{e}_3$, we obtain with $\boldsymbol{v} = v_1\boldsymbol{e}_1 + v_2\boldsymbol{e}_2$:

$$J = \sigma(E + v_1 B_2 - v_2 B_1) = \sigma(E - \boldsymbol{v}\cdot\nabla A) \qquad \text{in } \Lambda_1. \tag{9.12}$$

Moreover, we have from (3.3) the existence of constants C_k, $k = 1, 2, 3$ such that

$$\mathrm{i}\omega A + E = C_k \qquad \text{in } \Lambda_k,\ k = 1, 2, 3.$$

The Ohm's law (9.12) combined with (3.1) gives

$$\begin{aligned}\sigma(E - \boldsymbol{v}\cdot\nabla A) &= \operatorname{curl}\boldsymbol{H}\\ &= \operatorname{curl}(\mu^{-1}\,\mathbf{curl}\,A)\\ &= -\operatorname{div}(\mu^{-1}\nabla A).\end{aligned}$$

Then we have

$$\mathrm{i}\omega\sigma A - \operatorname{div}(\mu^{-1}\nabla A) + \sigma\boldsymbol{v}\cdot\nabla A = \sigma C_k \qquad \text{in } \Lambda_k. \tag{9.13}$$

where the constants C_k are calculated in the same way as for (3.44). More precisely, by using (9.12), and (3.34), (3.35):

$$\int_{\Lambda_2} J\,d\boldsymbol{x} = -\int_{\Lambda_3} J\,dx = I, \qquad \int_{\Lambda_1} J\,d\boldsymbol{x} = 0,$$

where I is the prescribed total current in the inductor, we obtain

$$\begin{aligned}C_1 &= \frac{\mathrm{i}\omega M_1(\sigma A) + M_1(\sigma\boldsymbol{v}\cdot\nabla A)}{M_1(\sigma)},\\ C_2 &= \frac{I + \mathrm{i}\omega|\Lambda_2|M_2(\sigma A)}{|\Lambda_2|M_2(\sigma)},\\ C_3 &= -C_2,\end{aligned}$$

with

$$M_k(\phi) := \frac{1}{|\Lambda_k|} \int_{\Lambda_k} \phi \, dx.$$

The instantaneous Lorentz force expressed in terms of A and $\boldsymbol{v}$ is given by

$$\begin{aligned} \boldsymbol{f}_{\mathrm{L}} &= \sigma\,(C_1 - \mathrm{i}\omega A - \boldsymbol{v}\cdot\nabla A)\,\nabla A \qquad \text{in } \Lambda_1 \\ &= \sigma\Big(\frac{\mathrm{i}\omega M_1(\sigma A) + M_1(\sigma \boldsymbol{v}\cdot\nabla A)}{M_1(\sigma)} - \boldsymbol{v}\cdot\nabla A - \mathrm{i}\omega A\Big)\nabla A. \end{aligned}$$

Averaging this expression on one time period, we obtain according to (9.1),

$$\begin{aligned} \boldsymbol{f}_{\mathrm{L}}^* = \frac{\sigma\omega}{2}\,\mathrm{Im}\Big(\Big(A - \frac{M_1(\sigma A)}{M_1(\sigma)}\Big)\nabla\overline{A}\Big) \\ + \frac{\sigma}{2}\,\mathrm{Re}\Big(\Big(\frac{M_1(\sigma\boldsymbol{v}\cdot\nabla A)}{M_1(\sigma)} - \boldsymbol{v}\cdot\nabla A\Big)\nabla\overline{A}\Big). \end{aligned}$$

The derivation of the final system of equations is now straightforward. To simplify the presentation, let us give this model for the particular case where the conductivity is constant in each conductor Λ_k, $k = 1, 2, 3$. Let us first notice that we have in this case, by using the Green formula and the fact that $\boldsymbol{v}$ will be required as to be divergence free and to vanish on γ_1:

$$\begin{aligned} \int_{\Lambda_1} \sigma\boldsymbol{v}\cdot\nabla A\,d\boldsymbol{x} &= \sigma_{|\Lambda_1}\int_{\Lambda_1}\mathrm{div}(A\boldsymbol{v})\,d\boldsymbol{x} - \sigma_{|\Lambda_1}\int_{\Lambda_1} A\,\mathrm{div}\,\boldsymbol{v}\,d\boldsymbol{x} \\ &= \sigma_{|\Lambda_1}\int_{\gamma_1} A\,\boldsymbol{v}\cdot\boldsymbol{n}\,ds \\ &= 0. \end{aligned}$$

The Lorentz force becomes then in this case:

$$\boldsymbol{f}_{\mathrm{L}}^* = \frac{\sigma}{2}\big(\omega\,\mathrm{Im}\,(A - M_1(A)) - \mathrm{Re}\,(\boldsymbol{v}\cdot\nabla A)\big)\nabla\overline{A}.$$

We can now deduce the final system of equations for $k = 1, 2, 3$:

$$-\,\mathrm{div}(\mu^{-1}\nabla A) + \mathrm{i}\omega\sigma(A - M_k(A)) + \sigma\boldsymbol{v}\cdot\nabla A = \frac{I_k}{|\Lambda_k|} \qquad \text{in } \bigcup_{k=1}^{3}\Lambda_k, \tag{9.14}$$

$$\Delta A = 0 \qquad \text{in } \Lambda_{\mathrm{ext}}, \tag{9.15}$$

$$- 2\eta \, \mathbf{div}\, \mathbf{D}(\boldsymbol{v}) + \varrho\, (\boldsymbol{v} \cdot \nabla)\boldsymbol{v} + \nabla p - \frac{\sigma\omega}{2} \operatorname{Im}(A\, \nabla \overline{A}) + \frac{\sigma}{2} \operatorname{Re}((\boldsymbol{v} \cdot \nabla A)\, \nabla \overline{A}) = 0 \qquad \text{in } \Lambda_1, \tag{9.16}$$

$$\operatorname{div} \boldsymbol{v} = 0 \qquad \text{in } \Lambda_1, \tag{9.17}$$

$$M_1(A) = 0, \tag{9.18}$$

$$[A] = \left[\mu^{-1} \frac{\partial A}{\partial n}\right] = 0 \qquad \text{on } \gamma, \tag{9.19}$$

$$A(\boldsymbol{x}) = \beta + \mathcal{O}(|\boldsymbol{x}|^{-1}) \qquad |\boldsymbol{x}| \to \infty, \tag{9.20}$$

$$\boldsymbol{v} = 0 \qquad \text{on } \gamma_1, \tag{9.21}$$

where the constants I_k are given by

$$I_1 = 0,\ I_2 = I,\ I_3 = -I,$$

the constant I denoting the imposed total current in the generator device. Note here that without Condition (9.18) the previous problem would have a solution $(A, \boldsymbol{v}, p)$ with A known up to an additive constant. Condition (9.18) is then helpful to fix the value of this constant.

The mathematical model (9.14)–(9.20) was studied in [149] where existence of a solution $(A, \boldsymbol{v}, p)$ is obtained and where uniqueness of this solution is ensured if the viscosity η is "large enough". In addition, a numerical method is established in [150]. It consists in coupling the use of a standard finite element method in the conductors and a boundary element method to represent the external potential A on the boundary γ. The convergence of the constructed numerical method is then proved.

More precisely we have the following result (cf. Rappaz–Touzani [149]).

Theorem 9.1.1. *Problem* (9.14)–(9.20) *has at least one solution*

$$(A, \boldsymbol{v}, p) \in \mathcal{W}^1(\mathbb{R}^2) \times \boldsymbol{\mathcal{W}}^{1,4}(\Lambda_1) \times \mathcal{L}^2_0(\Lambda_1).$$

As far as uniqueness of solutions is concerned, we have the same uniqueness result type as for the stationary Navier-Stokes equations. We quote here a result in [149] where the proof can be found.

Theorem 9.1.2. *Assume that the ratio $I\eta^{-\frac{1}{2}}$ is small enough, then Problem* (9.14)–(9.21) *has a unique solution.*

9.1.3 Bibliographical Comments

All the results we have mentioned here concern the coupling of stationary incompressible Navier-Stokes equations with the eddy current models presented in Chaps. 3 and 4. Let us mention some related works concerning either mathematical or numerical analysis of coupling between incompressible Navier-Stokes and eddy current equations.

- In [99] R.V. Hernandez considers a general purpose model that takes into account in addition to magnetohydrodynamics, heat transfer, i.e., induction heating. He studies existence and possible uniqueness of solution of the obtained system. This analysis concerns the case of steady state solutions, i.e., only magnetostatics are handled in this case.
- In [22] A. Bermúdez et al. develop a model that couples a quasi-static eddy current model with hydrodynamics and heat transfer. Numerical solution of the resulting system is obtained by a finite element method.
- A. Meir and P.G. Schmidt [129, 130] have considered the coupling of the stationary Navier-Stokes equations with magnetostatics, the problem being formulated with realistic boundary conditions, i.e. in the whole space. For this system they have shown well–posedness of the problem. Moreover, in [129], additional coupling with the heat equation is considered.
- In [91] J.-F. Gerbeau, C. Le Bris and T. Lelièvre derive and analyze various models for magnetohydrodynamics. In particular, they consider first a one-fluid time dependent model in which either magnetostatics or evolution eddy current equations (i.e. no time harmonic hypothesis is assumed) are considered. The also study a two-fluid model in which the multifluid feature is handled by a density dependent electric conductivity. This adds a mass conservation equation to the system. Existence of solutions of various models is proved. Numerical analysis of the equations is also considered. Note that, in all these cases, electromagnetic equations are formulated in a bounded domain by assuming a perfectly conducting boundary, i.e. by prescribing

$$\boldsymbol{B} \cdot \boldsymbol{n} = 0, \qquad \mathbf{curl}\, \boldsymbol{B} \times \boldsymbol{n} = \mathbf{0}$$

  on the conductor boundary.

The above list is not exhaustive. We have mainly quoted the works that are close to the models presented here.

9.2 Eddy Current Free Boundary Problems

Most of melt flow applications involve free boundaries, where a part of the fluid domain boundary is in contact with the free space and has unknown shape. This shape is entirely determined by a balance between the fluid pressure and the free space pressure. The mathematical formulation of the problem involves then the so-called *Laplace-Young* equation (see [116] for instance):

$$\tau\kappa = p_{\mathrm{F}} - p_{\mathrm{A}} \tag{9.22}$$

where p_{F} and p_{A} denote the fluid and air pressures respectively, τ is the fluid *surface tension* and κ is the curvature (mean Gauss curvature in the three-dimensional case) of the free boundary. Let us present some models that have been studied by some authors where mathematical and/or numerical approaches have been considered:

1. A first family presents simple models in which no fluid flow is available in the fluid part. In addition, the problem is simplified by assuming that the current density vector is given. This is equivalent to say that the induced current is neglected when compared to the source current. For these cases we first consider 2-D models where we assume in addition that the magnetic field does not penetrate the conductor. This yields a condition on the conductor boundary which results in a problem in a bounded domain. A 3-D model is also presented. For all these models, we present the mathematical setting as well as mathematical results when they are available.
2. In the second family we reconsider the two-dimensional magnetohydrodynamic problem studied in Sect. 9.1.2 and add the fact that a part of the boundary γ is unknown.

9.2.1 A 2-D Simple Magnetic Shaping Model

Magnetic shaping is the making of metal workpieces by solidifying molten metal, the shape being a result of a chosen configuration of inductors. A relevant issue in this domain is to consider the inverse problem consisting in prescribing the desired shape of the workpiece and trying to determine the shape and position of the inductor(s) that produce this shape. Some authors have studied, as we shall outline, this issue.

The term 'simple model' means here that a series of simplifying assumptions are made:

- The current density is given and its support lies outside the workpiece.
- No fluid flow is present, i.e. the metal workpiece shape is determined by the hydrostatic pressure only.
- The boundary of the molten metal domain is perfectly electrically conducting.

We consider the making of a cylindrical tube $\Omega = \Lambda \times \mathbb{R}$ by magnetic shaping. Due to this geometry, the current density vector is assumed to be parallel to the Ox_3–axis and to be independent of x_3:

Fig. 9.1 Setup for a 2-D magnetic shaping problem

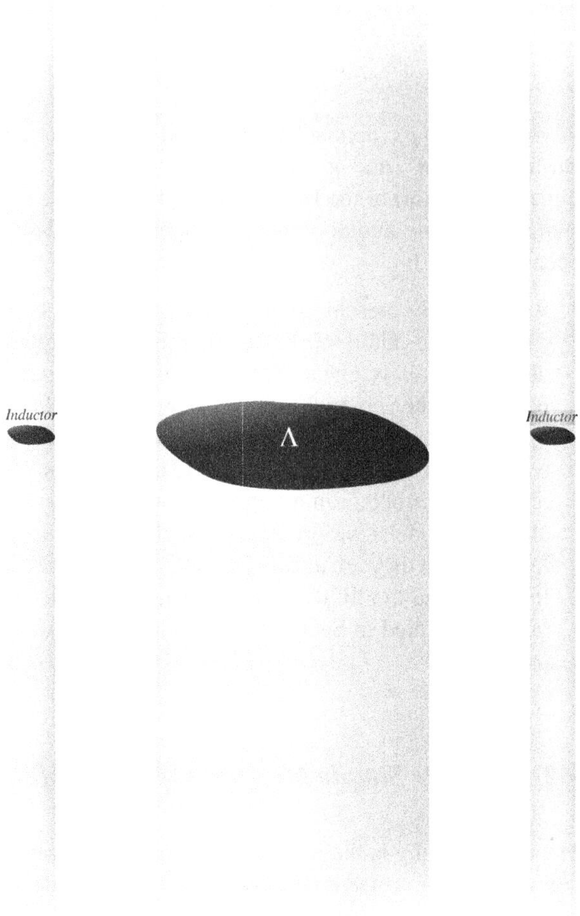

$$J(x_1, x_2, x_3) = J(x_1, x_2)\, e_3.$$

Figure 9.1 shows a typical section Λ of the conductor as well as the inductors.

Using the same technique as in Sect. 3.3, the magnetic induction has the form

$$B(x_1, x_2, x_3) = B_1(x_1, x_2)\, e_1 + B_2(x_1, x_2)\, e_2,$$

and we have the existence of a potential A such that (see (3.36))

$$-\Delta A = \mu_0 J \quad \text{in } \mathbb{R}^2.$$

Considering boundary conditions, the assumption of a perfectly electrically conducting boundary γ of Λ means that the magnetic induction $\boldsymbol{B}$ does not penetrate the conductor, i.e.

$$\boldsymbol{B} \cdot \boldsymbol{n} = 0 \qquad \text{on } \gamma.$$

This assumption implies first that the above partial differential equation can be reduced to the exterior domain $\Lambda_{\text{ext}} := \mathbb{R}^2 \setminus \overline{\Lambda}$, and second that the potential A is constant along γ. Since this one can be chosen up to an additive constant, we enforce

$$A = 0 \qquad \text{on } \gamma.$$

In addition, we have, according to (2.26)–(2.28) the condition at the infinity:

$$A(\boldsymbol{x}) = \beta + \mathcal{O}(|\boldsymbol{x}|^{-1}) \qquad |\boldsymbol{x}| \to \infty,$$

where $\beta \in \mathbb{R}$. These conditions are sufficient to solve the problem in the conductor if the domain Λ is known. Since this one is unknown, an additional condition is necessary. This is obtained by expressing a pressure balance equation. More precisely, we have for the Lorentz force the expression

$$\begin{aligned} \boldsymbol{f}_L &= \boldsymbol{J} \times \boldsymbol{B} \\ &= \frac{1}{\mu_0} \textbf{curl}\, \boldsymbol{B} \times \boldsymbol{B} \\ &= \frac{1}{2\mu_0} \nabla |\boldsymbol{B}|^2 - \frac{1}{\mu_0} (\boldsymbol{B} \cdot \nabla) \boldsymbol{B}. \end{aligned}$$

We note that no time averaging is involved here like in (9.1) since we deal with real valued fields, the current density being given.

It is classical in magnetohydrodynamics (see [165] for instance) to assume that for large frequencies, the fluid flow is mainly due to the gradient part of the Lorentz force, and therefore the pressure on the free boundary can be reasonably approximated by the term $\frac{1}{2\mu_0}|\boldsymbol{B}|^2$. Now, since the pressure of the ambient air is constant, we have the balance equation

$$\frac{1}{2\mu_0} |\boldsymbol{B}|^2 = \frac{1}{2\mu_0} |\nabla A|^2 = \text{Const.} \qquad \text{on } \gamma. \tag{9.23}$$

We can now summarize the obtained free boundary problem as follows: We seek a function $A : \Lambda_{\text{ext}} \to \mathbb{R}$ and a closed curve γ which is the boundary of Λ such that:

$$-\Delta A = \mu_0 J \qquad \text{in } \Lambda_{\text{ext}}, \tag{9.24}$$

$$A = 0 \qquad \text{on } \gamma, \tag{9.25}$$

$$\frac{1}{2\mu_0}|\nabla A|^2 = \lambda \qquad \text{on } \gamma, \tag{9.26}$$

$$A(\boldsymbol{x}) = \beta + \mathcal{O}(|\boldsymbol{x}|^{-1}) \qquad |\boldsymbol{x}| \to \infty, \tag{9.27}$$

where the positive constant λ is given. We note that since $A = 0$ on γ then its tangential derivative is also null and then (9.26) can be replaced by the Neumann boundary condition:

$$\frac{\partial A}{\partial n} = \sqrt{2\mu_0\lambda} \qquad \text{on } \gamma. \tag{9.28}$$

Remark 9.2.1. Problem (9.24)–(9.27) is known as the *Bernoulli* problem. For the mathematical analysis (existence, uniqueness and non-uniqueness, stability, qualitative properties), we refer to [63, 64, 76, 98] for instance and the references therein. For numerical aspects of the problem and the design of efficient solvers, we refer to [37, 114].

Remark 9.2.2. An interesting variant of (9.24)–(9.26) consists in taking into account surface tension. For this, (9.23) is replaced by the Laplace–Young equation

$$\frac{1}{2\mu_0}|\nabla A|^2 + \sigma\kappa = \lambda \qquad \text{on } \gamma,$$

where κ is the curvature of γ and $\sigma > 0$ is the surface tension of the liquid metal (see [117] and [91]). In this case, additional conditions called *wetting angle conditions* have to be added. In fact, any parameterization of the boundary γ involves for the expression of the curvature a second order derivative (with respect to curvilinear coordinates). The above equation appears then as a second order ordinary differential equation that requires for its solution, boundary conditions for the curvilinear abscissa.

Remark 9.2.3. Condition (9.26) is valid if the problem is formulated in the Ox_1x_2–plane. Otherwise, a gravity potential term must be added to the left-hand side of the equation, i.e.

$$\frac{1}{2\mu_0}|\nabla A|^2 + \varrho g x_3 = \lambda \qquad \text{on } \gamma,$$

where g is the gravity acceleration.

Without further precision, we mention here existence results in Henrot–Pierre [98] and studies on stability of the obtained shapes in Descloux [63, 64].

9.2.2 A Level Set Formulation of the Magnetic Shaping Model

Let us briefly present a formulation of (9.24)–(9.26) that is more adapted to its numerical solution.

Traditional methods for the numerical solution of (9.24)–(9.26) consist in the so-called *Front Tracking Methods*, i.e. the unknown boundary γ is considered as a perturbation of a reference boundary γ_0 following its outward unit normal. We then have to look for a closed curve

$$\gamma_t = \gamma_0 + t\boldsymbol{n} := \{\boldsymbol{x} + t\boldsymbol{n}(\boldsymbol{x});\ \boldsymbol{x} \in \gamma_0\}, \qquad t > 0, \tag{9.29}$$

where $\boldsymbol{n}$ is the outward unit normal to γ_0. Once γ_0 is parameterized, Eq. (9.29) is expressed as a first-order ordinary differential equation since the normal vector involves the derivative of the parameterization function. The variable t drives the iteration process. This approach yields a simple formulation that we do not intend to detail here, but it requires sufficient regularity of the boundary and furthermore the computation of the normal vector $\boldsymbol{n}$ which needs much care in the discrete case.

An alternative approach, referred to as *level set formulation*, has been successfully used in [37, 114] with accurate results even for non regular situations.

Like for the Front Tracking method, we deal here with a parameter that drives the iterations when numerical solution is involved. This parameter plays the role of a pseudo-time. Let us consider various boundaries $\gamma(t)$ obtained for various "time" values $t > 0$ with given "initial" boundary γ_0 and let us define $\gamma(t)$ as the 0-level set of a function ϕ, i.e.

$$\gamma(t) = \{\boldsymbol{x} \in \mathbb{R}^2;\ \phi(\boldsymbol{x}, t) = 0\}, \qquad t > 0.$$

We have for ϕ (see [164]) the level set equation

$$\frac{\partial \phi}{\partial t} + F|\nabla \phi| = 0 \qquad \text{in } \mathbb{R}^2, \quad t > 0, \tag{9.30}$$

where the initial condition is computed from the reference boundary γ_0 and F is the "propagation speed" of the front $\gamma(t)$. In general the level set equation is defined on the boundary and needs then to be extended to a fixed domain that contains

$$\bigcup_{t>0} \Lambda(t).$$

The propagation speed function can be defined in an adequate way. For instance, in [37, 114] Problem (9.24), (9.25), and (9.28) is solved by defining the following iterative procedure: Given a boundary γ^k of the exterior domain Λ^k_{ext} at the k-iteration, we solve the Neumann problem:

$$
\begin{cases}
-\Delta A_k = \mu_0 J & \text{in } \Lambda^k_{\text{ext}}, \\
\dfrac{\partial A_k}{\partial n_k} = \sqrt{2\mu_0\lambda} & \text{on } \gamma^k, \\
A_k(\boldsymbol{x}) = \beta + \mathcal{O}(|\boldsymbol{x}|^{-1}) & |\boldsymbol{x}| \to \infty.
\end{cases}
\tag{9.31}
$$

Problem (9.31) has a unique solution, up to an additive constant, given by the representation (see the proof of Theorem 1.4.5):

$$
\begin{aligned}
\frac{1}{2}A_k(\boldsymbol{x}) - \int_{\gamma^k} \frac{\partial}{\partial \boldsymbol{n}_y} G(\boldsymbol{x},\boldsymbol{y})A_k(\boldsymbol{y})\,ds(\boldsymbol{y}) &= \mu_0 \int_{\Lambda^k_{\text{ext}}} J(\boldsymbol{y})G(\boldsymbol{x},\boldsymbol{y})\,d\boldsymbol{y} \\
&\quad - \sqrt{2\mu_0\lambda}\int_{\gamma^k} G(\boldsymbol{x},\boldsymbol{y})\,ds(\boldsymbol{y}) + \beta \qquad \boldsymbol{x} \in \gamma^k,
\end{aligned}
\tag{9.32}
$$

where $\beta \in \mathbb{R}$, and G is the Green function given by (1.50). Note that the integral on Λ^k_{ext} reduces actually to the support of J which is a bounded set.

The propagation speed function at current iteration is then defined by $F_k = A_k$. To extend the level set function, the function F_k is extended by a Fast Marching technique (see [164] for instance). The updated function ϕ is defined by using a Forward Euler scheme to the Eq. (9.30), i.e.

$$
\phi_{k+1} = \phi_k - \tau F_k |\nabla\phi_k|,
$$

where $\tau > 0$ is a pseudo-time step that has to be chosen carefully (see [37]). The new boundary is naturally defined by

$$
\gamma_{k+1} := \{\boldsymbol{x};\ \phi_{k+1}(\boldsymbol{x}) = 0\}.
$$

Implementation details of this method can be found in [37] or [114] for a fast solver version.

Remark 9.2.4. One of the main features of level set formulations is that the geometric properties can be intrinsically deduced from the level set function ϕ. The unit normal to γ is defined by

$$
\boldsymbol{n}(\boldsymbol{x}) = \frac{\nabla\phi(\boldsymbol{x})}{|\nabla\phi(\boldsymbol{x})|}.
$$

Moreover, if the surface tension formulation is used, the curvature can be obtained from the formula:

$$
\kappa(\boldsymbol{x}) = \nabla \cdot \Big(\frac{\nabla\phi(\boldsymbol{x})}{|\nabla\phi(\boldsymbol{x})|}\Big).
$$

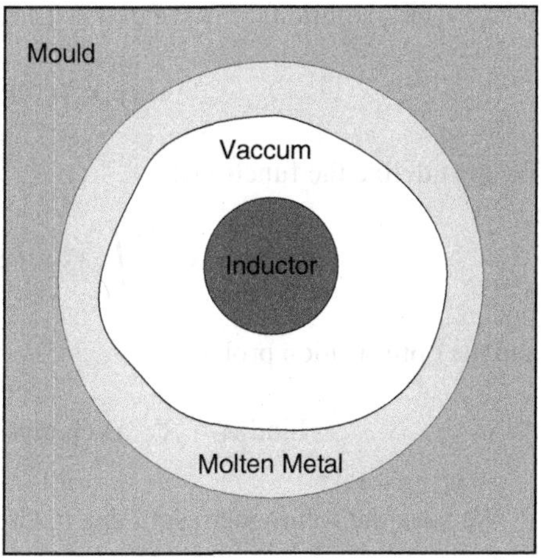

Fig. 9.2 Setup for a 2-D magnetic shaping problem

9.2.3 A Variant of the 2-D Model

We consider here a different configuration from the previous one, where the molten metal is confined in a mould. Figure 9.2 gives a schematic representation where we have a solid mould and the molten metal occupies a domain where a part of it adheres to the mould and another part is assumed to be in contact with a free space that separates it from the inductor.

We denote by Λ the domain consisting of the inductor and the vacuum between this one and the molten metal domain, and by $\Lambda_{\text{ext}} = \mathbb{R}^2 \setminus \overline{\Lambda}$. We also denote by γ the (unknown) boundary of Λ. Clearly the molten metal domain has a known part of its boundary that it shares with the mould and an unknown one that shares with Λ.

We use here the same hypotheses than for the first variant and add the assumption that the current density has its support contained in the inductor. Furthermore, we assume, as it may seem physically reasonable that the constant λ is not known but the area of Λ is known. This is equivalent to say that the area of the molten metal domain is known. We obtain by the same approach as before the following system of equations:

$$-\Delta A = \mu_0 J \qquad \text{in } \mathbb{R}^2, \tag{9.33}$$

$$A = 0 \qquad \text{on } \gamma, \tag{9.34}$$

$$|\Lambda| = m, \tag{9.35}$$

where $m > 0$ is the prescribed measure (area) of Λ.

Let us now give a mathematical setting for this problem and quote an existence result. We follow for this the formulation proposed by Crouzeix [60] which is based on the definition of (9.33)–(9.35) as an optimization problem.

We seek a solution (A, γ) of (9.33)–(9.35) such that the potential A lies in the set:

$$\mathcal{X} := \{v \in \mathcal{H}^1(\mathbb{R}^2);\ |\Lambda| = m\},$$

We next define the functional

$$W(v) := \frac{1}{2}\int_{\mathbb{R}^2} |\nabla v|^2\,d\boldsymbol{x} - \mu_0 \int_{\mathbb{R}^2} J\,v\,d\boldsymbol{x},$$

and the optimization problem:

$$\text{Find } A \in \mathcal{X} \quad \text{such that} \quad W(A) = \inf_{v\in\mathcal{X}} W(v), \tag{9.36}$$

We have the following result, due to Crouzeix [60]:

Theorem 9.2.1. *Assume that the given current density J is a bounded function with a support contained in the inductor. Then, Problem* (9.36) *admits at least one solution.*

Numerical solution of (9.36) can be naturally tackled by optimization techniques.

9.2.4 A 3–D Magnetic Shaping Problem

Let us apply the same approach for the three-dimensional case. For more details, the reader is referred to Pierre and Roche [144]. We consider the same setting as in Sect. 9.2.1. In other words, Ω will stand for a conductor made of an electrically conducting liquid metal with boundary Γ, and Ω_{ext} is the exterior domain. Adopting again the hypothesis of nonpenetrating magnetic field, we obtain the set of equations:

$$\textbf{curl}\,\boldsymbol{B} = \mu_0 \boldsymbol{J} \qquad \text{in } \Omega_{\text{ext}}, \tag{9.37}$$

$$\operatorname{div}\boldsymbol{B} = 0 \qquad \text{in } \Omega_{\text{ext}}, \tag{9.38}$$

$$\boldsymbol{B}\cdot\boldsymbol{n} = 0 \qquad \text{on } \Gamma, \tag{9.39}$$

$$\frac{1}{2\mu_0}|\boldsymbol{B}|^2 + \tau\kappa + \varrho g x_3 = \lambda \qquad \text{on } \Gamma. \tag{9.40}$$

where again, the magnetic permeability is assumed constant and equal to μ_0. Note that (9.40) involves the potential due to the gravity since this one is a part of the external pressure. A usual difficulty in the three-dimensional modelling is that the potential is vector-valued unlike the two-dimensional case where this one is scalar valued. We shall adopt hereafter an approach similar to the one described in Sect. 2.3.2 that consists in splitting the potential into a vector part that can be explicitly expressed and the gradient of a scalar field.

Let us define the vector field

$$\boldsymbol{B}_0(\boldsymbol{x}) := \frac{\mu_0}{4\pi} \int_{\mathbb{R}^3} \frac{\boldsymbol{J}(\boldsymbol{y}) \times (\boldsymbol{x} - \boldsymbol{y})}{|\boldsymbol{x} - \boldsymbol{y}|^3} \, ds(\boldsymbol{y}) \qquad \boldsymbol{x} \in \mathbb{R}^3.$$

According to the calculations made in Sect. 2.3.2, we have:

$$\mathbf{curl}\, \boldsymbol{B}_0 = \mu_0 \boldsymbol{J}, \tag{9.41}$$

$$\operatorname{div} \boldsymbol{B}_0 = 0. \tag{9.42}$$

From (9.37) and (9.41), we deduce that $\mathbf{curl}\,(\boldsymbol{B} - \boldsymbol{B}_0) = 0$, and then owing to Theorem 1.3.4 there exists a scalar field φ such that

$$\boldsymbol{B} = \boldsymbol{B}_0 + \nabla\varphi \qquad \text{in } \Omega. \tag{9.43}$$

Moreover, (9.38) and (9.42) imply

$$\Delta\varphi = 0 \qquad \text{in } \Omega. \tag{9.44}$$

A boundary condition for the potential φ is obtained from (9.43) and (9.39):

$$\frac{\partial\varphi}{\partial n} = -\boldsymbol{B}_0 \cdot \boldsymbol{n} \qquad \text{on } \Gamma. \tag{9.45}$$

Note that using Theorem 1.3.7, we have for the solution φ the expression

$$\varphi(\boldsymbol{x}) = \frac{1}{4\pi} \int_\Gamma \frac{p(\boldsymbol{y})}{|\boldsymbol{x} - \boldsymbol{y}|} \, ds(\boldsymbol{y}) \qquad \forall\, \boldsymbol{x} \in \Gamma,$$

where p is the unique solution in $\mathcal{H}^{\frac{1}{2}}(\Gamma)$ of the integral equation:

$$\frac{1}{2} p(\boldsymbol{x}) + \frac{1}{4\pi} \int_\Gamma p(\boldsymbol{y}) \frac{\boldsymbol{n}(\boldsymbol{x}) \cdot (\boldsymbol{x} - \boldsymbol{y})}{|\boldsymbol{x} - \boldsymbol{y}|^3} \, ds(\boldsymbol{y}) = \boldsymbol{B}_0(\boldsymbol{x}) \cdot \boldsymbol{n}_0(\boldsymbol{x}) \qquad \forall\, \boldsymbol{x} \in \Gamma.$$

Completing the system of equations with a prescribed volume constraint, the magnetic shaping problem can then be summarized as follows:

Given $m > 0$, $\boldsymbol{J} : \mathbb{R}^3 \to \mathbb{R}^3$,

Find $\boldsymbol{B} : \Omega \to \mathbb{R}^3$, $\Omega \subset \mathbb{R}^3$, $\lambda \in \mathbb{R}$ such that:

$$\boldsymbol{B} = \boldsymbol{B}_0 + \nabla\varphi \qquad \text{in } \Omega, \tag{9.46}$$

$$B_0(x) := \frac{\mu_0}{4\pi} \int_{\mathbb{R}^3} \frac{J(y) \times (x - y)}{|x - y|^3}\, dy \qquad \forall\, x \in \Omega, \tag{9.47}$$

$$\varphi(x) = \frac{1}{4\pi} \int_\Gamma \frac{p(y)}{|x - y|}\, ds(y) \qquad \forall\, x \in \Omega, \tag{9.48}$$

$$\frac{1}{2} p(x) + \frac{1}{4\pi} \int_\Gamma p(y) \frac{n(x) \cdot (x - y)}{|x - y|^3}\, ds(y) = B_0(x) \cdot n(x) \qquad \forall\, x \in \Gamma, \tag{9.49}$$

$$\frac{1}{2\mu_0} |B|^2 + \varrho g x_3 = \lambda \qquad \text{on } \Gamma. \tag{9.50}$$

$$|\Omega| = m, \tag{9.51}$$

where $m > 0$ is the prescribed volume of Ω. More clearly, if the domain Ω (and then its boundary Γ) is given, then the solution of (9.46)–(9.49) is obtained by performing the following steps:

(i) The magnetic induction B_0 in Ω is computed by (9.47).
(ii) We compute p by solving the integral equation (9.49) on Γ.
(iii) We compute the scalar potential φ on Ω by (9.48).
(iv) Finally the magnetic induction B is updated by (9.46).

9.3 An Electromagnetic Casting Problem

Let us consider (9.14)–(9.20) and assume that a part γ_F of γ is unknown. This situation is encountered in various applications like in Electromagnetic Casting (see [26] for instance). Figure 9.3 presents a schematic representation of an electromagnetic caster.

Electromagnetic casting (also called EMC) consists in solidifying liquid metals by making use of electromagnetic field. This field is generated by an alternating current which flows in the inductor. The Lorentz force maintains the melt flow in levitation and consequently avoids using sand moulds. The main advantage of such technology—in contrast with classical casting technologies—is the presence of stirring in the melt, which results in a reduction of the grain size in the solidified workpiece. More detailed descriptions of the process can be found for instance in [21, 161].

From a mathematical viewpoint simulation of an EMC process requires taking into account at least electromagnetic and hydrodynamic phenomena. We omit here to study the solidification process, this one being considered as independent of the others. The model we need here was already presented in Sect. 9.1.2 in the case where the ingot is assumed infinite in the Ox_3 direction, the only remaining

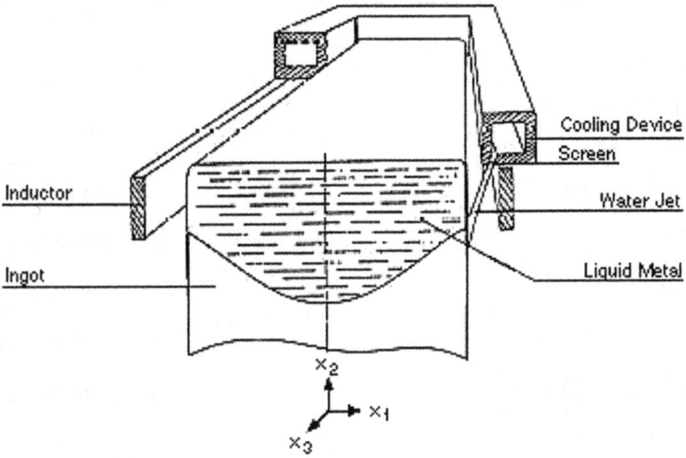

Fig. 9.3 An electromagnetic casting setup

difficulty is that the domain occupied by the fluid is a priori not known. The equation (9.22) has then to be added to describe the whole process.

Let us now translate (9.22) in terms of the unknowns involved in (9.14)–(9.20). The fluid pressure is defined by the normal traction on the free boundary γ_F and the external pressure is constant by assuming that the ambient air is static. This implies

$$\tau\kappa = p - 2\,\eta\,\boldsymbol{n}^{\mathsf{T}}\mathbf{D}(\boldsymbol{v})\,\boldsymbol{n} - p_A \qquad \text{on } \gamma_F, \tag{9.52}$$

where the tensor $\mathbf{D}(\boldsymbol{v})$ is defined by (9.4) and $\boldsymbol{n}$ is the outward unit normal to γ_F. Note that the value of the pressure p_A is unknown and, due to this, an additional condition must be supplied. For this, the value of the "height" of the liquid zone is imposed in [26]. Another possible choice is to prescribe the area of this zone.

At last, since we are dealing with a free boundary problem, the boundary condition on γ_F for $\boldsymbol{v}$ and p must be relaxed and replaced by a slip boundary condition. To summarize, the problem consists in looking for the potential $A : \mathbb{R}^2 \to \mathbb{C}$, the velocity $\boldsymbol{v} : \Lambda \to \mathbb{R}^2$, the pressure $p : \Lambda \to \mathbb{R}$, the free boundary γ_F and a real number p_A such that:

$$-\operatorname{div}(\mu^{-1}\nabla A) + \mathrm{i}\omega\sigma(A - M_k(A)) + \sigma\boldsymbol{v}\cdot\nabla A = \frac{I_k}{|\Lambda_k|} \qquad \text{in } \bigcup_{k=1}^{3}\Lambda_k, \tag{9.53}$$

$$\Delta A = 0 \qquad \text{in } \Lambda_{\text{ext}}, \tag{9.54}$$

$$-2\eta\,\mathbf{div}\,\mathbf{D}(\boldsymbol{v}) + \varrho\,(\boldsymbol{v}\cdot\nabla)\boldsymbol{v} + \nabla p - \frac{\sigma\omega}{2}\,\mathrm{Im}\,(A\,\nabla\overline{A}) + \frac{\sigma}{2}\,\mathrm{Re}\,((\boldsymbol{v}\cdot\nabla A)\,\nabla\overline{A}) = 0 \qquad \text{in } \Lambda_1, \tag{9.55}$$

$$\operatorname{div} \boldsymbol{v} = 0 \qquad \text{in } \Lambda_1, \tag{9.56}$$

$$M_1(A) = 0, \tag{9.57}$$

$$[A] = \left[\mu^{-1} \frac{\partial A}{\partial n}\right] = 0 \qquad \text{on } \gamma, \tag{9.58}$$

$$A(\boldsymbol{x}) = \beta + \mathcal{O}(|\boldsymbol{x}|^{-1}) \qquad |\boldsymbol{x}| \to \infty, \tag{9.59}$$

$$\boldsymbol{v} \cdot \boldsymbol{n} = 0 \qquad \text{on } \gamma_F, \tag{9.60}$$

$$\boldsymbol{n}^{\mathsf{T}} \mathbf{D}(\boldsymbol{v})\, \boldsymbol{t} = 0 \qquad \text{on } \gamma_F, \tag{9.61}$$

$$\boldsymbol{v} = \mathbf{0} \qquad \text{on } \gamma_1 \setminus \gamma_F, \tag{9.62}$$

$$\tau\kappa + p - 2\,\eta\, \boldsymbol{n}^{\mathsf{T}} \mathbf{D}(\boldsymbol{v})\, \boldsymbol{n} = p_A \qquad \text{on } \gamma_F, \tag{9.63}$$

$$|\Lambda_1| = m, \tag{9.64}$$

where $m > 0$ is given, $\boldsymbol{t}$ is the unit tangent vector to γ_F such that the pair $(\boldsymbol{t}, \boldsymbol{n})$ is positively oriented. Here again, if the boundary γ_F is known, then (9.53)–(9.62) is well defined. This remark suggests an iterative procedure for the numerical solution of (9.53)–(9.64).

To end, let us refer for a practical implementation of this model and its application in the aluminium industry to Besson et al. [26].

Remark 9.3.1. The interest of (9.53)–(9.64) goes beyond electromagnetic casting and defines a general setting for free boundary magnetohydrodynamic flows.

Chapter 10
Inductively Coupled Plasma Torches

We investigate in this chapter another type of application using eddy currents. An Inductively Coupled Plasma torch (commonly referred to as ICP) is a technical device used to analyze a given sample (gas, solid or liquid prepared as an aerosol) by injecting it in a plasma (generally made of argon) (see Fig. 10.1). The sample atoms are thus ionized thanks to the high temperature of the plasma. In such devices, energy is supplied by Joule heating to maintain a plasma source to a given temperature. This source is useful to dissolve, vaporize and ionize gas and a sample to analyze. An ICP generally includes a sample introduction system (generally a nebulizer), an ICP torch, a radio frequency generator and a spectrometer. More detailed description and applications of ICP torches can be found for instance in [1, 25, 133].

From a modelling point of view we are in presence of a compressible fluid flow where the energy is maintained by Joule heating. In addition, a closer look to an ICP experiment shows that the process can be considered as stationary.

10.1 The Model

Mathematical modelling of ICP devices requires the coupling of induction heating with the plasma and gas flow. A specific difficulty arises in this application in the fact that the electrically conducting region of the computational domain is not a priori known. This region consists in the inductors, any metallic component in the ICP device and the part of the gas which is transformed by heating into plasma. This part is characterized by its (unknown) temperature (or enthalpy or internal energy) when it exceeds a known critical value which depends on the chosen gas.

Let Ω stand for the open set in $\mathbb{R}^3$ made of the union of conducting parts with boundary Γ and let Ω_{ext} stand, as usual, for the nonconducting parts, including the free space ($\Omega_{\text{ext}} := \mathbb{R}^3 \setminus \overline{\Omega}$). As explained before, one should refer to these domains as $\Omega(e)$ and $\Omega_{\text{ext}}(e)$ respectively, where e is the internal energy. We shall

R. Touzani and J. Rappaz, *Mathematical Models for Eddy Currents and Magnetostatics: With Selected Applications*, Scientific Computation, DOI 10.1007/978-94-007-0202-8_10,

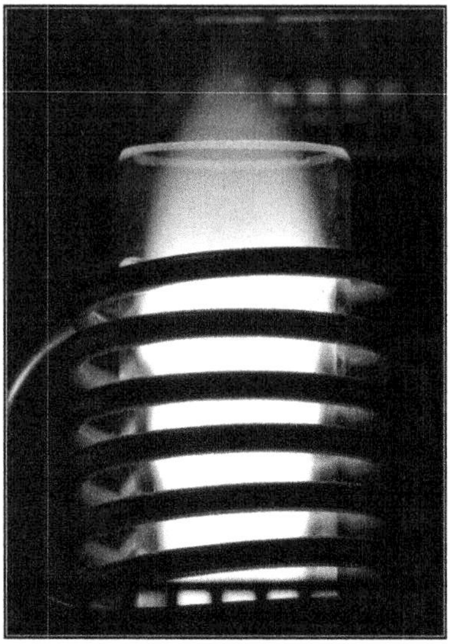

Fig. 10.1 An ICP device (Courtesy of the *Laboratoire de l'arc électrique et plasmas thermiques, Université Blaise Pascal, France*)

however avoid this notation for conciseness. In the sequel, we shall need to mention the domain that contains the fluid (gas), denoted by Ω_F, with its boundary Γ_F. This domain is given once for all since the gas is confined in a quartz container. Figure 10.2 illustrates the different parts of the conducting and nonconducting domains in the ICP setup. Let us now write down the model by distinguishing electromagnetic and hydrodynamic phenomena.

10.1.1 Eddy Currents

The main issue in choosing an eddy current model is the requirement that such a model should degenerate naturally from the equations in the conductors to those of the vacuum. More clearly, the choice of $\boldsymbol{H}$–model in Sect. 4.2 does not fulfill this condition since the term involving σ^{-1} becomes singular in the vacuum. The $\boldsymbol{E}$–model presented in Sect. 4.3 seems to be the most adapted to this situation. We note that models based on the vector potential $\boldsymbol{A}$ are also used for this purpose in some commercial codes.

Let us start by recalling a typical conductor setup where Ω_k is collection of conductors, which are connected and bounded domaines in $\mathbb{R}^3$, with respective boundaries Γ_k and defining

$$\Omega := \bigcup_i \Omega_i, \quad \Omega_{\text{ext}} := \mathbb{R}^3 \setminus \overline{\Omega}.$$

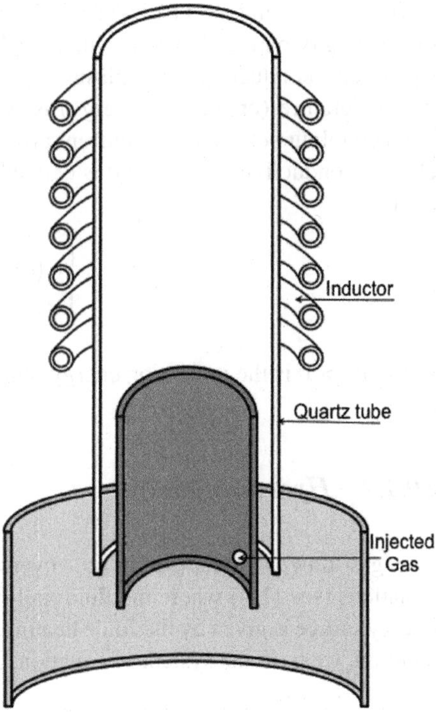

Fig. 10.2 Typical geometry of an ICP setup

We have the set of equations:

$$\mathbf{curl}\,(\mu^{-1}\,\mathbf{curl}\,\boldsymbol{E}) + \mathrm{i}\omega\sigma\boldsymbol{E} = -\mathrm{i}\omega\boldsymbol{J}_0 \qquad \text{in } \mathbb{R}^3, \tag{10.1}$$

$$\operatorname{div}\boldsymbol{E} = 0 \qquad \text{in } \Omega_{\text{ext}}, \tag{10.2}$$

$$\int_{\Gamma_k} \boldsymbol{E}\cdot\boldsymbol{n}\,ds = 0 \qquad k = 0, 1, \ldots \tag{10.3}$$

$$|\boldsymbol{E}(\boldsymbol{x})| = \mathcal{O}(|\boldsymbol{x}|^{-1}) \qquad |\boldsymbol{x}| \to \infty, \tag{10.4}$$

where $\boldsymbol{J}_0$ is a source current. Note that we have neglected convection contribution to the Ohm's law (2.10), where $\boldsymbol{v}$ is the fluid velocity. This hypothesis is generally made in ICP applications. Without this assumption (10.1) should be replaced by

$$\mathbf{curl}\,(\mu^{-1}\,\mathbf{curl}\,\boldsymbol{E}) - \sigma\boldsymbol{v}\times\mathbf{curl}\,\boldsymbol{E} + \mathrm{i}\omega\sigma\boldsymbol{E} = -\mathrm{i}\omega\boldsymbol{J}_0 \qquad \text{in } \mathbb{R}^3, \tag{10.5}$$

where the velocity field $\boldsymbol{v}$ is extended by $\mathbf{0}$ outside the fluid domain.

A crucial point in (10.1)–(10.4) is that the electric conductivity is null in the pure gas part and positive in the plasma, this one being ionized and then electrically conducting. This plasma region can be characterized by the fact that its temperature (or internal energy) is beyond the ionization energy. This is the main coupling between eddy current problem and the fluid flow. Consequently, the electric conductivity in the fluid is defined here as a function of the internal energy e with

$$\sigma = \begin{cases} \sigma(e) & \text{if } e \geq e_I, \\ 0 & \text{if } e < e_I, \end{cases} \tag{10.6}$$

where $e_I > 0$ is the ionization energy. This value is generally typical of gases.

10.1.2 Hydrodynamics

The gas flow is assumed to be governed by the compressible Navier-Stokes equations (see [16]) where the fluid motion is driven by the Lorentz force and the energy source is given by the Joule heating power density. Looking for a steady state solution, we have the system of equations:

$$-2\eta\big(\mathbf{div}\,\mathsf{D}(\boldsymbol{v}) - \frac{1}{3}\nabla \operatorname{div} \boldsymbol{v}\big) + \mathbf{div}(\varrho\boldsymbol{v} \otimes \boldsymbol{v}) + \nabla p = \boldsymbol{f}_L, \tag{10.7}$$

$$\operatorname{div}(\varrho\boldsymbol{v}) = 0, \tag{10.8}$$

$$-\operatorname{div}\big(k\nabla\theta\big) + \operatorname{div}((\varrho E + p)\boldsymbol{v}) = \frac{\sigma}{2}\boldsymbol{E}\cdot\overline{\boldsymbol{E}} - R(e), \tag{10.9}$$

$$E = e + \frac{1}{2}|\boldsymbol{v}|^2, \tag{10.10}$$

$$\theta = \Theta(\varrho, e), \tag{10.11}$$

$$p = P(\varrho, e), \tag{10.12}$$

in Ω_F, i.e. the domain occupied by the fluid. Here θ is the temperature, E and e are respectively the total and internal energies, p is the pressure, ϱ is the fluid density. The coefficients η and k stand respectively for the molecular viscosity and the thermal conductivity. For a realistic model, η and k depend strongly on the temperature. External sources are the Lorentz force density $\boldsymbol{f}_L$, the Joule heating density S_J and the radiation dissipation density $R(e)$ assumed to be dependent on the internal energy. Equations (10.11) and (10.12) are state equations of the used gas. Finally D is the symmetric deformation rate tensor defined by

$$\mathsf{D}(\boldsymbol{v}) := \frac{1}{2}\,(\nabla\boldsymbol{v} + (\nabla\boldsymbol{v})^{\mathsf{T}}).$$

We note that in (10.7)–(10.12) the following approximations have been adopted:

1. In (10.9) dissipation terms (due to viscosity) expressed by

$$2\eta\,(\mathbf{D}(\boldsymbol{v}) : \mathbf{D}(\boldsymbol{v}) - \frac{1}{3}(\operatorname{div}\boldsymbol{v})^2) + \boldsymbol{v}\cdot\nabla p$$

   (see [16], p. 153) are neglected.
2. The electric current due to fluid motion is neglected. Otherwise, one should use (2.10) to obtain (10.5) for the electric field and more complex expressions for the Lorentz force and the Joule heating power density.

Let us express the coupling terms in (10.7)–(10.10) in function of $\boldsymbol{E}$:

1. The instantaneous Lorentz force is given by $\boldsymbol{J}\times\boldsymbol{B}$. Since we are seeking a stationary solution, it is judicious to follow (9.1) and approximate this term by its average value over on time period, i.e., we define

$$\boldsymbol{f}_L := \frac{1}{2}\,\mathrm{Re}\,(\boldsymbol{J}\times\overline{\boldsymbol{B}}) = \frac{\sigma}{2\,\omega}\,\mathrm{Im}(\boldsymbol{E}\times\mathbf{curl}\,\overline{\boldsymbol{E}}).$$

2. The instantaneous Joule power density is given by $\boldsymbol{J}\cdot\boldsymbol{E}$. Proceeding in the same way as for the Lorentz force, we define

$$S_J := \frac{1}{2}\,\mathrm{Re}(\boldsymbol{J}\cdot\overline{\boldsymbol{E}}) = \frac{\sigma}{2}\boldsymbol{E}\cdot\overline{\boldsymbol{E}}.$$

3. The radiation dissipation term should be expressed by a radiative transfer equation. We avoid this difficulty by using *net emission approximation* models (see [94]) that involve an explicit dependency of $R = R(e)$ on the internal energy, where R is a nondecreasing function of e.

10.1.3 The Complete Model

We can now write down the complete set of equations that govern the ICP torch process:

$$\mathbf{curl}\,(\mu^{-1}\,\mathbf{curl}\,\boldsymbol{E}) + \mathrm{i}\omega\sigma\boldsymbol{E} = -\mathrm{i}\omega\boldsymbol{J}_0 \qquad \text{in } \mathbb{R}^3, \qquad (10.13)$$

$$\operatorname{div}\boldsymbol{E} = 0 \qquad \text{in } \Omega_{\text{ext}}, \qquad (10.14)$$

$$\begin{aligned} &-2\eta\big(\mathbf{div}\,\mathbf{D}(\boldsymbol{v}) - \frac{1}{3}\nabla\operatorname{div}\boldsymbol{v}\big) + \operatorname{div}(\varrho\boldsymbol{v}\otimes\boldsymbol{v}) + \nabla p \\ &\qquad = \frac{\sigma}{2\omega}\,\mathrm{Im}(\boldsymbol{E}\times\mathbf{curl}\,\overline{\boldsymbol{E}}) \qquad \text{in } \Omega_F, \qquad (10.15)\end{aligned}$$

$$\operatorname{div}(\varrho \boldsymbol{v}) = 0 \qquad \text{in } \Omega_F, \tag{10.16}$$

$$-\operatorname{div}\big(k\nabla\theta\big) + \operatorname{div}((\varrho E + p)\,\boldsymbol{v}) = \frac{\sigma}{2}\boldsymbol{E}\cdot\overline{\boldsymbol{E}} - R(e) \qquad \text{in } \Omega_F, \tag{10.17}$$

$$E = e + \frac{1}{2}|\boldsymbol{v}|^2 \qquad \text{in } \Omega_F, \tag{10.18}$$

$$\theta = \Theta(\varrho, e) \qquad \text{in } \Omega_F, \tag{10.19}$$

$$p = P(\varrho, e) \qquad \text{in } \Omega_F. \tag{10.20}$$

To these equations are associated the following conditions:

$$|\boldsymbol{E}(\boldsymbol{x})| = \mathcal{O}(|\boldsymbol{x}|^{-1}) \qquad |\boldsymbol{x}| \to \infty, \tag{10.21}$$

$$\int_{\Gamma_k} \boldsymbol{E}\cdot\boldsymbol{n}\, ds = 0 \qquad k = 0, 1, \ldots \tag{10.22}$$

$$\boldsymbol{v} = \boldsymbol{0} \qquad \text{on } \Gamma_{FC}, \tag{10.23}$$

$$\boldsymbol{v} = \boldsymbol{v}_I \qquad \text{on } \Gamma_{FI}, \tag{10.24}$$

$$\big((p + \frac{2}{3}\eta \operatorname{div}\boldsymbol{v})\,\mathbf{I} - 2\eta\mathbf{D}(\boldsymbol{v})\big)\boldsymbol{n} = \boldsymbol{0} \qquad \text{on } \Gamma_{FO}, \tag{10.25}$$

$$p = p_{\text{atm}} \qquad \text{on } \Gamma_{FO}, \tag{10.26}$$

$$\theta = \theta_0 \qquad \text{on } \Gamma, \tag{10.27}$$

where $\mathbf{I}$ is the identity tensor. In (10.23)–(10.26) the boundary Γ_F is partitioned into an input part Γ_{FI} in which the gas is injected with velocity $\boldsymbol{v}_I$, an output part Γ_{FO} from which the gas escapes and then the pressure p_{atm} (pressure of the atmosphere) and null traction are imposed, and finally the remaining part on which a null velocity is prescribed. In (10.27), a Dirichlet boundary condition for the temperature is prescribed for simplicity. Of course, other types of boundary conditions can be imposed on any part of the boundary.

Problem (10.13)–(10.26) is a nonlinear system of stationary partial differential equations. For this system no mathematical analysis is yet available.

In the sequel, we shall define a numerical procedure to solve the problem (10.13)–(10.26) and then give some numerical results for an application of the ICP process.

10.2 Numerical Approximation

In order to tackle the numerical solution of (10.13)–(10.26), two main approaches can be made:

1. The first one consists in defining an iterative procedure to solve the resulting nonlinear system of equations, once space discretization is applied. Using such a procedure can result in a large size system of equations and, moreover the choice of an initial guess to start the iterations can be a serious difficulty. Due to the strong nonlinearities that appear in this system, one generally resort to the Newton's method for its solution. This approach is used in works like [24, 176].
2. The second approach consists in considering the time dependent problem and performing a time stepping scheme where convergence toward a stationary solution is sought. This choice would lead to a more reliable solution technique where, an explicit scheme can be used for the solution of fluid flow equations, viscosity and thermal conductivity coefficients (η and k) being small. The drawback is that this needs a large number of time steps to converge to a stationary solution. Moreover it is well known that the use of an explicit scheme results in a restrictive time stepping condition (CFL condition).

We have chosen to present the second approach for its simplicity and because our numerical experiments are based on it.

Let us then consider Eqs. (10.13)–(10.26) where the fluid flow equations are replaced by their time dependent version, i.e.

$$\frac{\partial}{\partial t}(\varrho \boldsymbol{v}) - 2\eta\left(\mathbf{div}\,\mathsf{D}(\boldsymbol{v}) - \frac{1}{3}\nabla\,\mathrm{div}\,\boldsymbol{v}\right) + \mathbf{div}(\varrho\boldsymbol{v}\otimes\boldsymbol{v}) + \nabla p = \frac{\sigma}{2\omega}\,\mathrm{Im}(\boldsymbol{E}\times\mathbf{curl}\,\overline{\boldsymbol{E}}), \tag{10.28}$$

$$\frac{\partial\varrho}{\partial t} + \mathrm{div}(\varrho\boldsymbol{v}) = 0, \tag{10.29}$$

$$\frac{\partial}{\partial t}(\varrho E) - \mathrm{div}\left(k\nabla\theta\right) + \mathrm{div}((\varrho E + p)\,\boldsymbol{v}) = \frac{\sigma}{2}\boldsymbol{E}\cdot\overline{\boldsymbol{E}} - R(e), \tag{10.30}$$

$$E = e + \frac{1}{2}|\boldsymbol{v}|^2, \tag{10.31}$$

$$\theta = \Theta(\varrho, e), \tag{10.32}$$

$$p = P(\varrho, e), \tag{10.33}$$

in Ω_F, for all times $t > 0$. The system (10.28)–(10.33) is to be supplemented with the conditions (10.23)–(10.26) and with adequate initial conditions on the unknowns ϱ, $\varrho\boldsymbol{v}$ and ϱE.

We do not intend to present in detail the numerical approximation of the above problem. It is clear indeed that numerical approximation of compressible

Navier-Stokes equations presents some specific aspects that would drive us out of the scope of this book. For this we refer the reader to [56, 57]. Let us present the time stepping procedure that enables decoupling the equations in this system. This procedure was tested with realistic data and has given satisfactory results as it is shown in the following section.

Let us consider a maximal time value T and define a time step $\delta t > 0$ and time values $t^n = n\,\delta t$ for $n = 0, 1, \ldots, N$.

To start the time stepping procedure we define the initial values of ϱ, $\varrho\boldsymbol{v}$, and ϱE denoted respectively by ϱ^0, $(\varrho\boldsymbol{v})^0$, and ϱE^0. In particular, we initialize with a no flow configuration $(\varrho\boldsymbol{v})^0 = \mathbf{0}$, $\varrho^0 = \text{Const}$. This yields a starting value for the electric field $\boldsymbol{E}^0$ by solving (10.13), (10.14), (10.21), (10.22). Note that it is necessary to initialize a plasma zone where electric current flows. It is noteworthy that this is also required for the experimental setup.

We can now define the following algorithm: Assuming that the approximate values of $\boldsymbol{E}$, $\varrho\boldsymbol{v}$, ϱ, ϱE are given at time $t = t^n$ and denoted respectively by

$$\boldsymbol{E}^n,\ (\varrho\boldsymbol{v})^n,\ \varrho^n,\ (\varrho E)^n,$$

the computation of the other unknowns at the same time value is defined as follows:

- The velocity

$$\boldsymbol{v}^n = \frac{(\varrho\boldsymbol{v})^n}{\varrho^n}. \tag{10.34}$$

- We update the electric conductivity by setting $\sigma^n = \sigma(e^n)$, where e^n is obtained from (10.31) by

$$e^n := \frac{(\varrho E)^n}{\varrho^n} - \frac{1}{2}|\boldsymbol{v}^n|^2. \tag{10.35}$$

- We update the pressure by the state equation:

$$p^n = P(\varrho^n, e^n). \tag{10.36}$$

- The temperature is obtained from the state equation:

$$\theta^n = \Theta(\varrho^n, e^n). \tag{10.37}$$

- We compute the external force and heat source and radiation terms by

$$\boldsymbol{f}_L^n = \frac{\sigma^n}{2\omega}\,\mathrm{Im}(\boldsymbol{E}^n \times \mathbf{curl}\,\overline{\boldsymbol{E}}^n),$$

$$S_J^n = \frac{\sigma^n}{2} \boldsymbol{E}^n \cdot \overline{\boldsymbol{E}}^n,$$
$$R^n = R(e^n).$$

The computation of the values at $t = t^{n+1}$ are performed through the following steps:

- We obtain the approximate values of $\varrho \boldsymbol{v}$, ϱ, ϱE, $\boldsymbol{v}$, e, p and θ by performing one step of the forward Euler scheme:

$$(\varrho \boldsymbol{v})^{n+1} = (\varrho \boldsymbol{v})^n + 2\delta t\, \eta \left(\mathbf{div}\, \mathsf{D}(\boldsymbol{v}^n) - \frac{1}{3} \nabla \operatorname{div} \boldsymbol{v}^n \right)$$
$$- \delta t\, \mathbf{div}((\varrho \boldsymbol{v})^n \otimes \boldsymbol{v}^n) - \delta t\, \nabla p^n + \delta t\, \boldsymbol{f}_L^n,$$
$$\varrho^{n+1} = \varrho^n - \delta t\, \operatorname{div}(\varrho \boldsymbol{v})^n,$$
$$\boldsymbol{v}^{n+1} = \frac{(\varrho \boldsymbol{v})^{n+1}}{\varrho^{n+1}},$$
$$(\varrho E)^{n+1} = (\varrho E)^n + \delta t\, \operatorname{div}(k \nabla \theta^n) - \delta t\, \operatorname{div}((\varrho^n E^n + p^n)\, \boldsymbol{v}^{n+1})$$
$$+ \delta t\, (S_J^n - R^n).$$

- The fields e^{n+1}, p^{n+1} and θ^{n+1} are then computed using (10.34)–(10.37).

10.3 A Numerical Simulation

Let us report here some numerical simulations made in collaboration with D. Rochette in the framework of his habilitation thesis [156] and with S. Clain. Due to the geometry, we assume rotational symmetry thus using cylindrical coordinates with axisymmetry. This assumes that the inductor which is a coil formed of a few turns is actually approximated by a series of circular toroidal conductors.

For this, we use for eddy currents the model described in Chap. 5 and for fluid flow an axisymmetric version of (10.28)–(10.29). Numerical approximation uses the finite element method for the electric field, which is then piecewise linear while a finite volume method is used to solve the fluid flow problem that provides piecewise constant velocity, density, pressure and energy fields.

An important issue in the simulation of ICP torches is the presence of different space scales of electromagnetics and hydrodynamics: While for eddy currents, the mesh is rather coarse with local refinement in the inductors and their vicinity in order to capture the skin effect, a finer mesh is needed in the plasma region where an accurate knowledge of the Lorentz force and the dissipated Joule power is required. Figure 10.3 shows both meshes for a configuration with seven toroidal inductors. Naturally, the use of these meshes requires a strategy for information transmission from one mesh to another. For this, we use an intermediate very fine mesh for

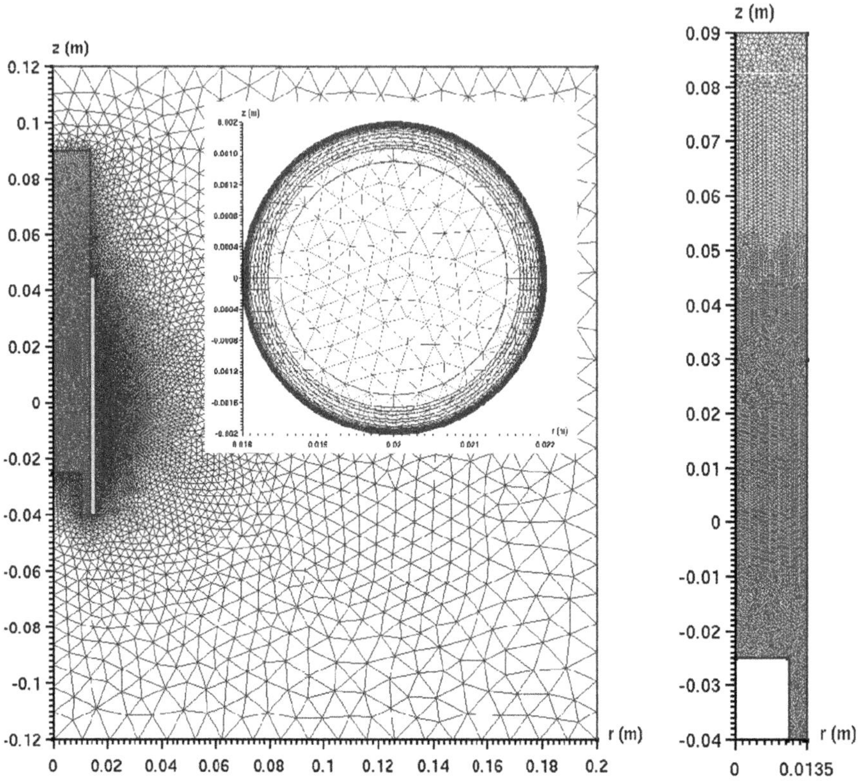

Fig. 10.3 Meshes for the ICP, *Left*: eddy currents (Zoom: mesh of an inductor), *Right*: hydrodynamics

which obtained fields are computed at nodes or element centers. These fields are then deduced for a mesh target simply by using piecewise linear approximation.

The simulation of the transient process of the ICP torch is made in two phases. In the first step, a constant rate of argon gas is injected at a part of the lower boundary of the setup. Initial conditions for hydrodynamics are given by the ambient temperature, atmospheric pressure and a zero-velocity, with no current alimentation. This phase takes 1 s. In a second phase a plasma zone with arbitrary size is created and a current voltage is prescribed. The argon plasma develops then in the torch and a stationary regime is obtained at a time of 400 ms.

We present hereafter in Fig. 10.4 the contour of the norm of the electric field (left) and the magnetic induction (right) around the inductor.

Figure 10.5 displays contours of the temperature and velocity vectors.

This numerical experiment shows that it is possible to obtain a stationary solution by using a time dependent model. Our hope is to model more properly the ignition phase and the starting of the torch. We mention that in other works (see [1, 25, 133] for instance), it is possible to obtain this solution from a stationary model where an iterative scheme is required to handle the nonlinearities of the model.

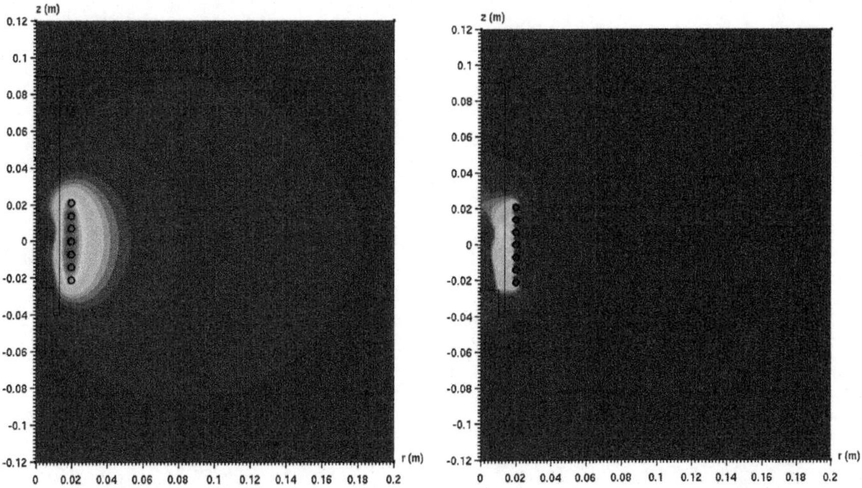

Fig. 10.4 Norm of the electric field (*left*) and the magnetic induction (*right*)

Fig. 10.5 Isothermal lines (*left*) and velocity field (*right*)

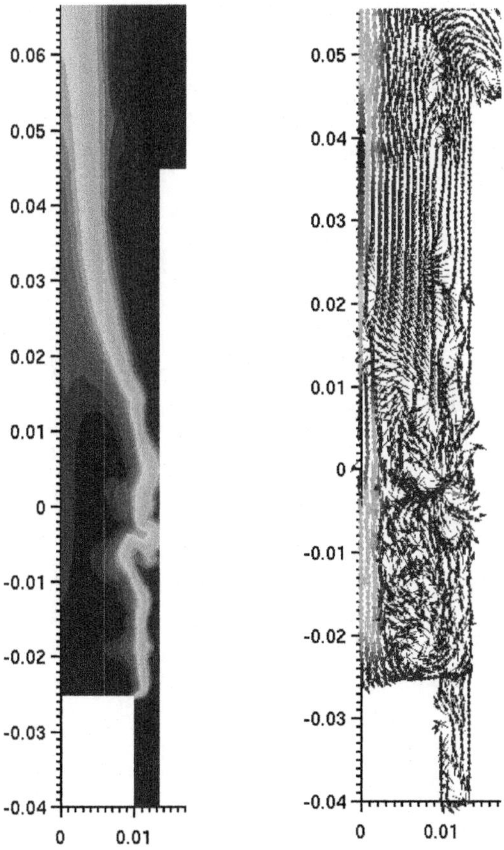

Chapter 11
Ferromagnetic Shielding

We present in this chapter an application of stationary electromagnetics in the case of ferromagnetic materials related to magnetic shielding.

It is well known that in a ferromagnetic conductor, the dependency of the magnetic induction $\boldsymbol{B}$ in function of the magnetic field $\boldsymbol{H}$ is nonlinear, and that $\boldsymbol{B}$ and $\boldsymbol{H}$ are often not collinear thus resulting in hysteresis phenomena. However, in most industrial applications, it is a good approximation to consider that $\boldsymbol{B}$ is collinear to $\boldsymbol{H}$, especially when the euclidean norm of $\boldsymbol{H}$ is large and in the stationary situation. In this case we have the relation

$$\boldsymbol{B} = \mu_0 \mu_r(|\boldsymbol{H}|)\boldsymbol{H},$$

where μ_r is the relative magnetic permeability which is a positive, decreasing and convex function of the euclidean norm of $\boldsymbol{H}$. The nonlinear mapping $\boldsymbol{B} \mapsto \boldsymbol{H}$ results in limiting the penetration of electromagnetic fields into a space, that is often referred to as *electromagnetic shielding* or *screening* of the ferromagnetic conductor.

A typical example takes place in the screen effect of the steel shell supporting a cell for aluminium production. Let us describe this case: In the aluminium industry, pure metal is produced in electrolytic cells by electrolytic processing of alumina by using very strong stationary electric currents (several hundreds of thousands of Amperes). The liquid aluminium produced at cell cathodes, and the electrolytic bath in which are placed the anodes, are submitted to intensive electromagnetic forces that produce motion of the liquids. To ensure strength of the device, lower and lateral cell faces are covered by plates in steel called "shell of the cell". The magnetic induction is shielded by these shells and has as effect to minimize Lorentz forces and consequently liquid motion in the cells. In order to compute the fluid flow and analyze the cell stability, it is important to predict this screen effect.

Let us consider a bounded domain Ω in the three-dimensional space occupied by a ferromagnetic material with relative magnetic permeability μ_r and let $\Omega_{\text{ext}} := \mathbb{R}^3 \setminus \overline{\Omega}$ stand for the exterior domain. We assume that a stationary electric current with density $\boldsymbol{J}$ flows in a conductor Λ surrounding Ω and creates a magnetic field $\boldsymbol{H}$

R. Touzani and J. Rappaz, *Mathematical Models for Eddy Currents and Magnetostatics: With Selected Applications*, Scientific Computation, DOI 10.1007/978-94-007-0202-8_11,

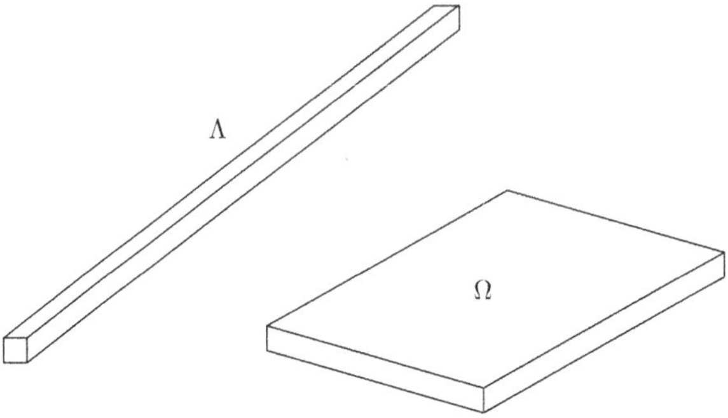

Fig. 11.1 A schematic ferromagnetic shielding setup

and a magnetic induction $\boldsymbol{B}$ (see Fig. 11.1). Following [49, 65] we start by deriving an equation that governs the electromagnetic induction in terms of the magnetic field $\boldsymbol{H}$. An existence and uniqueness result for the obtained model is then given following the approach in [67]. The numerical solution by an iterative procedure is then addressed: We first derive a fixed point algorithm to solve the nonlinear problem by a sequence of linear problems to solve at each iteration. The analysis of the method is handled for the continuous problem and a domain decomposition technique is presented and analyzed. This technique avoids solving the problem outside a bounded region. Finally we present a finite element method to solve the problem. We analyze convergence of the iterative procedure for the discrete problem and then present some numerical experiments performed on a realistic setup. Let us mention that this chapter is published in the paper [67].

A large number of publications is devoted to the mathematical modelling and numerical solution techniques for absorbing ferromagnetic materials with Landau-Lifschitz law for propagation and scattering of electromagnetic waves. Let us mention [109, 132] and the references therein. Other papers concern magnetostatics in thin plates [65, 95] for which the thickness of the plate tends to zero when μ_r tends to the infinity.

We have the magnetostatic equations (see Sect. 2.4.2):

$$\mathbf{curl}\,\boldsymbol{H} = \boldsymbol{J} \qquad \text{in } \mathbb{R}^3, \tag{11.1}$$

$$\operatorname{div}\boldsymbol{B} = 0 \qquad \text{in } \mathbb{R}^3, \tag{11.2}$$

$$\boldsymbol{B} = \mu_0\mu_r\boldsymbol{H} \qquad \text{in } \mathbb{R}^3, \tag{11.3}$$

$$|\boldsymbol{B}(\boldsymbol{x})| = \mathcal{O}(|\boldsymbol{x}|^{-2}) \qquad |\boldsymbol{x}| \to \infty, \tag{11.4}$$

with $\mu_r = \mu_r(|\boldsymbol{H}|)$ in Ω and $\mu_r = 1$ in Ω_{ext}.

The main simplifying assumption is that the source current density $\boldsymbol{J}$ is given in the wire Λ. To obtain a well posed set of equations for this problem, we denote by

$\boldsymbol{H}_0$ the magnetic field obtained in the absence of the plate Ω. Following the model presented in Sect. 2.4.2, we have for $\boldsymbol{H}_0$ the integral expression:

$$\boldsymbol{H}_0(\boldsymbol{x}) = \int_{\Lambda} \nabla_x G(\boldsymbol{x}, \boldsymbol{y}) \times \boldsymbol{J}(\boldsymbol{y})\, d\boldsymbol{y} \qquad \boldsymbol{x} \in \mathbb{R}^3, \tag{11.5}$$

where we recall that G is the Green function associated to the operator $-\Delta$ in the three-dimensional space, that is

$$G(\boldsymbol{x}, \boldsymbol{y}) = \frac{1}{4\pi} \frac{1}{|\boldsymbol{x} - \boldsymbol{y}|} \qquad \boldsymbol{x}, \boldsymbol{y} \in \mathbb{R}^3,\ \boldsymbol{x} \neq \boldsymbol{y}.$$

Note that, in practice, the support of the current density $\boldsymbol{J}$ is a compact set with a positive measure. The case of a current supported by a wire with null measure is not covered by the present analysis. Consequently we assume that $\boldsymbol{H}_0 \in \boldsymbol{\mathcal{L}}^2(\mathbb{R}^3)$ and

$$|\boldsymbol{H}_0(\boldsymbol{x})| = \mathcal{O}(|\boldsymbol{x}|^{-2}) \qquad \text{for } |\boldsymbol{x}| \to \infty. \tag{11.6}$$

The magnetic induction resulting from $\boldsymbol{H}_0$ is then obtained by:

$$\mathbf{curl}\, \boldsymbol{H}_0 = \boldsymbol{J} \qquad \text{in } \mathbb{R}^3, \tag{11.7}$$

$$\operatorname{div} \boldsymbol{B}_0 = 0 \qquad \text{in } \mathbb{R}^3. \tag{11.8}$$

The ferromagnetic property of the conductor Ω is generally expressed by the fact that the relative permeability depends on the norm of the magnetic field, that is $\mu_r = \mu_r(|\boldsymbol{H}|)$. We can then extend the function μ_r to $\mathbb{R}^3$ by defining

$$\tilde{\mu}_r(\boldsymbol{x}, s) = \begin{cases} \mu_r(s) & \text{if } \boldsymbol{x} \in \Omega, \\ 1 & \text{if } \boldsymbol{x} \in \Omega_{\text{ext}}. \end{cases}$$

Hence, $\tilde{\mu}_r$ is a positive function defined on $\mathbb{R}^3$. It maybe noticed that the extension $\tilde{\mu}_r$ depends on the position $\boldsymbol{x}$ and that it may exhibit a jump on the boundary of Ω.

Combining (11.1) and (11.7) and assuming that the presence of the plate Ω does not change the electric current in the wire Λ, we obtain $\mathbf{curl}\,(\boldsymbol{H} - \boldsymbol{H}_0) = 0$ and then using Theorem 1.3.6 we deduce the existence of a scalar field ψ such that

$$\boldsymbol{H} - \boldsymbol{H}_0 = \nabla\psi \qquad \text{in } \mathbb{R}^3.$$

Taking into account (11.2) and (11.3), we obtain the equation

$$\operatorname{div}\big(\tilde{\mu}_r(\cdot, |\boldsymbol{H}_0 + \nabla\psi|)(\boldsymbol{H}_0 + \nabla\psi)\big) = 0 \qquad \text{in } \mathbb{R}^3. \tag{11.9}$$

Let us give some remarks about this problem:

1. The assumption (11.6) and a similar behaviour for $\boldsymbol{H}$ imply

$$\psi(\boldsymbol{x}) = \mathcal{O}(|\boldsymbol{x}|^{-1}) \qquad |\boldsymbol{x}| \to \infty.$$

2. The relative magnetic permeability depends on $|\boldsymbol{H}| = |\boldsymbol{H}_0 + \nabla\psi|$. This implies that the obtained Eq. (11.9) is nonlinear. In the external domain Ω_{ext} we have, since $\mu_r = 1$ and $\operatorname{div} \boldsymbol{H}_0 = 0$ the Laplace equation:

$$\Delta\psi = 0 \qquad \text{in } \Omega_{\text{ext}}.$$

11.1 Mathematical Analysis

We consider a variational formulation of (11.9) in the space $\mathcal{W}^1(\mathbb{R}^3)$. This is clearly motivated by the fact that the problem is stated in the whole space $\mathbb{R}^3$. We obtain the weak formulation:

$$\begin{cases} \text{Find } \psi \in \mathcal{W}^1(\mathbb{R}^3) \text{ such that} \\ \displaystyle\int_{\mathbb{R}^3} \tilde{\mu}_r(\cdot, |\boldsymbol{H}_0 + \nabla\psi|)(\boldsymbol{H}_0 + \nabla\psi) \cdot \nabla\varphi \, d\boldsymbol{x} = 0 \quad \forall\, \varphi \in \mathcal{W}^1(\mathbb{R}^3). \end{cases} \tag{11.10}$$

Let us give an alternative formulation of this problem by considering the right-hand side. For any $\varphi \in \mathcal{W}^1(\mathbb{R}^3)$, we have

$$\begin{aligned} -\int_{\mathbb{R}^3} \tilde{\mu}_r(\cdot, |\boldsymbol{H}_0 + \nabla\psi|)\, \boldsymbol{H}_0 \cdot \nabla\varphi \, d\boldsymbol{x} &= -\int_{\Omega} \mu_r(|\boldsymbol{H}_0 + \nabla\psi|)\, \boldsymbol{H}_0 \cdot \nabla\varphi \, d\boldsymbol{x} \\ &\quad - \int_{\Omega_{\text{ext}}} \boldsymbol{H}_0 \cdot \nabla\varphi \, d\boldsymbol{x} \\ &= \int_{\Omega} \big(1 - \mu_r(|\boldsymbol{H}_0 + \nabla\psi|)\big)\, \boldsymbol{H}_0 \cdot \nabla\varphi \, d\boldsymbol{x} \\ &\quad - \int_{\mathbb{R}^3} \boldsymbol{H}_0 \cdot \nabla\varphi \, d\boldsymbol{x} \\ &= \int_{\Omega} \big(1 - \mu_r(|\boldsymbol{H}_0 + \nabla\psi|)\big)\, \boldsymbol{H}_0 \cdot \nabla\varphi \, d\boldsymbol{x} \\ &\quad + \int_{\mathbb{R}^3} \varphi \operatorname{div} \boldsymbol{H}_0 \, d\boldsymbol{x} \\ &= \int_{\Omega} \big(1 - \mu_r(|\boldsymbol{H}_0 + \nabla\psi|)\big)\, \boldsymbol{H}_0 \cdot \nabla\varphi \, d\boldsymbol{x}. \end{aligned}$$

This means that the a priori calculation of the magnetic field $\boldsymbol{H}_0$ can be made in the conductor Ω only.

Problem (11.10) can then be rewritten in the following form:

$$\begin{cases} \text{Find } \psi \in \mathcal{W}^1(\mathbb{R}^3) \text{ such that} \\ \displaystyle\int_\Omega \mu_r(|\boldsymbol{H}_0 + \nabla\psi|)\, \nabla\psi \cdot \nabla\varphi \, d\boldsymbol{x} + \int_{\Omega_{\text{ext}}} \nabla\psi \cdot \nabla\varphi \, d\boldsymbol{x} = \\ \qquad \displaystyle -\int_\Omega \big(\mu_r(|\boldsymbol{H}_0 + \nabla\psi|) - 1\big)\, \boldsymbol{H}_0 \cdot \nabla\varphi \, d\boldsymbol{x} \qquad \forall\, \varphi \in \mathcal{W}^1(\mathbb{R}^3). \end{cases} \tag{11.11}$$

We remark that if $\mu_r = 1$ everywhere, i.e., if the conductor Ω is not ferromagnetic, we obtain

$$\int_{\mathbb{R}^3} \nabla\psi \cdot \nabla\varphi \, d\boldsymbol{x} = 0 \qquad \forall\, \varphi \in \mathcal{W}^1(\mathbb{R}^3),$$

which implies $\psi = 0$ and then $\boldsymbol{H} = \boldsymbol{H}_0$.

Let us define, for $\psi \in \mathcal{W}^1(\mathbb{R}^3)$, the bilinear and linear forms:

$$\mathscr{B}_\psi(\phi, \varphi) = \int_\Omega \mu_r(|\boldsymbol{H}_0 + \nabla\psi|)\, \nabla\phi \cdot \nabla\varphi \, d\boldsymbol{x} + \int_{\Omega_{\text{ext}}} \nabla\psi \cdot \nabla\varphi \, d\boldsymbol{x},$$

$$\mathscr{L}_\psi(\varphi) = \int_\Omega \big(1 - \mu_r(|\boldsymbol{H}_0 + \nabla\psi|)\big)\, \boldsymbol{H}_0 \cdot \nabla\varphi \, d\boldsymbol{x}.$$

Then (11.10) or (11.11) is equivalent to finding a function $\psi \in \mathcal{W}^1(\mathbb{R}^3)$ that satisfies

$$\mathscr{B}_\psi(\psi, \varphi) = \mathscr{L}_\psi(\varphi) \qquad \forall\, \varphi \in \mathcal{W}^1(\mathbb{R}^3).$$

This formulation will be useful for the numerical approximation of (11.10).

We make the following assumption on the function μ_r:

Hypothesis 11.1.1. *The function $\mu_r : \mathbb{R}^+ \to \mathbb{R}^+$ is a C^1 decreasing function on $[0, \infty)$ that satisfies the following properties:*

$$1 \le \mu_r(s) \le \beta, \quad \mu_r(s) + s\mu_r'(s) \ge 1 \qquad \forall\, s \in \mathbb{R}^+,$$

where β is a positive constant.

This hypothesis is reasonable from a physical point of view since it means in particular that the magnetic permeability is bounded and that the magnetic energy density is strictly convex. This property will be clarified in the proof of the following theorem.

Theorem 11.1.1. *Assume that Hypothesis 11.1.1 holds. Then* (11.10) *(or* (11.11)*) admits a unique solution $\psi \in \mathcal{W}^1(\mathbb{R}^3)$.*

Proof. The complete proof can be found in [67]. The main argument is the following one:

Let $g : \mathbb{R}^+ \to \mathbb{R}^+$ be the function defined by

$$g(s) = \int_0^s t\mu_r(t)\,dt, \quad s \geq 0.$$

Then the functional $\mathscr{K}$ given by

$$\mathscr{K}(\varphi) = \int_\Omega g(|\boldsymbol{H}_0 + \nabla\varphi|)\,d\boldsymbol{x} + \frac{1}{2}\int_{\Omega_{\text{ext}}} |\boldsymbol{H}_0 + \nabla\varphi|^2\,d\boldsymbol{x}$$

is well defined for $\varphi \in \mathcal{W}^1(\mathbb{R}^3)$. In [67], it is proved that under Hypothesis 11.1.1:

1. The functional $\mathscr{K} : \mathcal{W}^1(\mathbb{R}^3) \to \mathbb{R}$ is $\mathcal{C}^1$, strictly convex and coercive.
2. The Gâteaux derivative $d\mathscr{K}(\psi;\xi)$ for $\xi \in \mathcal{W}^1(\mathbb{R}^3)$ is

$$d\mathscr{K}(\psi;\xi) = \int_{\mathbb{R}^3} \tilde{\mu}_r(\cdot, |\boldsymbol{H}_0 + \nabla\psi|)(\boldsymbol{H}_0 + \nabla\psi)\cdot\nabla\xi\,d\boldsymbol{x}.$$

The Euler equation is then given by

$$\int_{\mathbb{R}^3} \tilde{\mu}_r(\cdot, |\boldsymbol{H}_0 + \nabla\psi|)(\boldsymbol{H}_0 + \nabla\psi)\cdot\nabla\xi\,d\boldsymbol{x} = 0 \qquad \forall\,\xi \in \mathcal{W}^1(\mathbb{R}^3).$$

By using ([61], Chap. 3, Theorem 1.1) the functional $\mathscr{K}$ admits a unique minimum $\psi \in \mathcal{W}^1(\mathbb{R}^3)$ which is the unique solution to (11.11). □

11.2 An Iterative Procedure

Problem (11.10) is a nonlinear problem and its numerical solution requires constructing an iterative procedure. We present here an iterative algorithm to compute the potential ψ, each iteration consisting in solving a linear variational problem that can be solved by the finite element method. This algorithm is given in [67].

We define a mapping $F : \mathcal{W}^1(\mathbb{R}^3) \to \mathcal{W}^1(\mathbb{R}^3)$ by the variational identity:

$$\int_{\mathbb{R}^3} \tilde{\mu}_r(\cdot, |\boldsymbol{H}_0 + \nabla\xi|)(\boldsymbol{H}_0 + \nabla F(\xi))\cdot\nabla\varphi\,d\boldsymbol{x} = 0 \qquad \forall\,\varphi,\xi \in \mathcal{W}^1(\mathbb{R}^3).$$

The mapping F is well defined since Hypothesis 11.1.1 implies that the bilinear form

$$\mathscr{B}_\xi(\varphi,\zeta) := \int_{\mathbb{R}^3} \tilde{\mu}_r(\cdot, |\boldsymbol{H}_0 + \nabla\xi|)\nabla\varphi\cdot\nabla\zeta\,d\boldsymbol{x}$$

is continuous and coercive on $\mathcal{W}^1(\mathbb{R}^3)$ for every $\xi \in \mathcal{W}^1(\mathbb{R}^3)$. Moreover, we have the estimate

$$\|\nabla F(\xi)\|_{\mathcal{L}^2(\mathbb{R}^3)} \leq \beta \, \|\boldsymbol{H}_0\|_{\mathcal{L}^2(\mathbb{R}^3)} \qquad \forall\, \xi \in \mathcal{W}^1(\mathbb{R}^3).$$

Clearly, $\psi \in \mathcal{W}^1(\mathbb{R}^3)$ is solution of Problem (11.10) if and only if ψ is a fixed point of F in $\mathcal{W}^1(\mathbb{R}^3)$, i.e. $\psi = F(\psi)$.

In order to compute the solution of Problem (11.10) we construct a fixed point method that involves a relaxation parameter:

(i) We choose a function $\psi_0 \in \mathcal{W}^1(\mathbb{R}^3)$ and a relaxation parameter $\varepsilon > 0$.
(ii) For $k = 0, 1, \ldots$ we compute:

$$\tilde{\psi}_{k+1} = F(\psi_k), \tag{11.12}$$

$$\psi_{k+1} = (1 - \varepsilon)\, \psi_k + \varepsilon\, \tilde{\psi}_{k+1}. \tag{11.13}$$

(The case $\varepsilon = 1$ corresponds to the fixed point algorithm).

Note that solving (11.12) is equivalent to solving the linear variational problem (see (11.11)):

$$\int_\Omega \mu_r(|\boldsymbol{H}_0 + \nabla\psi_k|)\, \nabla\tilde{\psi}_{k+1} \cdot \nabla\varphi \, d\boldsymbol{x} + \int_{\Omega_{\text{ext}}} \nabla\tilde{\psi}_{k+1} \cdot \nabla\varphi \, d\boldsymbol{x}$$
$$= \int_\Omega (1 - \mu_r(|\boldsymbol{H}_0 + \nabla\psi_k|)\, \boldsymbol{H}_0 \cdot \nabla\varphi \, d\boldsymbol{x} \quad \forall\, \varphi \in \mathcal{W}^1(\mathbb{R}^3) \tag{11.14}$$

for $k = 0, 1, \ldots$

The following convergence result is proved in [67].

Theorem 11.2.1. *Assume that Hypothesis 11.1.1 holds. Then, if $\varepsilon < 2/\beta$, where β is the upper bound of μ_r, the iterative procedure* (11.12)–(11.13) *converges toward the unique solution ψ of* (11.10), *i.e.*

$$\lim_{k\to\infty} \|\nabla(\psi_k - \psi)\|_{\mathcal{L}^2(\mathbb{R}^3)} = 0.$$

11.3 Solution of the Linear Problem by a Domain Decomposition Method

Numerical solution of (11.14) requires solving a partial differential equation in an unbounded domain. Let us write this problem simply:

$$\int_{\mathbb{R}^3} \mu_k \nabla\tilde{\psi}_{k+1} \cdot \nabla\varphi \, d\boldsymbol{x} = -\int_\Omega (\mu_k - 1)\, \boldsymbol{H}_0 \cdot \nabla\varphi \, d\boldsymbol{x} \qquad \forall\, \varphi \in \mathcal{W}^1(\mathbb{R}^3), \tag{11.15}$$

with

$$\mu_k := \tilde{\mu}_r(\cdot, |\boldsymbol{H}_0 + \nabla\psi_k|).$$

Since $\mu_k = 1$ in Ω_{ext}, we have that $\tilde{\psi}_{k+1}$ is harmonic in Ω_{ext}. We shall use the same technique as in Chap. 3: Considering an open ball B_r containing $\overline{\Omega}$ centered at the origin, with radius r, and boundary ∂B_r, we have by the Poisson formula (see [62], Vol. 1, p. 249):

$$\tilde{\psi}_{k+1}(\boldsymbol{x}) = \frac{|\boldsymbol{x}|^2 - r^2}{4\pi r} \int_{\partial B_r} \frac{\tilde{\psi}_{k+1}(\boldsymbol{y})}{|\boldsymbol{x}-\boldsymbol{y}|^3}\, ds(\boldsymbol{y}) \qquad \forall\, \boldsymbol{x} \in \mathbb{R}^3 \setminus B_r. \tag{11.16}$$

Hence, if B_R is a ball centered at the origin with radius $R > r$, Problem (11.15) is equivalent to finding $\tilde{\psi}_{k+1} \in \mathcal{H}^1(B_R)$ satisfying (11.16) on the boundary ∂B_R of B_R, and for every $\varphi \in \mathcal{H}^1_0(B_R)$:

$$\int_{B_R} \mu_k \nabla\tilde{\psi}_{k+1} \cdot \nabla\varphi\, d\boldsymbol{x} = \int_{\Omega} (1-\mu_k)\, \boldsymbol{H}_0 \cdot \nabla\varphi\, d\boldsymbol{x}.$$

We want now to construct an alternating Schwarz domain decomposition method for the computation of $\tilde{\psi}_{k+1}$. We define for this the following algorithm:

(a) Define an initial guess $\psi^{(0)} \in \mathcal{H}^1_0(B_R)$ by solving the problem:

$$\int_{B_R} \mu_k \nabla\psi^{(0)} \cdot \nabla\varphi\, d\boldsymbol{x} = \int_{\Omega} (1-\mu_k)\, \boldsymbol{H}_0 \cdot \nabla\varphi\, d\boldsymbol{x} \qquad \forall\, \varphi \in \mathcal{H}^1_0(B_R).$$

(b) For $n = 0, 1, 2, \dots$ we compute $\psi^{(n+1)} \in \mathcal{H}^1(B_R)$ satisfying

$$\begin{cases} \displaystyle\int_{B_R} \mu_k \nabla\psi^{(n+1)} \cdot \nabla\varphi\, d\boldsymbol{x} = \int_{\Omega} (1-\mu_k)\, \boldsymbol{H}_0 \cdot \nabla\varphi\, d\boldsymbol{x} \\ \qquad\qquad\qquad\qquad \forall\, \varphi \in \mathcal{H}^1_0(B_R), \\ \displaystyle\psi^{(n+1)}(\boldsymbol{x}) = \frac{R^2 - r^2}{4\pi r} \int_{\partial B_r} \frac{\psi^{(n)}(\boldsymbol{y})}{|\boldsymbol{x}-\boldsymbol{y}|^3}\, ds(\boldsymbol{y}) \qquad \boldsymbol{x} \in \partial B_R. \end{cases} \tag{11.17}$$

We have the following convergence result:

Theorem 11.3.1. *Let* $\tilde{\psi}_{k+1} \in \mathcal{W}^1(\mathbb{R}^3)$ *denote the solution of* (11.15). *We have the estimate:*

$$\|\tilde{\psi}_{k+1} - \psi^{(n)}\|_{\mathcal{L}^\infty(\mathcal{B}_R)} \le \left(\frac{r}{R}\right)^n \|\tilde{\psi}_{k+1} - \psi^{(0)}\|_{\mathcal{L}^\infty(\mathcal{B}_R)}.$$

In particular,

$$\lim_{n\to\infty} \|\tilde{\psi}_{k+1} - \psi^{(n)}\|_{\mathcal{L}^\infty(\mathcal{B}_\mathcal{R})} = 0.$$

Proof. Let us a give here a sketch of the proof of this result. The complete proof can be found in [67].

Let us define $e^{(n)} = \tilde{\psi}_{k+1} - \chi^{(n)}$ on B_R. We have

$$e^{(n)}(\boldsymbol{x}) = \frac{R^2 - r^2}{4\pi r} \int_{\partial B_r} \frac{e^{(n-1)}(\boldsymbol{y})}{|\boldsymbol{x} - \boldsymbol{y}|^3}\, ds(\boldsymbol{y}) \qquad \text{for } \boldsymbol{x} \in \partial B_R. \tag{11.18}$$

It follows that

$$\|e^{(n)}\|_{\mathcal{L}^\infty(\partial\mathcal{B}_\mathcal{R})} \leq \left(\frac{R^2 - r^2}{4\pi r} \max_{\boldsymbol{x}\in\partial B_R} \int_{\partial B_r} \frac{1}{|\boldsymbol{x} - \boldsymbol{y}|^3}\, ds(\boldsymbol{y})\right) \|e^{(n-1)}\|_{\mathcal{L}^\infty(\partial\mathcal{B}_\triangledown)}.$$

By using spherical coordinates we obtain

$$\|e^{(n)}\|_{\mathcal{L}^\infty(\partial B_R)} \leq \frac{r}{R} \|e^{(n-1)}\|_{\mathcal{L}^\infty(\partial B_R)}.$$

The maximum principle for a harmonic function between B_r and B_R (see for instance [92]) implies

$$\|e^{(n)}\|_{\mathcal{L}^\infty(\partial B_R)} \leq \frac{r}{R} \|e^{(n-1)}\|_{\mathcal{L}^\infty(\partial B_r)} \leq \frac{r}{R} \|e^{(n-1)}\|_{\mathcal{L}^\infty(\partial B_R)}.$$

The conclusion of the theorem follows then. □

Remark 11.3.1. Using (11.18) we prove that $\|e^{(n)}\|_{\mathcal{H}^{\frac{\infty}{\in}}(\partial B_R)}$ is bounded by $\|e^{(n-1)}\|_{\mathcal{L}^\infty(\partial\mathcal{B}_\triangledown)}$ and consequently there exists a constant C such that

$$\|e^{(n)}\|_{\mathcal{H}^1(B_R)} \leq C\, \|e^{(n-1)}\|_{\mathcal{L}^\infty(\partial\mathcal{B}_\triangledown)}.$$

It follows that

$$\|e^{(n)}\|_{\mathcal{H}^1(B_R)} \leq C\, \|e^{(n-1)}\|_{\mathcal{L}^\infty(\mathcal{B}_\mathcal{R})}.$$

From Theorem 11.3.1 and from the definition of $e^{(n)}$, we deduce that there is a constant, also denoted by C, such that

$$\|\tilde{\psi}_{k+1} - \psi^{(n)}\|_{\mathcal{H}^1(B_R)} \leq C \left(\frac{r}{R}\right)^n.$$

11.4 An Iterative Procedure for the Discrete Nonlinear Problem

Let $\mathcal{W}_h$ be a finite dimensional subspace of $\mathcal{W}^1(\mathbb{R}^3) \cap \mathcal{W}^{1,\infty}(\Omega)$. By using the same arguments used for proving the existence and uniqueness of the solution $\psi \in \mathcal{W}^1(\mathbb{R}^3)$ of the exact problem, it is possible to prove that there exists a unique solution $\psi_h \in \mathcal{W}_h$ satisfying:

$$\int_{\mathbb{R}^3} \tilde{\mu}_r(\cdot, |\boldsymbol{H}_0 + \nabla\psi_h|)\,(\boldsymbol{H}_0 + \nabla\psi_h)\cdot\nabla\varphi_h\,d\boldsymbol{x} = 0 \qquad \forall\ \varphi_h \in \mathcal{W}_h. \quad (11.19)$$

Let us consider the mapping $F_h : \mathcal{W}_h \to \mathcal{W}_h$ defined by

$$\int_{\mathbb{R}^3} \tilde{\mu}_r(\cdot, |\boldsymbol{H}_0 + \nabla z_h|)\,(\boldsymbol{H}_0 + \nabla F_h(z_h))\cdot\nabla\varphi_h\,d\boldsymbol{x} = 0 \qquad \forall\ \varphi_h, z_h \in \mathcal{W}_h.$$

Clearly F_h is well defined since Hypothesis 11.1.1 implies that the bilinear form $\mathscr{B}_\psi$ is continuous and coercive on $\mathcal{W}^1(\mathbb{R}^3)$ for every $\psi \in \mathcal{W}^1(\mathbb{R}^3)$ and the linear form $\mathscr{L}_\psi$ is continuous for every $\psi \in \mathcal{W}^1(\mathbb{R}^3)$. Moreover we have

$$\|\nabla F_h(\psi)\|_{\mathcal{L}^2(\mathbb{R}^3)} \le \beta\,\|\boldsymbol{H}_0\|_{\mathcal{L}^2(\mathbb{R}^3)} \qquad \forall\ \psi \in \mathcal{W}^1(\mathbb{R}^3),$$

and ψ_h is solution of (11.19) if and only if ψ_h is a fixed point of F_h in $\mathcal{W}_h$ i.e. $\psi_h = F_h(\psi_h)$.

In order to compute the solution ψ_h of (11.19), we will use the following fixed point method

$$\psi_h^{(k+1)} = F_h(\psi_h^{(k)}), \quad k = 0, 1, \ldots$$

which is equivalent to solve at each iteration k the finite dimensional linear problem:

$$\begin{cases} \text{Find } \psi_h^{(k+1)} \in \mathcal{W}_h \text{ such that} \\ \mathscr{B}_{\psi_h^{(k)}}(\psi_h^{(k+1)}, \varphi) = \mathscr{L}_{\psi_h^{(k)}}(\varphi) \qquad \forall\ \varphi \in \mathcal{W}_h, \end{cases}$$

the initial guess $\psi_h^{(0)} \in \mathcal{W}_h$ being given.

Let us introduce the inner product in $\mathcal{W}_h$

$$((u, w))_{\psi_h} = \int_{\mathbb{R}^3} \tilde{\mu}_r(\cdot, |\boldsymbol{H}_0 + \nabla\psi_h|)\,\nabla u\cdot\nabla w\,d\boldsymbol{x},$$

and its associated norm

$$\|u\|_{\psi_h} := ((u, u))_{\psi_h}^{\frac{1}{2}}.$$

Note that, since $\mathcal{W}_h$ has a finite dimension, this norm is equivalent to all norms in $\mathcal{W}_h$.

We reinforce Hypothesis 11.1.1 by setting:

Hypothesis 11.4.1. *The function μ_r is C^2 decreasing on $[0,\infty)$ and fulfills for all $s \geq 0$ the property:*

$$|\mu_r''(s)| \leq \gamma, \quad \mu_r'(0) = 0,$$

where γ is a nonnegative constant.

Remark that Hypothesis 11.1.1 implies

$$\lim_{s\to\infty} |\mu_r'(s)| = \lim_{s\to\infty} (-\mu_r'(s)) \leq \lim_{s\to\infty} \frac{\mu_r(s)-1}{s} = 0,$$

and Hypothesis 11.4.1 gives

$$\lim_{s\to 0} \frac{|\mu_r'(s)|}{s} = \lim_{s\to 0} \frac{|\mu_r'(s) - \mu_r'(0)|}{s} = \mu_r''(0).$$

As a consequence, there exists a positive constant λ such that

$$\frac{|\mu_r'(s)|}{s} \leq \lambda \qquad \forall\, s \in [0,\infty).$$

Theorem 11.4.1. *Assume that $\boldsymbol{H}_0 \in \mathcal{L}^2(\mathbb{R}^3) \cap \mathcal{L}^\infty(\mathbb{R}^3)$ and that Hypotheses 11.1.1 and 11.4.1 hold. Let $\psi_h \in \mathcal{W}_h$ be the unique solution of* (11.19)*. Then there exists $\kappa > 0$ such that if $\|\psi_h - \psi_h^{(0)}\|_{\psi_h} \leq \kappa$, then we have*

$$\lim_{k\to\infty} \|\psi_h - \psi_h^{(k)}\|_{\psi_h} = 0.$$

The proof of this theorem is based on two technical lemmas that are easy to prove.

Lemma 11.4.1. *Let $f : \mathbb{R}^3 \to \mathbb{R}$ be the function defined by $f(\boldsymbol{\xi}) = \mu_r(|\boldsymbol{\xi}|)$. If $\boldsymbol{\xi}, \boldsymbol{\eta} \in \mathbb{R}^3$, $\boldsymbol{\xi} \neq \boldsymbol{0}$, then we have:*

$$f'(\boldsymbol{\xi})\,\boldsymbol{\eta} = \mu_r'(|\boldsymbol{\xi}|)\,\frac{\boldsymbol{\xi}\cdot\boldsymbol{\eta}}{|\boldsymbol{\xi}|},$$

$$f''(\boldsymbol{\xi})(\boldsymbol{\eta},\boldsymbol{\zeta}) = \mu_r''(|\boldsymbol{\xi}|)\frac{\boldsymbol{\xi}\cdot\boldsymbol{\eta}}{|\boldsymbol{\xi}|}\frac{\boldsymbol{\xi}\cdot\boldsymbol{\zeta}}{|\boldsymbol{\xi}|} + \mu_r'(|\boldsymbol{\xi}|)\left(\frac{\boldsymbol{\zeta}\cdot\boldsymbol{\eta}}{|\boldsymbol{\xi}|} - \frac{\boldsymbol{\xi}\cdot\boldsymbol{\eta}}{|\boldsymbol{\xi}|^2}\frac{\boldsymbol{\xi}\cdot\boldsymbol{\zeta}}{|\boldsymbol{\xi}|}\right).$$

Lemma 11.4.2. *Under the same hypotheses as in Lemma 11.4.1, we have the bound*

$$|f''(\boldsymbol{\xi})(\boldsymbol{\eta},\boldsymbol{\zeta})| \leq (\gamma + 2\lambda)|\boldsymbol{\zeta}|\,|\boldsymbol{\eta}| \qquad \forall\, \boldsymbol{\xi},\boldsymbol{\eta},\boldsymbol{\zeta} \in \mathbb{R}^3.$$

In order to give a sketch of the proof to Theorem 11.4.1 (The complete proof can be found in [67]), we define

$$\phi_h = F_h(\psi_h + \varphi_h) - F_h(\psi_h) = F_h(\psi_h + \varphi_h) - \psi_h \qquad \text{for } \varphi_h \in \mathcal{W}_h.$$

Using the Taylor expansion and Lemma 11.4.2, we can prove that

$$\begin{aligned}\|\phi_h\|_{\psi_h}^2 \le \int_\Omega &\Big(|\mu_r'(|\boldsymbol{H}_0 + \nabla\psi_h|)|\,|\nabla\varphi_h| + \frac{1}{2}(\gamma + 2\lambda)|\nabla\varphi_h|^2\Big)\\ &\Big(|\nabla\phi_h|^2 + |\boldsymbol{H}_0 + \nabla\psi_h|\,|\nabla\phi_h|\Big)\,d\boldsymbol{x}.\end{aligned}$$

From Hypothesis 11.1.1 we have

$$0 \le -s\mu_r'(s) \le \mu_r(s) - 1,$$

and by setting

$$M = \sup_{s\in[0,\infty)} |\mu'(s)|,$$

$$\begin{aligned}\|\phi_h\|_{\psi_h}^2 \le\ &\int_\Omega \big(\mu_r(|\boldsymbol{H}_0 + \nabla\psi_h|) - 1\big)|\nabla\varphi_h|\,|\nabla\phi_h|\,d\boldsymbol{x}\\ &+ M\int_\Omega |\nabla\varphi_h|\,|\nabla\phi_h|^2\,d\boldsymbol{x}\\ &+ \frac{1}{2}(\gamma + 2\lambda)\int_\Omega |\nabla\varphi_h|^2|\nabla\phi_h|^2\,d\boldsymbol{x}\\ &+ \frac{1}{2}(\gamma + 2\lambda)\int_\Omega |\nabla\varphi_h|^2|\boldsymbol{H}_0 + \nabla\psi_h|\,|\nabla\phi_h|\,d\boldsymbol{x}.\end{aligned}$$

From this, we can deduce (see [67]) that there exists $\kappa > 0$ such that if $\|\varphi_h\|_{\psi_h} \le \kappa$ then

$$\|\phi_h\|_{\psi_h} \le \big(1 - \frac{1}{2\beta}\big)\,\|\varphi_h\|_{\psi_h}.$$

Setting $\varphi_h = \psi_h^{(k)} - \psi_h$, we obtain $\phi_h = \psi_h^{(k+1)} - \psi_h$ and the above inequality implies that if we choose $\psi_h^{(0)} \in \mathcal{W}_h$ such that $\|\psi_h^{(0)} - \psi_h\|_{\psi_h} < \kappa$, then

$$\|\psi_h^{(k)} - \psi_h\|_{\psi_h} \le \big(1 - \frac{1}{2\beta}\big)^k\,\kappa.$$

This achieves the proof. □

11.5 Numerical Results

We end this chapter by reporting some numerical experiments that were made on the presented ferromagnetic shielding model in [67]. The goal is mainly to test the features of the iterative solution algorithm presented in Sect. 11.2.

In Sect. 11.2 we have seen that the relaxed fixed point method converges when the relaxation parameter $\varepsilon < 2/\beta$ independently of the initial guess $\psi^{(0)} \in \mathcal{W}^1(\mathbb{R}^3)$ (see Theorem 11.3.1). In our physical applications we have $\beta = 4{,}000$ (see Fig. 11.3) which implies very small values for ε. This choice of ε would lead to a poor convergence behaviour, the relaxation parameter being small. However for the numerical approximation (see Sect. 11.4), the convergence of $\psi_h^{(k)}$ to $\psi_h \approx \psi$ when k tends to the infinity is faster since we can take $\varepsilon = 1$ (see Theorem 11.4.1), but in this case $\psi_h^{(0)}$ must be chosen close enough to ψ_h. Actually our convergence result holds when $\|\psi_h - \psi_h^{(0)}\|_{\psi_h} \leq \kappa$ but κ could depend on h. Theorems 11.1.1 and 11.4.1 claim that κ converges to zero when $\mathcal{W}_h$ becomes dense to the limit in $\mathcal{W}^1(\mathbb{R}^3)$, but numerical tests show that this is not the case. In all our tests, the iterative procedure presented in Sect. 11.2 with $\varepsilon = 1$ converges for any initial guess $\psi^{(0)}$:

$$\operatorname{div}\left(\tilde{\mu}_r(\cdot\,, |\boldsymbol{H}_0 + \nabla\psi^{(k)}|)\,(\boldsymbol{H}_0 + \nabla\psi^{(k+1)})\right) = 0.$$

In the following, we present an industrial setup with dimensions measured in meters and current intensities in Amperes. We consider a ferromagnetic rectangular plate $\Omega = (-0.01, 0.01) \times (-2.5, 2.5) \times (-2, 2)$ placed in front of an idealized

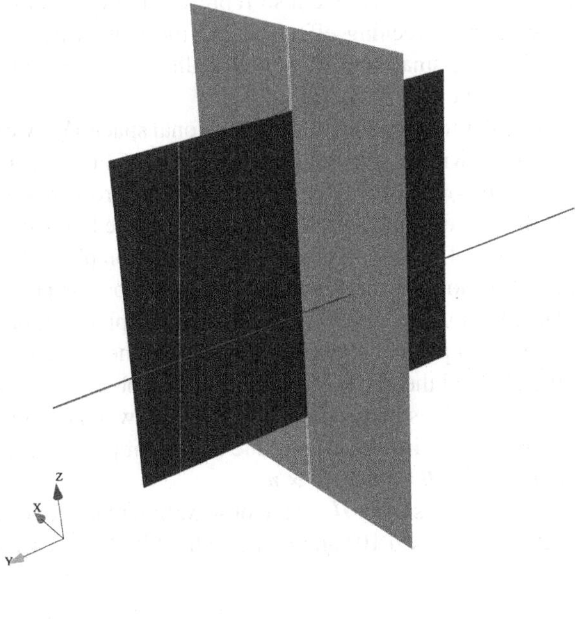

Fig. 11.2 Geometry of the test case with the plate in the Oyz plane, the observation plane in the Oxz plane

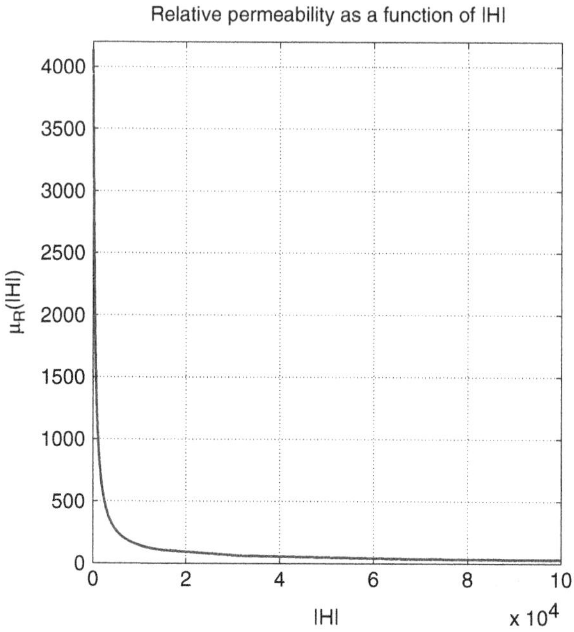

Fig. 11.3 Relative permeability (Curve $|\boldsymbol{H}| \mapsto \mu_r(|\boldsymbol{H}|)$)

infinitely long wire with zero section. The total electric current is equal to 1 and flows in the wire parallel to Oy and passes through the point $(1.01, 0, 0)$ (see Fig. 11.2). The relative magnetic permeability μ_r as a function of $|\boldsymbol{H}|$ is given in Fig. 11.3. In Fig. 11.2 we also represent the observation plane we have chosen to observe the screening effect behind the plate when we go away perpendicularly from it. The small ball B_r is with radius $r = 3.5$ centered at the origin O and the radius of the ball B_R is $R = 4.4$.

In order to build the finite dimensional space $\mathcal{W}_h$, we use a finite element method with piecewise polynomials of degree 1 on a tetrahedral mesh $\mathcal{T}_h$ discretizing the large ball B_R and we solve the approximate problem set in all the space $\mathbb{R}^3$ by using the domain decomposition algorithm presented in Sect. 11.3.

Figure 11.4 shows (left) on the observation plane the magnetic field $\boldsymbol{H}_0$ as if the plate was not present. Since the current support is perpendicular to the observation plane, the magnetic field is parallel to that plane. Figure 11.4 shows (right) on the observation plane the magnetic field $\boldsymbol{H}$ at the same scale. We can see the shielding effect behind the plate. Moreover, we can observe that near the plate boundary, the magnetic field is perpendicular to the plate when outside and parallel to the plane of the plate when inside. This is due to the jump of μ_r across the plate boundary with continuity of $\boldsymbol{B} \cdot \boldsymbol{n}$ and $\boldsymbol{H} \times \boldsymbol{n}$.

Figure 11.5 shows $\boldsymbol{H}$ on the observation plane for different values of the current: to the left we used 10^6 and to the right 10^8. We can see the saturation effect: with a

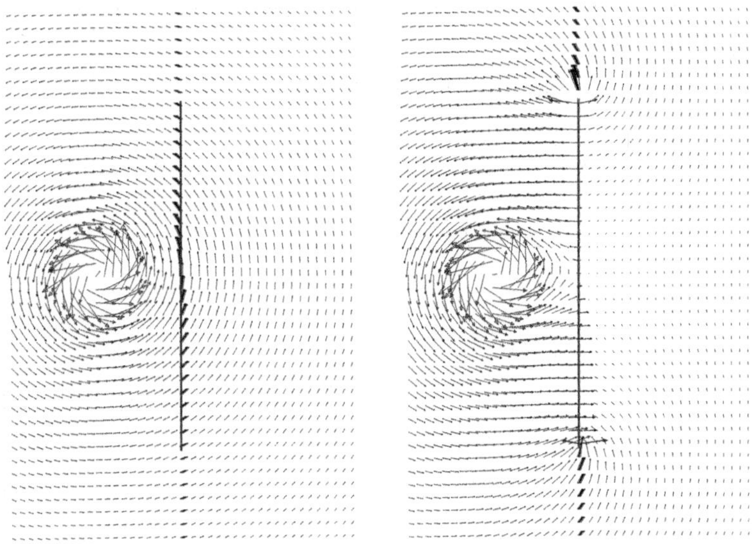

Fig. 11.4 Magnetic field in the observation plane without (*left*) and with the plate (*right*)

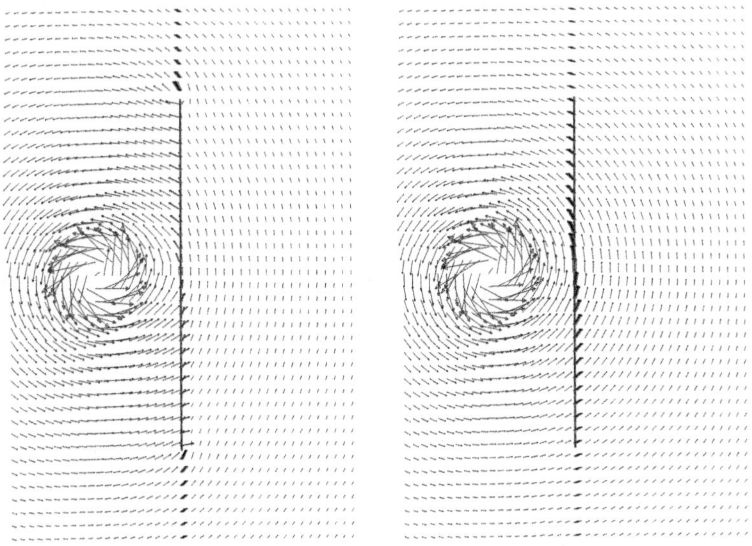

Fig. 11.5 Magnetic field $\boldsymbol{H}$ in the observation plane as for a current of 10^6 A (*left*) and of 10^8 A (*right*)

current of 10^8 the plate has no shielding effect anymore. This can also be seen with μ_r: for a current equal to 1, the relative permeability in the plate is almost constant to 4,136; for a current of 10^8, μ_r is almost constant to 106.

Chapter 12
The Electrolytic Process for Aluminium Production

12.1 Introduction

The industrial production of aluminium is based, since the end of the nineteenth century, on the electrolysis process. Aluminium is in fact not present in the nature in its pure metallic form but as an oxide called *Alumina* (Al_2O_3) and an adapted processing (electrolysis) is to be used to extract the metal. Aluminium electrolysis is performed in large Hall-Héroult cells, as illustrated for an industrial setting in Fig. 12.1. In the aluminium industry, these cells are connected electrically in series and supplied with direct current (DC). The electric current is supplied to the cells through metal busbars made of aluminium or copper. Figure 12.2 gives a schematic representation of one Hall-Héroult cell.

The procedure can be roughly outlined as the following: Alumina is dissolved in a bath essentially composed of molten cryolite (a double fluoride of aluminium and sodium). The resulting mixture is then electrolyzed by experiencing a strong static[1] electric current (current density J is about $10{,}000\,\mathrm{A\,m^{-2}}$) that deposits the aluminium liquid metal at the bottom and the bath while the liberated oxygen combines with the carbon of the anode to form carbon dioxide.

In this process the liquids (aluminium and electrolyte) are immiscible and have respective densities ϱ_{al} and ϱ_{el} with $\varrho_{\mathrm{el}} < \varrho_{\mathrm{al}}$. They occupy at time $t > 0$ respectively the domains $\Omega_{\mathrm{al}}(t)$ and $\Omega_{\mathrm{el}}(t)$. The union of these domains constitutes the fixed fluid domain, denoted by Ω where

$$\overline{\Omega} := \overline{\Omega}_{\mathrm{el}}(t) \cup \overline{\Omega}_{\mathrm{al}}(t) \qquad t > 0.$$

Since liquid aluminium is heavier than electrolyte, the domain $\Omega_{\mathrm{al}}(t)$ is below $\Omega_{\mathrm{el}}(t)$ as shown in Fig. 12.2.

[1]A precise definition of this term is given later

R. Touzani and J. Rappaz, *Mathematical Models for Eddy Currents and Magnetostatics: With Selected Applications*, Scientific Computation, DOI 10.1007/978-94-007-0202-8_12,

Fig. 12.1 View of a potroom. *Photo Rio Tinto Alcan* (Copied from website: http://www.lawyersweekly.ca)

The motion of both fluids is driven by the Lorentz force produced by the electric current density $\boldsymbol{J}$ and the induced magnetic induction $\boldsymbol{B}$, and by gravity acceleration. In addition, it is well known that this motion results in a deformation of the interface between the liquids. It is then essential to control as much as possible this motion in order to prevent shortcut damages caused by contact between the anode and the liquid aluminium which is a good electric conductor. Following these considerations, all the involved fields will be considered as time dependent. We note that this is not in contradiction with the fact that electromagnetic phenomena are static, since in a first approximation, time dependency involves the time variable as a parameter, and no time derivative in the electromagnetic equations is present.

Let us finally denote by $\tilde{\Omega}$ the union of all electrical conductors of the cell, and by Ω_{ext} the vacuum defined by $\Omega_{\text{ext}} = \mathbb{R}^3 \setminus \overline{\tilde{\Omega}}$.

In this chapter we present a mathematical model for the aluminium electrolysis, define a numerical method to solve the resulting system of equations and then present some numerical results that were obtained in an industrial environment.

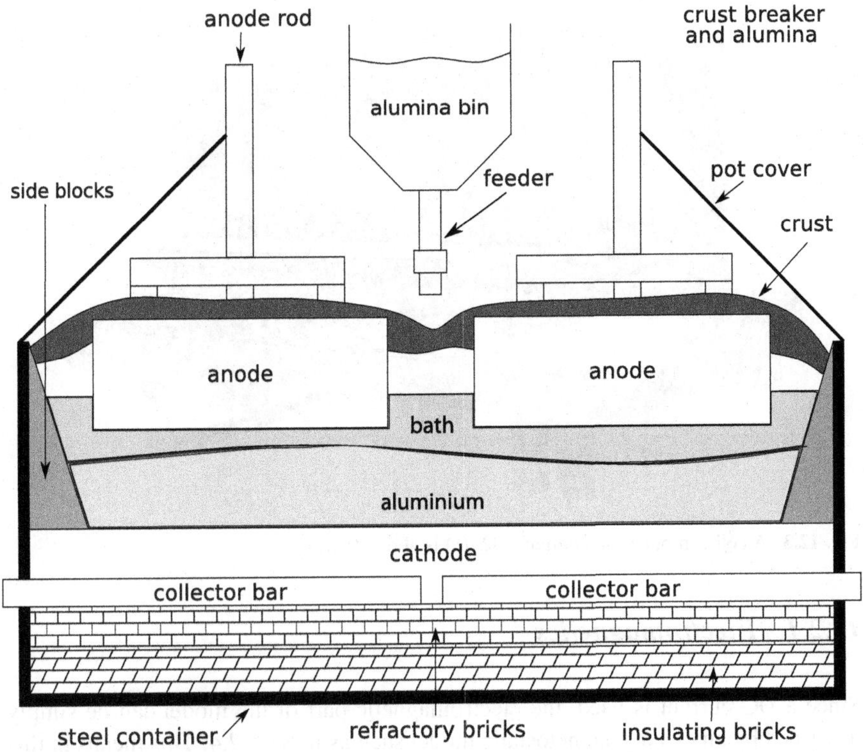

Fig. 12.2 A schematic Hall-Héroult cell (cf. Hofer [102])

12.2 The Model

From the preceding, we conclude that a reliable modeling of the electrolytic process for aluminium production has to take into account electromagnetics and hydrodynamics coupled with free surface description. Moreover, since we are interested in time evolution of the process and, in particular, in stability issues, the model we present is time dependent.

Let us note that it would be unfeasible to simulate the whole hall of cells. Instead, one generally assumes that a static current with given intensity is brought to each cell (by the busbars) and that this current is not modified by the electromagnetic phenomena in cells. This enables considering the cells individually. Figure 12.3 illustrates the computational domain, where each color corresponds to a different material.

We now describe each of the involved physical phenomena separately before stating the coupled system that governs the process.

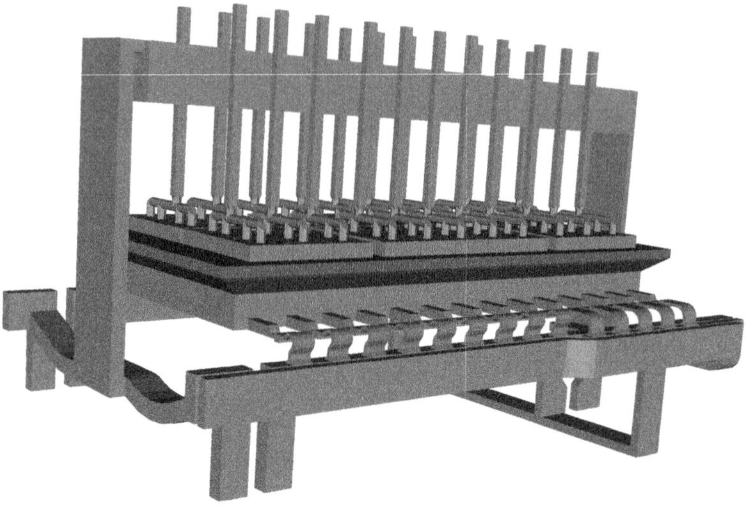

Fig. 12.3 A cell computational domain ($\Omega_{\rm el}(t)$ is above $\Omega_{\rm al}(t)$)

12.2.1 Electromagnetics

Since a DC current is used, the electromagnetic part of the model can be simply described by an electromagnetostatic model such as in Sect. 2.4. This means in this context that the electromagnetic induction term is neglected, i.e. all time derivatives are dropped. However, due to domain motions, the electric conductivity depends on time which implies that time is a parameter in the equations (see [80] and the references therein for more details). We obtain the set of equations:

$$\mathbf{curl}\,\boldsymbol{H} = \boldsymbol{J}, \tag{12.1}$$

$$\mathbf{curl}\,\boldsymbol{E} = 0, \tag{12.2}$$

$$\mathrm{div}\,\boldsymbol{B} = 0, \tag{12.3}$$

in $\mathbb{R}^3$. Now we assume that all conducting parts of the setup are not ferromagnetic, except in the steel shell described in Chap. 11 which is not taken into account in this chapter. This implies

$$\boldsymbol{B} = \mu_0 \boldsymbol{H}. \tag{12.4}$$

Moreover, due to the presence of the fluids, we have the Ohm's law for a moving electric conductor (see (2.10))

$$\boldsymbol{J} = \sigma\,(\boldsymbol{E} + \boldsymbol{v} \times \boldsymbol{B}), \tag{12.5}$$

where $\boldsymbol{v}$ is the fluid velocity extended by zero outside the fluid domain. Using Theorem 1.3.4, we deduce from (12.2) the existence of a scalar potential ϕ such that

$$\boldsymbol{E} = -\nabla\phi \qquad \text{in } \mathbb{R}^3. \tag{12.6}$$

Reporting (12.5) in (12.1) and taking the divergence of the resulting equation, we get

$$\operatorname{div}(\sigma\, \nabla\phi) = \operatorname{div}(\sigma \boldsymbol{v} \times \boldsymbol{B}) \qquad \text{in } \mathbb{R}^3. \tag{12.7}$$

The equation for the magnetic induction $\boldsymbol{B}$ is clearly given by (2.25) when $\boldsymbol{J}$ is known. Finally, we assume that a source current density is given at the boundary of $\tilde{\Omega}$, that is

$$\boldsymbol{J} \cdot \boldsymbol{n} = j_S \qquad \text{on } \partial\tilde{\Omega},$$

where the function j_S is given and fulfills the compatibility property

$$\int_{\partial\tilde{\Omega}} j_S \, ds = 0. \tag{12.8}$$

In practice the input current j_S is prescribed on boundary parts at which the current enters or leaves. Its value is null on any other part of the boundary.

This prescription can be written as a boundary condition for (12.7) by expressing the normal component of (12.5) on $\partial\tilde{\Omega}$, using (12.6) and the fact that the fluid velocity $\boldsymbol{v}$ is assumed to vanish on the boundary:

$$\sigma \frac{\partial\phi}{\partial n} = -j_S \qquad \text{on } \partial\tilde{\Omega}. \tag{12.9}$$

Hence, thanks to (12.8) for a given $\boldsymbol{v}$ and $\boldsymbol{B}$ and a fixed interface between aluminium and bath $\Gamma(t)$, by Theorem 1.2.1, (12.7) and (12.9) provide a well posed problem for a potential ϕ, up to an additive constant. It is, at this point, worth noting that σ admits a jump on the interface $\Gamma(t)$.

Let us now recall that we have assumed that the cells are decoupled. This implies in particular that the fluid motion, and then the interface, does not modify the electric current supplied by the busbars. This hypothesis can be implemented by assuming that the electric current density $\boldsymbol{J}$ is split as

$$\boldsymbol{J}(\boldsymbol{x},t) = \boldsymbol{J}_0(\boldsymbol{x}) + \boldsymbol{\delta J}(\boldsymbol{x},t) \qquad \boldsymbol{x} \in \mathbb{R}^3,\ t > 0, \tag{12.10}$$

where $\boldsymbol{J}_0$ is the electric current density which flows in all the busbars and cells of the hall and $\boldsymbol{\delta J}$ is a perturbation of $\boldsymbol{J}_0$ in the cell $\tilde{\Omega}$ when the interface between the liquids moves. We have

$$\operatorname{div} \boldsymbol{J}_0 = \operatorname{div} \boldsymbol{\delta J} = 0 \qquad \text{in } \mathbb{R}^3, \tag{12.11}$$

and the current perturbation $\delta \boldsymbol{J}$ has its support included in $\tilde{\Omega}$. Consequently, we can define a static magnetic induction $\boldsymbol{B}_0$ by

$$\begin{aligned} \mathbf{curl}\, \boldsymbol{B}_0 &= \mu_0 \boldsymbol{J}_0 && \text{in } \mathbb{R}^3, \\ \operatorname{div} \boldsymbol{B}_0 &= 0 && \text{in } \mathbb{R}^3. \end{aligned}$$

Using (2.25), we have for $\boldsymbol{x} \in \mathbb{R}^3$,

$$\begin{aligned} \boldsymbol{B}_0(\boldsymbol{x}) &= \mu_0 \int_{\mathbb{R}^3} \nabla_x G(\boldsymbol{x}, \boldsymbol{y}) \times \boldsymbol{J}_0(\boldsymbol{y})\, d\boldsymbol{y} \\ &= \frac{\mu_0}{4\pi} \int_{\mathbb{R}^3} \frac{(\boldsymbol{x} - \boldsymbol{y}) \times \boldsymbol{J}_0(\boldsymbol{y})}{|\boldsymbol{x} - \boldsymbol{y}|^3}\, d\boldsymbol{y}. \end{aligned} \tag{12.12}$$

Let now $\delta \boldsymbol{B}$ denote the perturbed magnetic induction due to $\delta \boldsymbol{J}$ defined by

$$\begin{aligned} \mathbf{curl}\, \delta \boldsymbol{B} &= \mu_0 \delta \boldsymbol{J}, \\ \operatorname{div} \delta \boldsymbol{B} &= 0 \end{aligned}$$

in $\mathbb{R}^3$. We have obviously

$$\boldsymbol{B}(\boldsymbol{x}, t) = \boldsymbol{B}_0(\boldsymbol{x}) + \delta \boldsymbol{B}(\boldsymbol{x}, t) \qquad \boldsymbol{x} \in \mathbb{R}^3,\ t > 0.$$

Note that $\delta \boldsymbol{B}$ can be obtained by using the vector potential $\boldsymbol{A}$ (see Sect. 2.2.4) that satisfies:

$$\begin{cases} \delta \boldsymbol{B} = \mathbf{curl}\, \boldsymbol{A} & \text{in } \mathbb{R}^3, \\ \operatorname{div} \boldsymbol{A} = 0 & \text{in } \mathbb{R}^3, \\ |\boldsymbol{A}(\boldsymbol{x}, t)| = \mathcal{O}(|\boldsymbol{x}|^{-1}) & |\boldsymbol{x}| \to \infty. \end{cases}$$

It follows that:

$$\begin{cases} -\Delta \boldsymbol{A} = \mu_0 \delta \boldsymbol{J} & \text{in } \mathbb{R}^3, \\ \operatorname{div} \boldsymbol{A} = 0 & \text{in } \mathbb{R}^3, \\ |\boldsymbol{A}(\boldsymbol{x}, t)| = \mathcal{O}(|\boldsymbol{x}|^{-1}) & |\boldsymbol{x}| \to \infty. \end{cases} \tag{12.13}$$

Since $\delta \boldsymbol{J}$ has a compact support in $\tilde{\Omega}$, the field $\boldsymbol{A}$ is harmonic outside $\tilde{\Omega}$. From the first equation of (12.13), we deduce

$$\begin{cases} -\Delta A_j = \mu_0 \delta J_j & \text{in } \mathbb{R}^3, \\ A_j(\boldsymbol{x}, t) = \mathcal{O}(|\boldsymbol{x}|^{-1}) & |\boldsymbol{x}| \to \infty, \end{cases} \tag{12.14}$$

for $j = 1, 2, 3$. Setting $\psi = \operatorname{div} \boldsymbol{A}$ and taking the divergence of the first equation in (12.14), we have for all $t > 0$,

$$\begin{cases} \Delta\psi = 0 & \text{in } \mathbb{R}^3, \\ \psi(\boldsymbol{x}, t) = \mathcal{O}(|\boldsymbol{x}|^{-2}) & |\boldsymbol{x}| \to \infty, \end{cases}$$

the condition at the infinity resulting also from the fact that $\boldsymbol{A}$ is harmonic outside $\tilde{\Omega}$. Therefore, trivially $\psi = \operatorname{div} \boldsymbol{A} = 0$ and the second equation in (12.13) is useless.

The main advantage of using the current splitting approximation (12.10) is that the current density $\boldsymbol{J}_0$, which has a wide support (in principle all conducting parts of the whole aluminium production hall) induces the static magnetic induction $\boldsymbol{B}_0$ which has to be computed once for all, whereas the time dependent perturbed magnetic induction $\boldsymbol{\delta B}$ involves a current density with reduced support Ω.

12.2.2 Hydrodynamics

The fluid flow is assumed governed by the incompressible Navier-Stokes equations, i.e.

$$\varrho\frac{\partial \boldsymbol{v}}{\partial t} + \varrho\,(\boldsymbol{v}\cdot\nabla)\,\boldsymbol{v} - 2\,\mathbf{div}(\eta\,\mathbf{D}(\boldsymbol{v})) + \nabla p = \boldsymbol{f} \qquad \text{in } \Omega_{\mathrm{el}}(t) \cup \Omega_{\mathrm{al}}(t), \quad (12.15)$$

$$\operatorname{div} \boldsymbol{v} = 0 \qquad \text{in } \Omega_{\mathrm{el}}(t) \cup \Omega_{\mathrm{al}}(t), \quad (12.16)$$

for $t > 0$, where p is the fluid pressure, η is the molecular viscosity, assumed constant in each fluid, ϱ is the density that equals ϱ_{al} in $\Omega_{\mathrm{al}}(t)$ and ϱ_{el} in $\Omega_{\mathrm{el}}(t)$. In addition, $\mathbf{D}(\boldsymbol{v})$ is the symmetric strain tensor defined by

$$\mathbf{D}(\boldsymbol{v}) = \frac{1}{2}(\nabla\boldsymbol{v} + (\nabla\boldsymbol{v})^{\mathsf{T}}),$$

and $\boldsymbol{f}$ is the sum of external forces given by

$$\boldsymbol{f} = \varrho\boldsymbol{g} + \boldsymbol{J}\times\boldsymbol{B}, \qquad (12.17)$$

where $\boldsymbol{g}$ is the gravity acceleration vector and the term $\boldsymbol{J}\times\boldsymbol{B}$ expresses the Lorentz force that drives the fluid motion. Note that in view of formulating the final problem in terms of ϕ and $\boldsymbol{B}$, the external forces can be expressed as

$$\boldsymbol{f}(\boldsymbol{v}, \boldsymbol{B}, \phi) := \varrho\boldsymbol{g} - \sigma\nabla\phi\times\boldsymbol{B} + \sigma\,(\boldsymbol{v}\times\boldsymbol{B})\times\boldsymbol{B}.$$

Considering boundary conditions, we have a homogeneous Dirichlet boundary condition for $\boldsymbol{v}$ on $\partial\Omega$. On the interface

$$\Gamma(t) := \overline{\Omega}_{\mathrm{al}}(t) \cap \overline{\Omega}_{\mathrm{el}}(t) \qquad t > 0,$$

that separates the two fluids, we have naturally the continuity of the velocity

$$[\boldsymbol{v}] = 0 \qquad \text{on } \Gamma(t),\ t > 0.$$

Here, the jump $[\cdot]$ stands for the value

$$[v] := v_{|\Omega_{\rm el}(t)} - v_{|\Omega_{\rm al}(t)}.$$

Next, taking into account surface tension effects we have the Laplace–Young equation

$$[(2\eta\,\mathbf{D}(\boldsymbol{v}) - p\,\mathbf{I})\boldsymbol{n}] = \gamma\kappa\boldsymbol{n} \qquad \text{on } \Gamma(t),\ t > 0, \tag{12.18}$$

where $\boldsymbol{n}$ is the unit normal to $\Gamma(t)$ pointing from $\Omega_{\rm al}(t)$ to $\Omega_{\rm el}(t)$, γ is the surface tension coefficient that depends on the fluids and κ is the Gauss curvature of the surface $\Gamma(t)$ positively counted with respect to the normal $\boldsymbol{n}$. Such a condition requires enforcing a *wetting angle* condition (see [91]). We restrict however ourselves in this presentation to the case where the surface tension can be neglected. In this case, (12.15)–(12.16) with interface conditions reduce to

$$\varrho\frac{\partial \boldsymbol{v}}{\partial t} + \varrho\,(\boldsymbol{v}\cdot\nabla)\,\boldsymbol{v} - 2\,\mathbf{div}(\eta\,\mathbf{D}(\boldsymbol{v})) + \nabla p = \boldsymbol{f} \qquad \text{in } \Omega, \tag{12.19}$$

$$\operatorname{div}\boldsymbol{v} = 0 \qquad \text{in } \Omega, \tag{12.20}$$

in the sense $\boldsymbol{v}(\cdot,t) \in \mathcal{H}^1(\Omega)$, $p(\cdot,t) \in \mathcal{L}^2(\Omega)$, for almost all $t > 0$.

Remark 12.2.1. For the sake of simplicity we do not consider here turbulence modelling. In practice, a mixing length model is used to derive a turbulent viscosity, and Navier's conditions are prescribed on the boundary $\partial\Omega$ instead of Dirichlet boundary conditions.

12.2.3 The Interface

The interface $\Gamma(t)$ that separates the aluminium and the electrolyte is described by a level set equation. Let φ denote a continuous function defined on $\Omega \times \mathbb{R}^+$ such that $\varphi > 0$ in $\Omega_{\rm al}(t)$ and $\varphi < 0$ in $\Omega_{\rm el}(t)$ for instance. The interface $\Gamma(t)$ is then the 0-level set for φ. We have the following transport equation:

$$\begin{cases} \dfrac{\partial\varphi}{\partial t} + \boldsymbol{v}\cdot\nabla\varphi = 0 & \text{in } \Omega,\ t > 0, \\ \varphi(\cdot,0) = \varphi_0 & \text{in } \Omega, \end{cases}$$

where the initial level set function φ_0 resulting from the initial configuration. For instance the signed distance function can be chosen, i.e., such that φ_0 satisfies the eikonal equation

$$|\nabla \varphi_0| = 1.$$

12.2.4 The Complete Model

Let us now summarize the complete set of equations and boundary conditions after adding an initial condition on the velocity and on φ. Setting

$$\boldsymbol{B}_0(\boldsymbol{x}) = \mu_0 \int_{\mathbb{R}^3} \nabla_{\boldsymbol{x}} G(\boldsymbol{x}, \boldsymbol{y}) \times \boldsymbol{J}_0(\boldsymbol{y})\, d\boldsymbol{y} \qquad \boldsymbol{x} \in \Omega, \tag{12.21}$$

we have the following equations:

$$\varrho \frac{\partial \boldsymbol{v}}{\partial t} + \varrho\,(\boldsymbol{v} \cdot \nabla)\,\boldsymbol{v} - 2\,\mathbf{div}(\eta\,\mathbf{D}(\boldsymbol{v})) + \nabla p = \boldsymbol{f}(\boldsymbol{v}, \boldsymbol{B}, \phi) \qquad \text{in } \Omega, \tag{12.22}$$

$$\operatorname{div} \boldsymbol{v} = 0 \qquad \text{in } \Omega, \tag{12.23}$$

$$\operatorname{div}(\sigma \nabla \phi) = \operatorname{div}(\sigma \boldsymbol{v} \times \boldsymbol{B}) \qquad \text{in } \tilde{\Omega}, \tag{12.24}$$

$$\boldsymbol{B}(\boldsymbol{x}, t) = \mathbf{curl}\,\boldsymbol{A}(\boldsymbol{x}, t) + \boldsymbol{B}_0(\boldsymbol{x}) \qquad \boldsymbol{x} \in \Omega, \tag{12.25}$$

$$-\,\boldsymbol{\Delta} \boldsymbol{A} = \mu_0 \boldsymbol{\delta} \boldsymbol{J} \qquad \text{in } \mathbb{R}^3, \tag{12.26}$$

$$\boldsymbol{\delta} \boldsymbol{J} = \sigma(\boldsymbol{v} \times \boldsymbol{B} - \nabla \phi) - \boldsymbol{J}_0 \qquad \text{in } \tilde{\Omega}, \tag{12.27}$$

$$\frac{\partial \varphi}{\partial t} + \boldsymbol{v} \cdot \nabla \varphi = 0 \qquad \text{in } \Omega, \tag{12.28}$$

for $t > 0$, with the infinity and boundary conditions:

$$|\boldsymbol{A}(\boldsymbol{x})| = \mathcal{O}(|\boldsymbol{x}|^{-1}) \qquad |\boldsymbol{x}| \to \infty, \tag{12.29}$$

$$\sigma \frac{\partial \phi}{\partial n} = -js \qquad \text{on } \partial\tilde{\Omega}, \tag{12.30}$$

$$\boldsymbol{v} = \boldsymbol{0} \qquad \text{in } \tilde{\Omega} \setminus \Omega, \tag{12.31}$$

for $t > 0$, and the initial conditions:

$$\boldsymbol{v} = \boldsymbol{v}_0 \qquad \text{in } \Omega,\ t = 0, \tag{12.32}$$

$$\varphi = \varphi_0 \qquad \text{in } \Omega,\ t = 0. \tag{12.33}$$

In (12.24) and (12.27) the velocity $\boldsymbol{v}$ is extended by $\mathbf{0}$ outside Ω. Problem (12.22)–(12.33) constitutes a system of equations and boundary and initial conditions that govern the electrolysis process for aluminium production. Note that we have omitted, for simplicity, to mention the dependency of the interface $\Gamma(t)$ on φ, being defined by $\varphi = 0$. In addition, the physical parameters ϱ, η, σ depend on $\Omega_{\rm el}(t)$ and $\Omega_{\rm al}(t)$, and consequently on φ.

Remark 12.2.2. A more complete model would take into account heat transfer in the cell by Joule heating. A discussion about this topic can be found in [79].

Remark 12.2.3. In the case where surface tension on $\Gamma(t)$ is taken into account, it can be shown that we have the following equivalent formulation:

$$\varrho\frac{\partial \boldsymbol{v}}{\partial t} + \varrho\,(\boldsymbol{v}\cdot\nabla)\,\boldsymbol{v} - 2\,\mathbf{div}(\eta\,\mathbf{D}(\boldsymbol{v})) + \nabla p = \boldsymbol{f}(\boldsymbol{v}, \boldsymbol{B}, \phi) + \gamma\kappa\delta_{\Gamma(t)} \quad \text{in } \Omega,$$

where $\delta_{\Gamma(t)}$ is the surface Dirac distribution defined by the duality pairing

$$< \delta_{\Gamma(t)}, \psi > := \int_{\Gamma(t)} \psi \, ds \qquad \forall\, \psi \in \mathscr{D}(\Omega), \quad t > 0.$$

Note that, after use of the Green formula on the interface $\Gamma(t)$, the contribution of the surface tension term involves the prescription of *wetting angles* where the interface is in contact with the domain boundary. This topic is developed in [91].

12.3 Numerical Approximation

Starting from the evolution equation (12.28), we construct a time integration scheme to discretize in time the system (12.22)–(12.33). Furthermore, in view of deriving an efficient scheme, in terms of computational time, we intend to decouple through this algorithm the involved fields.

In the following, we describe a time integration scheme that allows for such decoupling and present it step by step. Let us for this consider the time interval $(0, T)$ and let $\delta t = T/N$ denote the time step and consider time values $t^n = n\,\delta t$, for $n = 0, \ldots, N$, with $t^N = T$.

In the sequel, each step will be presented for the continuous problem and then space discretization is considered. For that end we define a finite element mesh $\tilde{\mathscr{T}}_h$ of the domain $\tilde{\Omega}$ into tetrahedra and assume that this mesh fulfills the regularity assumptions of Chap. 7. This mesh is constructed in the following way: We consider $\tilde{\mathscr{T}}_h$ as the union of the mesh $\mathscr{T}_h$ of Ω and the one of its complement $\tilde{\Omega} \setminus \Omega$. In addition, since we deal in the present numerical method with a lagrangian approach for interface description, $\mathscr{T}_h$ is constructed as the union of two meshes $\mathscr{T}^n_{h,\rm el}$ and $\mathscr{T}^n_{h,\rm al}$ of $\Omega_{\rm el}(t^n)$ and $\Omega_{\rm al}(t^n)$ respectively. Consequently, in this approach, the discretized interface Γ^n_h obtained at $t = t^n$ is defined as a union of triangles. For simplicity, we shall omit the subscript h when there is no confusion. For these reasons all used meshes depend on the time t^n and will thus have superscript n. The construction of these meshes will be described later.

In the sequel, we consider piecewise linear and constant finite elements by defining the following finite element spaces:

$$\mathcal{V}_h^n = \{v \in C^0(\overline{\Omega});\ v_{|T} \in \mathbb{P}_1\ \forall\, T \in \mathscr{T}_h^n\}, \quad \boldsymbol{\mathcal{V}}_h^n = (\mathcal{V}_h^n)^3, \tag{12.34}$$

$$\tilde{\mathcal{V}}_h^n = \{v \in C^0(\overline{\tilde{\Omega}});\ v_{|T} \in \mathbb{P}_1\ \forall\, T \in \tilde{\mathscr{T}}_h^n\}, \quad \tilde{\boldsymbol{\mathcal{V}}}_h^n = (\tilde{\mathcal{V}}_h^n)^3, \tag{12.35}$$

$$\tilde{\mathcal{X}}_h^n = \{v \in \mathcal{L}^2(\tilde{\Omega});\ v_{|T} \in \mathbb{P}_0\ \forall\, T \in \tilde{\mathscr{T}}_h^n\}, \quad \tilde{\boldsymbol{\mathcal{X}}}_h^n = (\tilde{\mathcal{X}}_h^n)^3. \tag{12.36}$$

12.3.1 Initialization

The initial conditions are given by:

$$\boldsymbol{v}^0 = \boldsymbol{v}_0,\ \varphi^0 = \varphi_0.$$

This means that initial domains Ω_{al}^0 and Ω_{el}^0 are given and therefore an initial electric conductivity σ^0, density ϱ^0 and viscosity η^0 are defined. A particularly interesting choice for our application is to take a no flow initial condition ($\boldsymbol{v}^0 = 0$). We set in addition at $t = 0$

$$\delta \boldsymbol{J}^0 = \delta \boldsymbol{B}^0 = 0.$$

The initial interface Γ^0 is assumed given by a flat horizontal surface for instance.

We have next to compute the static magnetic induction $\boldsymbol{B}_0$ by using (12.12). We consider for this a piecewise constant approximation of $\boldsymbol{B}_0$, i.e. we define $\boldsymbol{B}_{0,h} \in \tilde{\boldsymbol{\mathcal{X}}}_h$ by

$$\boldsymbol{B}_{0,h}(\boldsymbol{x}_T) = \frac{\mu_0}{4\pi} \int_{\mathbb{R}^3} \frac{\boldsymbol{x}_T - \boldsymbol{y}}{|\boldsymbol{x} - \boldsymbol{y}|^3} \times \boldsymbol{J}_0(\boldsymbol{y})\, d\boldsymbol{y}, \tag{12.37}$$

where $\boldsymbol{x}_T$ is the barycenter of the tetrahedron $T \in \mathscr{T}_h^0$.

12.3.2 Time Stepping

Now, we assume that at a time $t^n := n\,\delta t$, the following approximations are known:

$$\varphi^n \approx \varphi(\cdot, t^n),\ \boldsymbol{v}^n \approx \boldsymbol{v}(\cdot, t^n),\ \boldsymbol{J}^n \approx \boldsymbol{J}(\cdot, t^n),\ \boldsymbol{B}^n \approx \boldsymbol{B}(\cdot, t^n),\ \phi^n \approx \phi(\cdot, t^n).$$

This enables setting

$$\Gamma^n := \{\boldsymbol{x} \in \mathbb{R}^3;\ \varphi^n(\boldsymbol{x}) = 0\},$$

and the resulting subdomains $\Omega^n_{\text{al}} \approx \Omega_{\text{al}}(t^n)$ and $\Omega^n_{\text{el}} \approx \Omega_{\text{el}}(t^n)$. We emphasize that this implies a finite element mesh $\mathscr{T}^n_h$.

We compute the approximate solution at $t = t^{n+1}$ by performing the following steps:

1. We compute φ^{n+1} on the mesh $\mathscr{T}^n_h$ by running one time step of the backward Euler scheme for (12.28), i.e.

$$\frac{\varphi^{n+1} - \varphi^n}{\delta t} + \boldsymbol{v}^n \cdot \nabla\varphi^{n+1} = 0 \qquad \text{in } \Omega. \tag{12.38}$$

2. We look for the zeros of the function φ^{n+1}, which yield Γ^{n+1} and then the subdomains $\Omega_{\text{al}}(t^{n+1})$, $\Omega_{\text{el}}(t^{n+1})$. We obtain consequently a new mesh $\mathscr{T}^{n+1}_h$ and ϱ^{n+1}, σ^{n+1} and η^{n+1}.
3. We interpolate $\boldsymbol{v}^n$, $\boldsymbol{B}^n$ and ϕ^n on $\mathscr{T}^{n+1}_h$, which results in $\hat{\boldsymbol{v}}^n$, $\hat{\boldsymbol{B}}^n$ and $\hat{\phi}^n$ respectively, and compute $\boldsymbol{v}^{n+1}$ by solving the semi-discrete (by a semi-implicit Euler scheme) incompressible Navier-Stokes equations:

$$\begin{cases} \varrho^n \dfrac{\boldsymbol{v}^{n+1} - \hat{\boldsymbol{v}}^n}{\delta t} - 2\,\mathbf{div}(\eta^{n+1}\,\mathbf{D}(\boldsymbol{v}^{n+1})) & \\ \quad + \varrho^n\,(\hat{\boldsymbol{v}}^n \cdot \nabla)\,\boldsymbol{v}^{n+1} + \nabla p^{n+1} = \boldsymbol{f}(\hat{\boldsymbol{v}}^n, \hat{\boldsymbol{B}}^n, \hat{\phi}^n) & \text{in } \Omega, \\ \operatorname{div} \boldsymbol{v}^{n+1} = 0 & \text{in } \Omega, \\ \boldsymbol{v}^{n+1} = \boldsymbol{0} & \text{on } \partial\Omega. \end{cases} \tag{12.39}$$

4. We compute ϕ^{n+1} on the mesh $\mathscr{T}^{n+1}_h$ by solving the boundary value problem:

$$\begin{cases} \operatorname{div}(\sigma^{n+1}\nabla\phi^{n+1}) = \operatorname{div}(\sigma^{n+1}\boldsymbol{v}^{n+1} \times \hat{\boldsymbol{B}}^n) & \text{in } \tilde{\Omega}, \\ \sigma^{n+1}\dfrac{\partial\phi^{n+1}}{\partial n} = -j_S & \text{on } \partial\tilde{\Omega}. \end{cases} \tag{12.40}$$

5. We compute $\delta\boldsymbol{J}^{n+1}$ by

$$\delta\boldsymbol{J}^{n+1} = \sigma^{n+1}(\boldsymbol{v}^{n+1} \times \hat{\boldsymbol{B}}^n - \nabla\phi^{n+1}) - \boldsymbol{J}_0.$$

6. We compute components A^{n+1}_j, $j = 1, 2, 3$ of the vector potential $\boldsymbol{A}^{n+1}$ by solving the problems:

$$\begin{cases} -\Delta A_j^{n+1} = \mu_0\, \delta J_j^{n+1} & \text{in } \mathbb{R}^3, \\ A_j^{n+1} = \mathcal{O}(|\boldsymbol{x}|^{-1}) & |\boldsymbol{x}| \to \infty, \end{cases} \tag{12.41}$$

and update the magnetic induction by

$$\boldsymbol{B}^{n+1} = \hat{\boldsymbol{B}}^n + \mathbf{curl}\, \boldsymbol{A}^{n+1}.$$

Steps 1. to 6. are repeated for $n = 1, \ldots, N$ where $T = N\delta t$ is the final time value.

Let us now describe space discretization of each step of this time integration scheme.

12.3.3 Space Discretization

1. *Solution of the interface propagation equation.*
   Let us consider the space discretization of (12.38). We consider for this a finite element Streamline Upwind (SU) method in order to treat the convection term (see [42]). Let us introduce this through a variational formulation. The approximation of (12.38) consists in looking for $\varphi_h^{n+1} \in \mathcal{V}_h^n$ such that for all $\psi \in \mathcal{V}_h^n$,

$$\int_\Omega \Big(\frac{\varphi_h^{n+1} - \varphi_h^n}{\delta t} + \boldsymbol{v}_h^n \cdot \nabla \varphi_h^{n+1}\Big)\psi\, d\boldsymbol{x} \\ + \sum_{T\in\mathscr{T}_h} \frac{\beta h_T}{2\, \|\boldsymbol{v}_h^n\|_{\mathcal{L}^2(T)}} \int_T (\boldsymbol{v}_h^n \cdot \nabla\varphi_h^{n+1})\, (\boldsymbol{v}_h^n \cdot \nabla\psi)\, d\boldsymbol{x} = 0, \tag{12.42}$$

   the given functions $\boldsymbol{v}_h^n$ and φ_h^n being respectively in $\boldsymbol{\mathcal{V}}_{h,0}^n$ and $\mathcal{V}_h^n$. Here β is an upwinding parameter to choose independently of h. Note that the method used in (12.42) is a variant of the method developed in [42] that adds a streamwise artificial diffusion to stabilize the numerical solution.
2. *Interface updating*
   The discrete updated interface is defined by

$$\Gamma_h^{n+1} := \{\boldsymbol{x} \in \mathbb{R}^3;\ \varphi_h^{n+1}(\boldsymbol{x}) = 0\}.$$

   Since φ_h^{n+1} is a piecewise linear function, its trace on any straight line (e.g. vertical) connecting nodes on Γ_h^n can be computed. The new interface Γ_h^{n+1} is then determined by moving the nodes to the points where $\varphi_h^{n+1} = 0$.

This induces the updated mesh $\mathscr{T}_h^{n+1}$ and provides new physical properties σ^{n+1}, ϱ^{n+1} and η^{n+1}. We do not intend to present in detail the procedure that deduces $\mathscr{T}_h^{n+1}$ from $\mathscr{T}_h^n$. This is described in [166].

3. *Solution of fluid flow equations*
Space discretization of (12.39) makes use of the spaces:

$$\boldsymbol{\mathcal{V}}_{h,0}^{n+1} := \{\boldsymbol{v} \in \boldsymbol{\mathcal{V}}_h^{n+1};\ \boldsymbol{v} = \boldsymbol{0} \text{ on } \partial\Omega\},$$

$$\mathcal{Q}_h^{n+1} := \Big\{q \in \mathcal{V}_h^{n+1};\ \int_\Omega q\,d\boldsymbol{x} = 0\Big\}.$$

Note that we have used the same approximation degree ($\mathbb{P}_1$) for both velocity and pressure. This is allowable thanks to the following stabilized finite element method (cf. [81]):
Find $\boldsymbol{v}_h^{n+1} \in \boldsymbol{\mathcal{V}}_{h,0}^{n+1}$, $p_h^{n+1} \in \mathcal{Q}_h^{n+1}$ such that for all $(\boldsymbol{w}, q) \in \boldsymbol{\mathcal{V}}_{h,0}^{n+1} \times \mathcal{Q}_h^{n+1}$:

$$\begin{aligned}
&\int_\Omega \varrho^{n+1} \frac{\boldsymbol{v}_h^{n+1} - \hat{\boldsymbol{v}}_h^n}{\delta t} \cdot \boldsymbol{w}\,d\boldsymbol{x} + 2\int_\Omega \eta^{n+1}\,\mathbf{D}(\boldsymbol{v}_h^{n+1}):\mathbf{D}(\boldsymbol{w})\,d\boldsymbol{x} \\
&\quad + \int_\Omega \varrho^{n+1}\,(\hat{\boldsymbol{v}}_h^n \cdot \nabla)\,\boldsymbol{v}_h^{n+1} \cdot \boldsymbol{w}\,d\boldsymbol{x} - \int_\Omega p_h^{n+1}\,\mathrm{div}\,\boldsymbol{w}\,d\boldsymbol{x} \\
&\quad + \sum_{T \in \mathscr{T}_h} \frac{\alpha h_T^2}{\eta^n} \int_T \nabla p_h^{n+1} \cdot \nabla q\,d\boldsymbol{x} \\
&\quad + \int_\Omega q\,\mathrm{div}\,\boldsymbol{v}_h^{n+1}\,d\boldsymbol{x} = \int_\Omega \boldsymbol{f}(\hat{\boldsymbol{v}}_h^n, \hat{\boldsymbol{B}}_h^n, \hat{\phi}_h^n) \cdot \boldsymbol{w}\,d\boldsymbol{x},
\end{aligned} \tag{12.43}$$

where $\alpha > 0$ is a parameter to choose independently of h and h_T is the diameter of the tetrahedron T. We recall that $\hat{\boldsymbol{v}}_h^n$, $\hat{\phi}_h^n$ and $\hat{\boldsymbol{B}}_h^n$ are respective interpolations of $\boldsymbol{v}_h^n$, ϕ_h^n and $\boldsymbol{B}_h^n$ on the mesh $\mathscr{T}_h^{n+1}$.

4. *Solution of the potential equation*
Clearly (12.40) is an elliptic boundary value problem that can be solved using a standard finite element method. Using the finite element space $\tilde{\mathcal{V}}_h^{n+1}$ we have the discrete variational formulation that consists in seeking $\phi_h^{n+1} \in \tilde{\mathcal{V}}_h^{n+1}$ such that

$$\begin{aligned}
\int_{\tilde\Omega} \sigma^{n+1}\nabla\phi_h^{n+1} \cdot \nabla\psi\,d\boldsymbol{x} &= \int_{\tilde\Omega} \sigma^{n+1}(\boldsymbol{v}_h^{n+1} \times \hat{\boldsymbol{B}}_h^n) \cdot \nabla\psi\,d\boldsymbol{x} \\
&\quad - \int_{\partial\tilde\Omega} j s \psi\,ds \qquad \forall\,\psi \in \tilde{\mathcal{V}}_h^{n+1}.
\end{aligned}$$

Note that the right-hand side is obtained after the use of a Green formula that enables handling the Neumann boundary condition.

5. *Computation of the current density*
   Compute the electric current density $\boldsymbol{J}^{n+1}$ by using (12.5) and (12.6),

$$\boldsymbol{J}^{n+1} = \sigma^{n+1}\,(\boldsymbol{v}^{n+1} \times \hat{\boldsymbol{B}}^{n} - \nabla\phi^{n+1}),$$
$$\delta\boldsymbol{J}^{n+1} = \boldsymbol{J}^{n+1} - \boldsymbol{J}_0,$$

   in $\tilde{\Omega}$, where $\boldsymbol{J}_0$ is the given static current density.
   The evaluation of the current density by using (12.40) is rather simple and enables constructing a piecewise constant function $\boldsymbol{J}_h^{n+1}$, that we define elementwise by the following:

$$\boldsymbol{J}_{h|T}^{n+1} = \frac{1}{|T|}\int_T \sigma^{n+1}\,(\boldsymbol{v}_h^{n+1} \times \hat{\boldsymbol{B}}_h^{n} - \nabla\phi_h^{n+1})\,d\boldsymbol{x} \qquad \forall\, T \in \tilde{\mathscr{T}}_h^{n+1}.$$

6. *Computation of the magnetic induction*
   Update the magnetic induction $\boldsymbol{B}^{n+1}$ by setting

$$\boldsymbol{B}^{n+1}(\boldsymbol{x}) = \boldsymbol{B}_0(\boldsymbol{x}) + \delta\boldsymbol{B}^{n+1}(\boldsymbol{x}) \qquad \boldsymbol{x} \in \Omega. \tag{12.44}$$

   The perturbed magnetic induction $\delta\boldsymbol{B}^{n+1}$ is obtained by solving (12.14) for each component A_j of $\boldsymbol{A}$.
   Numerical solution of (12.14) can be carried out by using the same technique as in Sect. 3.3.4, i.e. by using the Poisson formula. This can be formulated as the following: Let Υ be a domain that contains $\overline{\tilde{\Omega}}$, with boundary $\partial\Upsilon$, and let B_r denote the ball with center $\boldsymbol{0}$ and radius r such that

$$\overline{\tilde{\Omega}} \subset B_r,\ \overline{B}_r \subset \Upsilon.$$

   Using ([62], Vol. 1, p. 249), we deduce that (12.14) is equivalent to:

$$-\Delta A_j^{n+1} = \mu_0\,\delta J_j^{n+1} \qquad \text{in } \Upsilon, \tag{12.45}$$

$$A_j^{n+1}(\boldsymbol{x}) = \frac{r^2 - |\boldsymbol{x}|^2}{4\pi r}\int_{\partial B_r} \frac{A_j^{n+1}(\boldsymbol{y})}{|\boldsymbol{x}-\boldsymbol{y}|^3}\,ds(\boldsymbol{y}) \qquad \boldsymbol{x} \in \partial\Upsilon, \tag{12.46}$$

   for $j = 1,2,3$. Problem (12.45)–(12.46) is solved by an iterative procedure that can be presented as a domain decomposition technique, and that can be outlined as follows:

   (1) *Initialization:* Define the j-th component $A_j^{n+1,0}$ of $\boldsymbol{A}^{n+1,0}$ as the unique solution of the Dirichlet problem:

$$\begin{cases} -\Delta A_j^{n+1,0} = \mu_0\,\delta J_j^{n+1} & \text{in } \Upsilon, \\ A_j^{n+1,0} = 0 & \text{on } \partial\Upsilon. \end{cases}$$

Here the second index in the superscript starts for the domain decomposition iteration index. Note that the boundary condition can be justified by the fact that if $\partial\Upsilon$ is *far enough* from $\partial\tilde{\Omega}$, then $A_j^{n+1,0}$ is a good approximation of the condition at the infinity in (12.14). If we consider the numerical solution of (12.14), it is necessary to extend the mesh $\tilde{\mathscr{T}}_h^{n+1}$ outside $\tilde{\Omega}$ in order to obtain a conforming mesh $\breve{\mathscr{T}}_h^n$ of Υ.

(2) Given the iterate $A_j^{n+1,k}$ for a $k \geq 0$, update the value on the boundary $\boldsymbol{g}$ by (12.46), i.e.

$$g^{k+1}(\boldsymbol{x}) = \frac{r^2 - |\boldsymbol{x}|^2}{4\pi r} \int_{\partial B_r} \frac{A_j^{n+1,k}(\boldsymbol{y})}{|\boldsymbol{x} - \boldsymbol{y}|^3} \, ds(\boldsymbol{y}) \qquad \forall\, \boldsymbol{x} \in \partial\Upsilon.$$

The above integral does not involve any singularity, the sphere ∂B_r being far from the nodes on the boundary of Υ.

Space discretization of this step uses standard finite element approximation with the mesh $\breve{\mathscr{T}}_h^{n+1}$.

(3) Solve, by finite elements, the boundary value problem:

$$\begin{cases} -\Delta A_j^{n+1,k+1} = \mu_0\, \delta J_{j,h}^{n+1} & \text{in } \Upsilon, \\ A_j^{n+1,k+1} = g_{j,h}^{k+1} & \text{on } \partial\Upsilon. \end{cases} \tag{12.47}$$

(4) Set $k := k + 1$ and repeat (2)–(3) until convergence

(5) The numerical solution of problems (12.47) provides the components of the approximate potential $\boldsymbol{A}_h$. The magnetic induction is then obtained by computing the piecewise constant vector

$$\boldsymbol{B}_h^{n+1} := \boldsymbol{B}_{0,h} + \textbf{curl}\, \boldsymbol{A}_h^{n+1} \qquad \text{in } \tilde{\Omega}.$$

where $\boldsymbol{B}_{0,h}$ is defined by (12.37).

Remark 12.3.1. The presented iterative scheme has some analogy with the one described in [18], where an integral representation of the exterior solution replaces the Poisson formula.

12.3.4 Lagrangian or Eulerian Approach

We have presented in this section a Lagrangian procedure to describe the interface motion. The finite element mesh is deformed according to the obtained interface displacement. This method is rather complex to implement but has the advantage of yielding a fitted mesh to the interface which enables accurate handling of the gaps in physical properties (mainly σ). More details about the implementation of this method can be found in [166]. An alternative consists in using an Eulerian procedure where the mesh is constant throughout time steps and the obtained interface, at each time step determines the values of σ, ϱ and η in the tetrahedra where their discontinuity appears. The advantage of this approach is that no remeshing is required each time step. However, it well known that the use of a non fitting mesh yields poor accuracy of the finite element method. To overcome this difficulty, tetrahedra in which the interface is localized must be split into tetrahedra such that the interface appears as union of boundaries of elements. This creates new nodes at each time step and requires an efficient technique to handle the resulting change of the matrix size and structure at each time step (See for this [75]).

12.4 Numerical Results

Let us present a numerical benchmark of the time stepping procedure introduced in the previous section. For this end, let us consider a simplified geometry that illustrates the main behaviour of Hall-Héroult cells. This geometry was mainly studied by Steiner [166] in his PhD thesis. The model geometry consists in a cylinder with a circular section of radius 3.5 cm and height 15 cm (See Fig. 12.4). An electric current flows in this tank entering from a cylindrical anode with radius 2.5 cm placed at the top of the tank and partially immersed in the bath. It leaves by a cathode that entirely covers the bottom. This experimental setup is relevant since it has been already considered in some studies (see [91] for instance). It enables observing typical MHD instabilities.

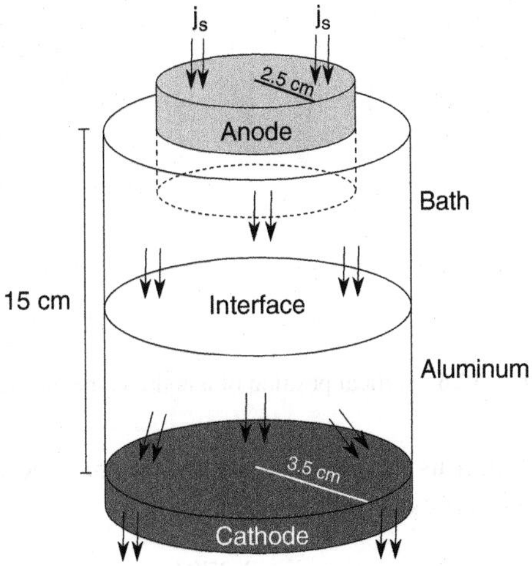

Fig. 12.4 A crucible: model for aluminium production

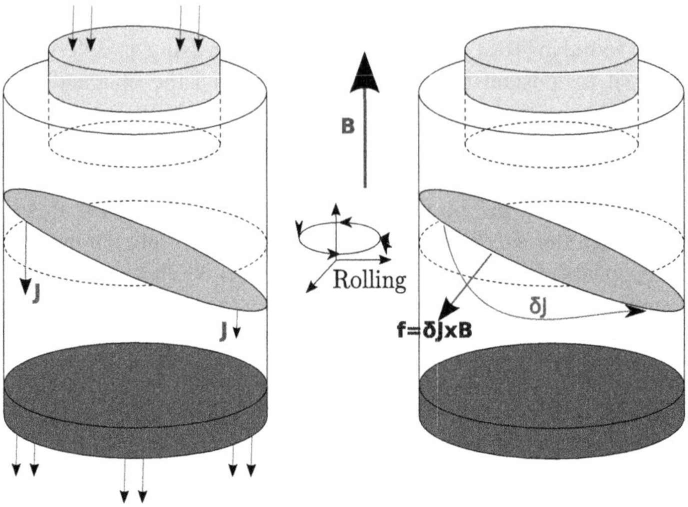

Fig. 12.5 Illustration of the rolling phenomenon in a model crucible

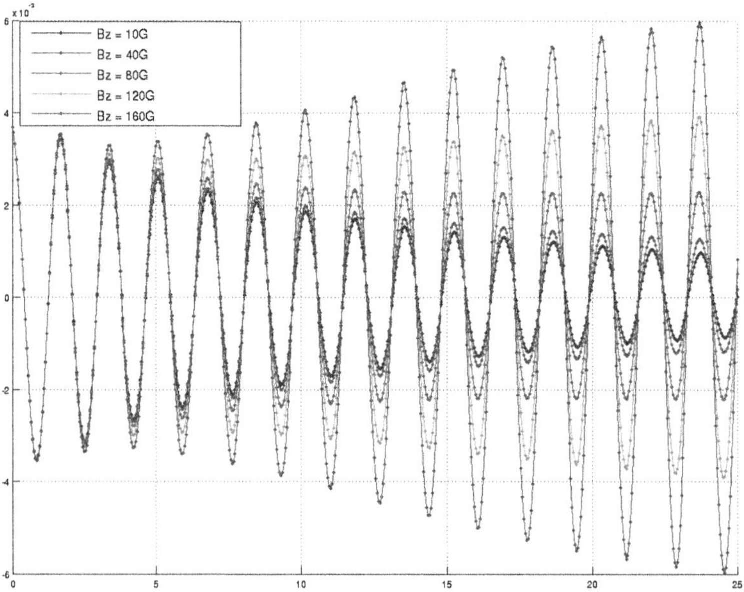

Fig. 12.6 Vertical position of a node on the interface

Let us now consider a numerical experiment. We consider for this the material properties

$$\varrho_{\mathrm{al}} = 2{,}270\,\mathrm{kg\,m^{-3}}, \qquad \varrho_{\mathrm{el}} = 2{,}130\,\mathrm{kg\,m^{-3}},$$

$$\sigma_{\mathrm{al}} = 3.33\times10^{6}\,\mathrm{S\,m^{-1}}, \qquad \sigma_{\mathrm{el}} = 210\,\mathrm{S\,m^{-1}}.$$

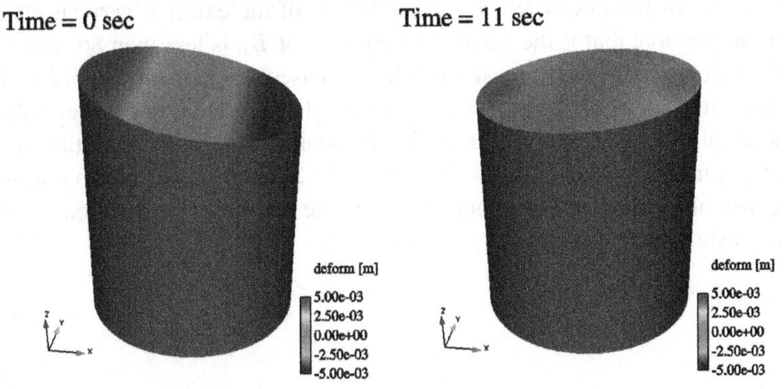

Fig. 12.7 Unstable rolling: Time values 0 and 11 s

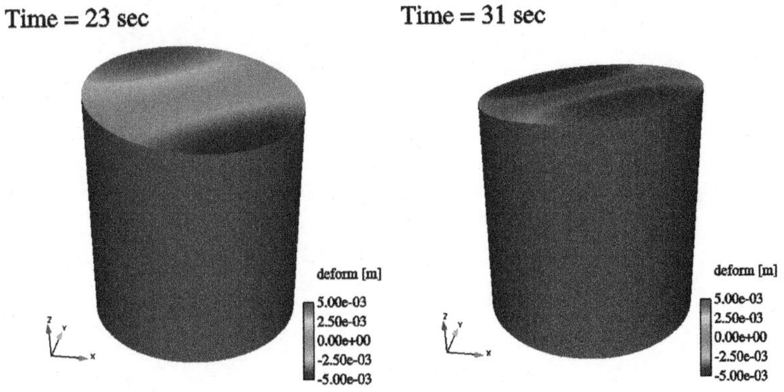

Fig. 12.8 Unstable rolling: Time values 23 and 31 s

We note that these data show a significant gap of the electric conductivity on the interface. This implies that particular care must be taken in the mesh motion procedure or if an eulerian technique is adopted.

We still note that when the source current is null ($j_S = 0$), this one being the unique source of motion in this setup, we have a stationary solution of the system (12.22)–(12.32) given by:

$$\boldsymbol{v} = \boldsymbol{0},\ \phi = \text{Const.},\ \boldsymbol{B} = \boldsymbol{J} = \boldsymbol{0}.$$

Moreover, it is easy to see that in this case the interface is flat.

Now, we prescribe a source electric current $j_S = 10\,\text{A}$ on the anode and increase an external vertical magnetic induction $\boldsymbol{B}_0$ in the range 10–160 Gauss. By taking a non flat interface Γ as initial condition with $\boldsymbol{v} = \boldsymbol{0}$, we can compute the vertical oscillations of the nodes on the interface. As an illustration, we represent in Fig. 12.6 the time history ($t = 0$ to $t = 25$ s.) of the vertical position of a node located near the boundary of the interface for various values of the external vertical magnetic induction. We note that if the vertical component of $\boldsymbol{B}_0$ is less than 80 Gauss, then the oscillation amplitude decreases as time increases which means stability. In this case, the interface tends asymptotically to the flat interface. For larger values of $\boldsymbol{B}_0$ the amplitude increases which results in instability. In both cases, the "rolling" phenomenon takes place as show Figs. 12.7–12.8. From a practical viewpoint, a severe instability can result in contact between the interface and the anode, and then in current shortcut.

Mathematical Symbols

Physical Fields and Quantities

$\boldsymbol{B}$ Magnetic induction,
$\boldsymbol{H}$ Magnetic field,
$\boldsymbol{D}$ Electric displacement field,
$\boldsymbol{E}$ Electric field,
$\boldsymbol{J}$ Electric current density,
$\boldsymbol{M}$ Magnetization field,
$\boldsymbol{v}$ Velocity of fluid,
p Pressure of fluid,
θ Temperature,
h Enthalpy,
V Voltage,
I Total electric current.

Physical Coefficients

ϱ_q Charge density,
σ Electric conductivity,
ε Electric permittivity,
μ_r Relative magnetic permeability,
μ_0 Magnetic permeability of vacuum,
μ Magnetic permeability: $\mu = \mu_r \mu_0$,
ω Angular frequency,
k Thermal conductivity,
ϱ Density,
η Molecular viscosity,

R. Touzani and J. Rappaz, *Mathematical Models for Eddy Currents and Magnetostatics: With Selected Applications*, Scientific Computation, DOI 10.1007/978-94-007-0202-8,

τ Surface tension,
κ Mean curvature of a surface.

Other Symbols

$\boldsymbol{e}_k$ k-th vector of the canonical basis of $\mathbb{R}^3$,
$\boldsymbol{x}$ Generic point of $\mathbb{R}^d$, $\boldsymbol{x} = x_1\,\boldsymbol{e}_1 + \ldots x_d\,\boldsymbol{e}_d$,
$G(\boldsymbol{x},\boldsymbol{y})$ Green function of the operator $-\Delta$ for both two-dimensional and three-dimensional situations depending on the context.

Differential Operators (Cartesian Coordinates)

$$\mathbf{curl}\,\boldsymbol{u} := \Big(\frac{\partial u_3}{\partial x_2} - \frac{\partial u_2}{\partial x_3}\Big)\boldsymbol{e}_1 + \Big(\frac{\partial u_1}{\partial x_3} - \frac{\partial u_3}{\partial x_1}\Big)\boldsymbol{e}_2 + \Big(\frac{\partial u_2}{\partial x_1} - \frac{\partial u_1}{\partial x_2}\Big)\boldsymbol{e}_3$$
(*curl* operator in 3-D),

$$\mathbf{curl}\,u := \frac{\partial u}{\partial x_2}\boldsymbol{e}_1 - \frac{\partial u_3}{\partial x_1}\boldsymbol{e}_2 \text{ (Vector } curl \text{ operator in 2-D)},$$

$$\mathrm{curl}\,\boldsymbol{u} := \frac{\partial u_2}{\partial x_1} - \frac{\partial u_1}{\partial x_2} \text{ (Scalar } curl \text{ operator in 2-D)},$$

$$\mathrm{div}\,\boldsymbol{u} := \sum_{i=1}^d \frac{\partial u_i}{\partial x_i} \text{ (Divergence operator)},$$

$$\mathbf{div}\,\mathbf{D} := \sum_{i,j=1}^d \frac{\partial D_{ij}}{\partial x_j}\boldsymbol{e}_i \text{ (Divergence of a second–order tensor } \mathbf{D}),$$

$$\nabla u := \sum_{i=1}^d \frac{\partial u}{\partial x_i}\boldsymbol{e}_i \text{ (Gradient operator for a scalar field)},$$

$$\nabla \boldsymbol{u} := \Big(\frac{\partial u_i}{\partial x_j}\Big)_{ij} \text{ (Gradient operator for a vector field)},$$

$$\Delta u := \sum_{i=1}^d \frac{\partial^2 u}{\partial x_i^2} \text{ (Laplace operator)}.$$

$$\boldsymbol{\Delta u} := \sum_{i=1}^d \frac{\partial^2 \boldsymbol{u}}{\partial x_i^2} \text{ (Laplace operator for a vector field)}.$$

Function Spaces

Let X denote an open subset of $\mathbb{R}^d$, $d = 2, 3$ with boundary ∂X.
$\mathscr{D}(X)$ Space of indefinitely differentiable functions on X with compact support,
$\boldsymbol{\mathscr{D}}(X)$ $:= \mathscr{D}(X)^d$,

$\mathcal{C}^k(X)$	Space of k-times continuously differentiable functions on X,
$\mathcal{C}^{0,\alpha}(X)$	Space of Hölder continuous functions with exponent α on X,
$\mathcal{L}^p(X)$	$(1 \le p < \infty)$ Space of measurable functions f such that $\|f\|^p$ is Lebesgue–integrable over X,
$\boldsymbol{\mathcal{L}}^p(X)$	$:= \mathcal{L}^p(X)^d$,
$\mathcal{L}^\infty(X)$	Space of measurable functions that are almost everywhere bounded on X,
$\mathcal{H}^1(X)$	$:= \{u \in \mathcal{L}^2(X);\ \nabla u \in \boldsymbol{\mathcal{L}}^2(X)\}$ with weak partial derivatives (in the sense of distributions),
$\boldsymbol{\mathcal{H}}^1(X)$	$:= \mathcal{H}^1(X)^d$,
$\mathcal{H}_0^1(X)$	$:= \{u \in \mathcal{H}^1(X);\ u = 0 \text{ on } \partial X\}$,
$\mathcal{W}^{m,p}(X)$	$:= \left\{ v \in \mathcal{L}^p(X);\ \dfrac{\partial^{\|\alpha\|} v}{\partial x_1^{\alpha_1} \ldots + \partial x_d^{\alpha_d}} \in \mathcal{L}^p(X), \|\alpha\| = \alpha_1 + \ldots + \alpha_d \le m \right\}$,
$\mathcal{H}^{\frac{1}{2}}(\partial X)$	Space of traces of functions of $\mathcal{H}^1(X)$ on the boundary ∂X of X,
$\mathcal{H}^{-\frac{1}{2}}(\partial X)$	Dual space of $\mathcal{H}^{\frac{1}{2}}(\partial X)$,
$\mathcal{W}^1(X)$	$:= \{v : X \to \mathbb{C};\ \theta v \in \mathcal{L}^2(X),\ \nabla v \in \boldsymbol{\mathcal{L}}^2(X)\}$, with $\theta(\boldsymbol{x}) = \begin{cases} \dfrac{1}{1 + \|\boldsymbol{x}\|} & \text{if } d = 3, \\ \dfrac{1}{(1 + \|\boldsymbol{x}\|)\ln(2 + \|\boldsymbol{x}\|)} & \text{if } d = 2, \end{cases}$
$\boldsymbol{\mathcal{W}}^1(X)$	$:= \mathcal{W}^1(X)^d$,
$\mathcal{W}_0^1(X)$	Closure of $\mathscr{D}(X)$ for the semi-norm $\varphi \mapsto \|\nabla \varphi\|_{\boldsymbol{\mathcal{L}}^2(X)}$, $:= \{v : X \to \mathbb{C};\ \nabla v \in \boldsymbol{\mathcal{L}}^2(X)\}$,
$\boldsymbol{\mathcal{H}}(\mathbf{curl}, X)$	$:= \{\boldsymbol{u} \in \boldsymbol{\mathcal{L}}^2(X);\ \mathbf{curl}\,\boldsymbol{u} \in \boldsymbol{\mathcal{L}}^2(X)\}$,
$\boldsymbol{\mathcal{H}}(\mathrm{div}, X)$	$:= \{\boldsymbol{u} \in \boldsymbol{\mathcal{L}}^2(X);\ \mathrm{div}\,\boldsymbol{u} \in \mathcal{L}^2(X)\}$.

References

1. Abeele, D.V., Degrez, G.: Efficient computational model for inductive plasma flows. AIAA J. **38**(2), 234–242 (2000)
2. Abramowitz, M., Stegun, I.: Handbook of Mathematical Functions. U.S. Government Printing Office, Washington (1964)
3. Adams, R.: Sobolev Spaces. Academic, New York (1975)
4. Ainsworth, M., McLean, W., Tran, T.: The conditioning of boundary element equations on locally refined meshes and preconditioning by diagonal scaling. SIAM J. Numer. Anal. **36**, 1901–1932 (1999)
5. Albanese, R., Rubinacci, G.: Finite element methods for the solution of 2-D eddy current problems. Adv. Imaging Electron Phys. **102**, 1–86 (1998)
6. Albanese, R., Rubinacci, G., Tamburrino, A., Ventre, S., Villone, F.: A fast 3-D eddy current integral formulation. COMPEL **20**(2), 317–331 (2001)
7. Amiez, G., Gremaud, P.A.: On a numerical approach to stefan like problems. Numer. Math. **59**, 71–89 (1991)
8. Amirat, Y., Touzani, R.: Self–inductance coefficient for toroidal thin conductors. Nonlinear Anal. B **131**, 233–240 (2001)
9. Amirat, Y., Touzani, R.: Asymptotic behavior of the inductance coefficient for thin conductors. Math. Models Methods Appl. Sci. **12**(2), 273–289 (2002)
10. Amirat, Y., Touzani, R.: A two-dimensional eddy current model using thin inductors. Asymptot. Anal. **58**(3), 171–188 (2008)
11. Amirat, Y., Touzani, R.: A singular perturbation problem in eddy current models (2013, submitted)
12. Ammari, H., Nédélec, J.C.: Propagation d'ondes électromagnétiques à basses fréquences. J. Math. Pures Appl. **77**, 839–849 (1998)
13. Ammari, H., Buffa, A., Nédélec, J.C.: A justification of eddy currents model for the Maxwell equations. SIAM J. Appl. Math. **60**(5), 1805–1823 (2000)
14. Amrouche, C., Bernardi, C., Dauge, M., Girault, V.: Vector potentials in three–dimensional non-smooth domains. Math. Methods Appl. Sci. **21**, 823–864 (1998)
15. Arnold, D.N., Wendland, W.L.: On the asymptotic convergence of collocation methods. Math. Comput. **41**(164), 349–381 (1983)
16. Batchelor, G.K.: An Introduction to Fluid Dynamics. Cambridge University, Cambridge (1967)
17. Bay, F., Labbé, V., Favennec, Y., Chenot, J.L.: A numerical model for induction heating processes coupling electromagnetism and thermomechanics. Int. J. Numer. Method Eng. **58**(6), 839–867 (2003)

R. Touzani and J. Rappaz, *Mathematical Models for Eddy Currents and Magnetostatics: With Selected Applications*, Scientific Computation, DOI 10.1007/978-94-007-0202-8,

18. Belgacem, F.B., Fournié, M., Gmati, N., Jelassi, F.: On the Schwarz algorithm for the elliptic exterior boundary value problems. Model. Math. Anal. Numer. **39**(4), 693–714 (2005)
19. Bénilan, P., Boccardo, L., Gallouët, T., Gariepy, R., Pierre, M., Vazquez, J.: An L^1–theory of existence and uniqueness of solutions of nonlinear elliptic equations. Annali della Scuola Normale Superiore di Pisa – Classe di Scienze, Ser. 4 **22**(2), 241–273 (1995)
20. Bermúdez, A., Rodríguez, R., Salgado, P.: A finite element method with Lagrange multipliers for low-frequency harmonic Maxwell equations. SIAM J. Numer. Anal. **40**(5), 1823–1849 (2002)
21. Bermúdez, A., Muñiz, M.C., Salgado, P.: Asymptotic approximation and numerical simulation of electromagnetic casting. Metall. Trans. B **34**(1), 83–91 (2003)
22. Bermúdez, A., Gómez, D., Muñiz, M.C., Salgado, P., Vázquez, R.: Numerical simulation of a thermo-electromagneto-hydrodynamic problem in an induction heating furnace. Appl. Numer. Math. **59**(1), 2082–2104 (2009)
23. Bernardi, C., Dauge, M., Maday, Y.: Spectral Methods for Axisymmetric Domains. Series in Applied Mathematics. Gauthier-Villars, Paris (1999)
24. Bernardi, D., Colombo, V., Ghedini, E., Mentrelli, A.: Three-dimensional modeling of inductively coupled plasma torches. Pure Appl. Chem. **77**(2), 359–372 (2005)
25. Bernardi, D., Colombo, V., Ghedini, E., Mentrelli, A., Trombetti, T.: 3-D numerical analysis of powder injection in inductively coupled plasma torches. IEEE Trans. Plasma Sci. **33**(2), 424–425 (2005)
26. Besson, O., Bourgeois, J., Chevalier, P.A., Rappaz, J., Touzani, R.: Numerical modelling of electromagnetic casting processes. J. Comput. Phys. **92**(2), 482–507 (1991)
27. Biro, O.: Edge element formulations of eddy current problems. Comput. Methods Appl. Mech. Eng. **169**, 391–405 (1999)
28. Biro, O., Preis, K.: An edge finite element eddy current formulation using a reduced magnetic and a current vector potential. IEEE Trans. Magn. **36**(5), 3128–3130 (2000)
29. Boccardo, L., Gallouët, T.: Nonlinear elliptic and parabolic equations involving measure data. J. Funct. Anal. **87**(1), 149–169 (1989)
30. Bodart, O., Boureau, A.V., Touzani, R.: Numerical investigation of optimal control of induction heating processes. Appl. Math. Model. **25**, 697–712 (2001)
31. Bonnans, J.F., Gilbert, J., Lemaréchal, C., Sagastizabal, C.: Numerical Optimization. Springer, New York (2006)
32. Bossavit, A.: On the numerical analysis of eddy current problems. Comput. Methods Appl. Mech. Eng. **27**, 303–318 (1981)
33. Bossavit, A.: Two dual formulations of the 3-D eddy currents problem. COMPEL **4**(2), 103–116 (1985)
34. Bossavit, A.: Computational Electromagnetism. Associated Press (1998)
35. Bossavit, A., Rodrigues, J.F.: On the electromagnetic induction heating problem in bounded domains. Adv. Math. Sci. Appl. **4**(1), 79–92 (1994)
36. Bossavit, A., Vérité, J.C.: The TRIFOU code: solving the 3-D eddy–current problem by using h as a state variable. IEEE Trans. Magn. (MAG–19) **6**, 2465–2470 (1983)
37. Bouchon, F., Clain, S., Touzani, R.: Numerical solution of the free boundary Bernoulli problem using a level set formulation. Comput. Methods Appl. Mech. Eng. **194**(36–38), 3934–3948 (2005)
38. Braess, D.: Finite Elements, Theory, Fast Solvers, and Applications in Solid Mechanics. Cambridge University Press, Cambridge (2001)
39. Brancher, J.P., Sero-Guilaume, O.: Sur l'équilibre des liquides magnétiques. Application à la magnétostatique. J. Mec. Theor. Appl. **2**(2), 265–283 (1983)
40. Brenner, S.C., Scott, L.R.: The Mathematical Theory of Finite Element Methods. Springer, New York (2002)
41. Brezis, H.: Analyse Fonctionnelle. Masson, Paris (1983)
42. Brooks, A.N., Hughes, T.J.R.: Streamline upwind/petrov-galerkin formulations for convection dominated flows with particular emphasis on the incompressible Navier-Stokes equations. Comput. Methods Appl. Mech. Eng. **32**, 199–259 (1982)

43. Buffa, A., Ciarlet, P. Jr.: On traces for functional spaces related to maxwell's equations. Part I: an integration by parts formula in Lipschitz polyhedra. Math. Methods Appl. Sci. **24**, 9–30 (2001)
44. Buffa, A., Ciarlet, P. Jr.: On traces for functional spaces related to Maxwell's equations. Part II: Hodge decompositions on the boundary of Lipschitz polyhedra and applications. Math. Methods Appl. Sci **24**, 31–48 (2001)
45. Casado-Díaz, J., Rebollo, T.C., Girault, V., Mármol, M.G., Murat, F.: Finite elements approximation of second order linear elliptic equations in divergence form with right-hand side in L^1. Numer. Math. **105**(3), 337–374 (2006)
46. Casas, E.: Pontryagin's principle for state-constraint boundary control problems of seminlinear parabolic equations. SIAM J. Control Optim. **35**, 1297–1327 (1997)
47. Chaboudez, C., Clain, S., Glardon, R., Rappaz, J., Swierkosz, M., Touzani, R.: Numerical modelling in induction heating of long workpieces. IEEE Trans. Magn. **30**(6), 5028–5037 (1994)
48. Chaboudez, C., Clain, S., Mari, D., Glardon, R., Swierkosz, M., Rappaz, J.: Numerical modelling in induction heating for axisymmetric geometries. IEEE Trans. Magn. **33**(1), 739–745 (1997)
49. Chadebec, O., Colomb, J.L., Janet, F.: A review of magnetostatic moment method. IEEE Trans. Magn. **42**(4), 515–520 (2006)
50. Ciarlet, P.G.: Finite element methods. In: Ciarlet, P.G., Lions, J.L. (eds.) Handbook of Numerical Analysis, vol. I, pp. 209–485. North-Holland, Amsterdam (1991)
51. Ciarlet, P.J., Sonnendrücker, E.: A decomposition of the electromagnetic field – application to the Darwin model. Math. Models Methods Appl. Sci. **7**(8), 1085–1120 (1997)
52. Clain, S.: Analyse mathématique et numérique d'un modèle de chauffage par induction. Ph.D. thesis, École Polytechnique Fédérale de Lausanne (1994)
53. Clain, S., Touzani, R.: Solution of a two–dimensional stationary induction heating problem without boundedness of the coefficients. Model. Math. Anal. Numer. **31**(7), 845–870 (1997)
54. Clain, S., Touzani, R.: A two–dimensional stationary induction heating problems. Math. Methods Appl. Sci. **20**, 759–766 (1997)
55. Clain, S., Rappaz, J., Swierkosz, M., Touzani, R.: Numerical modelling of induction heating for 2-D geometries. Math. Models Methods Appl. Sci. **3**(6), 805–822 (1993)
56. Clain, S., Rochette, D., Touzani, R.: A multislope MUSCL method on unstructured meshes applied to compressible euler equations for swirling flows. J. Comput. Phys. **229**, 4884–4906 (2010)
57. Clain, S., Touzani, R., Silva, M.L.D., Vacher, D., André, P.: A contribution on the numerical simulation of ICP torches. In: Fifth European Conference on Computational Fluid Dynamics, ECCOMAS CFD, Lisbon (2010)
58. Costabel, M.: Symmetric methods for the coupling of finite elements and boundary elements. In: Brebbia, C., Wendland, W., Kuhn, G. (eds.) Boundary Elements IX, pp. 411–420. Springer, Berlin (1987)
59. Coulaud, O., Henrot, A.: Numerical approximation of free boundary problem arising in electromagnetic shaping. Tech. Rep., INRIA (1992)
60. Crouzeix, M.: Variational approach of a magnetic shaping problem. Eur. J. Mech., B/Fluids **10**(5), 527–536 (1991)
61. Dacorogna, B.: Direct Methods in the Calculus of Variations. Applied Mathematical Sciences. Springer, Berlin (1989)
62. Dautray, R., Lions, J.L.: Mathematical Analysis and Numerical Methods for Science and Technology. Springer, Berlin (1990)
63. Descloux, J.: Stability of the solutions of the bidimensional magnetic shaping problem in absence of surface tension. Eur. J. Mech. B/Fluids **10**(5), 513–526 (1991)
64. Descloux, J.: A stability result for the magnetic shaping problem. Z. Angew. Math. Phys. **45**, 544–555 (1994)
65. Descloux, J., Flück, M., Rappaz, J.: A problem of magnetostatics related to thin plates. Model. Math. Anal. Numer. **32**, 859–876 (1998)

66. Descloux, J., Flück, M., Romerio, M.: A modelling of the stability of aluminium electrolytic cells. In: Nonlinear Partial Differential Equations and Their Applications, Collège de France Seminar, vol. XIII (Paris, 1994/1996). Volume 391 of Pitman Research Notes in Mathematics Series, pp. 117–133. Longman, Harlow (1998)
67. Descloux, J., Flück, M., Rappaz, J.: Modelling and mathematical results arising from ferromagnetic problems. Sci. China Math. **55**(5), 1053–1067 (2012)
68. Djaoua, M.: Équations intégrales pour un problème singulier dans le plan. Ph.D. thesis, École Polytechnique (1977)
69. Dreyfuss, P.: Analyse numérique d'une méthode intégrale sans singularité – application à l'électromagnétisme. Ph.D. thesis, École Polytechnique Fédérale de Lausanne (1999)
70. Dreyfuss, P., Rappaz, J.: Numerical analysis of a non singular boundary integral method. Part I: the circular case. Math. Methods Appl. Sci. **24**, 847–863 (2001)
71. Dreyfuss, P., Rappaz, J.: Numerical analysis of a non singular boundary integral method. Part II: the general case. Math. Methods Appl. Sci. **25**, 557–570 (2002)
72. Egan, L.R., Furlani, E.P.: A computer simulation of an induction heating system. IEEE Trans. Magn. **27**, 4343–4354 (1991)
73. Favennec, Y., Labbé, V., Bay, F.: Induction heating processes optimization: a general optimal control approach. J. Comput. Phys. **187**, 68–94 (2003)
74. Feynman, R.P., Leighton, R.B., Sands, M.: The Feynman Lectures on Physics. The Commemorative Ed. Addison Wesley, Redwood City (1989)
75. Flotron, S.: Simulations numériques de phénomènes MHD–thermiques avec interface libre dans l'électrolyse de l'aluminium. Ph.D. thesis, École Polytechnique Fédérale de Lausanne (2013)
76. Flück, M., Rumpf, M.: Bernoulli's free-boundary problem, qualitative theory and numerical approximation. J. Reine Angew. Math. **486**, 165–204 (1997)
77. Flück, M., Hofer, T., Picasso, M., Rappaz, J., Steiner, G.: Scientific computing for aluminium production. Int. J. Numer. Anal. Model. **1**(1), 1–20 (2008)
78. Flück, M., Janka, A., Laurent, C., Picasso, M., Rappaz, J., Steiner, G.: Some mathematical and numerical aspects in aluminium production. J. Sci. Comput. **1**(1), 1–20 (2009)
79. Flück, M., Rappaz, J., Safa, Y.: Numerical simulation of thermal problems coupled with magnetohydrodynamic effects in aluminum cells. Appl. Math. Model. **33**(3), 1479–1492 (2009)
80. Flück, M., Hofer, T., Janka, A., Rappaz, J.: Numerical methods for ferromagnetic plates. Appl. Numer. Partial Differ. Equ. Comput. Methods Appl. Sci. **15**, 169–182 (2010)
81. Franca, L.P., Muller, R.L., Hughes, T.J.R.: Convergence analyses of Galerkin least-squares methods for symmetric advective-diffusiv forms of the Stokes and incompressible Navier-stokes equations. Comput. Methods Appl. Mech. Eng. **105**(2), 285–298 (1993)
82. Friedman, M.J.: Mathematical study of the nonlinear singular integral magnetic field equation. I. SIAM J. Appl. Math. **39**(1), 14–20 (1980)
83. Friedman, M.J.: Mathematical study of the nonlinear singular integral magnetic field equation. II. SIAM J. Appl. Math. **18**(4), 644–653 (1981)
84. Friedman, M.J.: Mathematical study of the nonlinear singular integral magnetic field equation. III. SIAM J. Appl. Math. **12**(4), 536–540 (1981)
85. Gagnoud, A., Etay, J., Garnier: Le problème de lévitation en frontière libre électromagnétique. J. Mec. Theor. Appl. **5**(6), 911–925 (1986)
86. Gallouët, T., Herbin, R.: Existence of a solution to a coupled elliptic system. Appl. Math. Lett. **7**(2), 49–55 (1994)
87. Gatica, G.: An alternative variational formulation for the Johnson & Nédélec's coupling procedure. Rev. Math. Appl. **16**, 17–41 (1995)
88. Gauthier-Béchonnet, S.: Résolution et mise en œuvre d'un modèle tridimensionnal des courants de foucault. Ph.D. thesis, Université Blaise Pascal, Clermont-Ferrand (1998)
89. Gerbeau, J.F., Le Bris, C., Bercovier, M.: Existence of solution for a density-dependent magnetohydrodynamic equation. Adv. Differ. Equ. **2**(3), 427–452 (1997)

90. Gerbeau, J.F., Le Bris, C., Bercovier, M.: Spurious velocities in the steady flow of an incompressible fluid subjected to external forces. Int. J. Numer. Method Fluids **25**, 679–695 (1997)
91. Gerbeau, J.F., Le Bris, C., Le Lièvre, T.: Mathematical Methods for the Magnetohydrodynamics of Liquid Metals. Oxford University Press, Oxford (2006)
92. Gilbarg, D., Trudinger, N.: Elliptic Partial Differential Equations of Second Order. Springer, Berlin (1970)
93. Girault, V., Raviart, P.A.: Finite Element Approximation of the Navier-Stokes Equations. Springer, Berlin (1985)
94. Gleize, A., Gonzales, J., Freton, P.: Thermal plasma modelling. J. Phys. D Appl. Phys. **38**, R153–R183 (2005)
95. Haddar, H., Joly, P.: Effective boundary conditions for thin ferromagnetic layers: the one dimensional model. SIAM J. Appl. Math. **6**(4), 1386–1417 (2001)
96. Henneron, T.: Contribution à la prise en compte des grandeurs globales dans les problèmes d'électromagnétisme résolus avec la méthode des éléments finis. Ph.D. thesis, Université Lille I (2004)
97. Henrot, A., Pierre, M.: Un problème inverse en formage des métaux liquides. Model. Math. Anal. Numer. **23**(1), 155–177 (1989)
98. Henrot, A., Pierre, M.: Variation et optimisation de forme: une analyse géométrique. Springer, Berlin (2005)
99. Hernández, R.V.: Contributions to the mathematical study of some problems in magnetohydrodynamics and induction heating. Ph.D. thesis, Universidade de Santiago de Compostela (2008)
100. Hiptmair, R.: Symmetric coupling for eddy current problems. SIAM J. Numer. Anal. **40**(1), 41–65 (2002)
101. Hiptmair, R., Sterz, O.: Current and voltage excitations for the eddy current model. Int. J. Numer. Model. **18**, 1–21 (2005)
102. Hofer, T.: Numerical simulation and optimization of the alumina distribution in an aluminium electrolysis pot. Ph.D. thesis, École Polytechnique Fédérale de Lausanne (2011)
103. Hsiao, G.: On boundary integral equations of the first kind. J. Comput. Math. **7**(2), 121–131 (1989)
104. Hsiao, G.: Boundary element methods – an overview. Appl. Numer. Math. **56**, 1356–1369 (2006)
105. Hsiao, G., Wendland, W.: Boundary element methods: foundation and error analysis. In: Encyclopedia of Computational Mechanics, vol. 1, chap. 12, pp. 339–373. Wiley (2005)
106. Hughes, T.J.R.: The Finite Element Method, Linear Static and Dynamic Finite Element Analysis. Dover, Mineola (2000)
107. Jackson, J.: Classical Electrodynamics. Wiley, London (1965)
108. Johnson, C., Nédélec, J.C.: On the coupling of boundary integral and finite element methods. Math. Comput. **35**, 1063–1079 (1980)
109. Joly, P., Vacus, O.: Mathematical and numerical studies of nonlinear ferromagnetic material. Model. Math. Anal. Numer. **33**(3), 593–626 (1999)
110. Kanayama, H., Tagami, D., Saito, M., Kikuchi, F.: A numerical method for 3-D eddy current problems. Jpn. J. Ind. Appl. Math. **18**(2), 603–612 (2001)
111. Kim, D.H., Hahn, S.Y., Park, I.H., Cha, G.: Computation of three–dimensional electromagnetic field including moving media by indirect boundary integral equation method. IEEE Trans. Magn. **35**(3), 1932–1938 (1999)
112. Klein, O., Philip, P.: Correct voltage distribution for axisymmetric sinusoidal modelling of induction heating with prescribing current, voltage, or power. IEEE Trans. Magn. **38**(3), 1519–1523 (2002)
113. Kuhn, M., Steinbach, O.: Symmetric coupling of finite and boundary elements for exterior magnetic field problems. Math. Methods Appl. Sci. **25**, 357–371 (2002)
114. Kuster, C., Gremaud, P., Touzani, R.: Fast numerical methods for Bernoulli free boundary problems. SIAM J. Sci. Comput. **29**(2), 622–634 (2007)

115. Labridis, D., Dokopoulos, P.: Calculation of eddy current losses in nonlinear ferromagnetic materials. IEEE Trans. Magn. **25**, 2665–2669 (1989)
116. Landau, L., Lifshitz, E.: Electrodynamics of Continuous Media. Pergamon, London (1960)
117. Landau, L., Lifshitz, E.: Fluid Mechanics. Pergamon, London (1960)
118. Leray, J., Schauder, J.: Topologie et équations fonctionnelles. Ann. Sci. Ecole Norm. Sup. **51**, 45–78 (1934)
119. Leroux, M.N.: Résolution numérique du problème du potentiel dans le plan par une méthode variationnelle d'éléments finis. Ph.D. thesis, Université de Rennes (1974)
120. Leroux, M.N.: Méthode d'éléments finis pour la résolution numérique de problèmes extérieurs en dimension 2. R.A.I.R.O. Analyse Numérique **11**(1), 27–60 (1977)
121. Li, B.Q.: The fluid flow aspects of electromagnetic levitation processes. Int. J. Eng. Sci. **32**(1), 45–67 (1989)
122. Li, H.: Finite element analysis for the axisymmetric Laplace operator on polygonal domains. J. Comput. Appl. Math. **235**, 5155–5176 (2011)
123. Li, B.Q., Evans, J.W.: Computation of shapes of electromagnetically supported menisci in electromagnetic casters. Part I: calculations in two dimensions. IEEE Trans. Magn. **25**(6), 4443–4448 (1989)
124. Lions, J.L., Magenes, E.: Problèmes aux limites non homogènes et applications, Tome I. Dunod, Paris (1968)
125. Massé, P., Morel, B., Breville, T.: A finite element prediction correction scheme for magneto-thermal coupled problem during Curie transition. IEEE Trans. Magn. **25**, 181–183 (1989)
126. Masserey, A.: Optimisation et simulation numérique du chauffage par induction pour le procédé de thixoformage. Ph.D. thesis, École Polytechnique Fédérale de Lausanne (2002)
127. Masserey, A., Rappaz, J., Rozsnyo, R., Touzani, R.: Optimal control of an induction heating process for thixoforming. IEEE Trans. Magn. **40**(3), 1657–1663 (2004)
128. Masserey, A., Rappaz, J., Rozsnyo, R., Touzani, R.: Power formulation for the optimal control of an industrial induction heating process for thixoforming. Int. J. Appl. Electromagn. Mech. **19**, 51–56 (2004)
129. Meir, A.: Thermally coupled, stationary, incompressible MHD flow; existence, uniqueness, and finite element approximation. Numer. Methods PDE **11**, 311–337 (1995)
130. Meir, A., Schmidt, P.G.: Variational methods for stationary MHD flow under natural interface conditions. Nonlinear Anal. Theory Methods Appl. **24**(4), 659–689 (1996)
131. Monk, P.: Finite Element Methods for Maxwell's Equations. Oxford University Press, Oxford (2003)
132. Monk, P., Vacus, O.: Error estimates for a numerical scheme for ferromagnetic problems. SIAM J. Numer. Anal. **36**(3), 696–718 (1999)
133. Montaser, A., Golightly, D.W. (eds.): Inductively Coupled Plasmas in Analytical Atomic Spectrometry. VCH Publishers, Inc., New York (1992)
134. Natarajan, T., El-Kaddah, N.: A methodology for two-dimensional finite element analysis of electromagnetically driven flow in induction stirring systems. IEEE Trans. Magn. **35**(3), 1773–1776 (1999)
135. Nédélec, J.C.: Notions sur les équations intégrales de la physique. Centre de Mathématiques Appliquées, École Polytechnique, Palaiseau (1977)
136. Nédélec, J.C.: Mixed finite elements in $\mathbb{R}^3$. Numer. Math. **35**(3), 315–341 (1980)
137. Nédélec, J.C.: A new family of mixed finite elements in $\mathbb{R}^3$. Numer. Math. **50**, 57–81 (1986)
138. Nédélec, J.C.: Acoustic and Electromagnetic Equations. Integral Representations for Harmonic Problems. Springer, New York (2001)
139. Neff, H.: Introductory Electromagnetics. Wiley, New York (1991)
140. Parietti, C.: Modélisation mathématique et analyse numérique d'un problème de chauffage électromagnétique. Ph.D. thesis, École Polytechnique Fédérale de Lausanne (1998)
141. Parietti, C., Rappaz, J.: A quasi–static two–dimensional induction heating problem. Part I: modelling and analysis. Math. Models Methods Appl. Sci. **8**(6), 1003–1021 (1998)
142. Parietti, C., Rappaz, J.: A quasi–static two–dimensional induction heating problem. Part II: numerical analysis. Math. Models Methods Appl. Sci. **9**(9), 1333–1350 (1999)

143. Pierre, M., Roche, J.R.: Computation of free surfaces in the electromagnetic shaping of liquid metals by optimization algorithms. Eur. J. Mech. B/Fluids **10**(5), 489–500 (1991)
144. Pierre, M., Roche, J.R.: Numerical simulation of electromagnetic shaping of liquid metals. Tech. Rep., INRIA (1992)
145. Rapetti, F., Bouillaut, F., Santandrea, L., Buffa, A., Maday, Y., Razek, A.: Calculation of eddy currents with edge elements on non-matching grids in moving structures. IEEE Trans. Magn. **10**(5), 482–507 (1991)
146. Rappaz, J., Swierkosz, M.: Mathematical modeling and numerical simulation of induction heating process. Appl. Math. Comput. Sci. **6**(2), 207–221 (1996)
147. Rappaz, J., Swierkosz, M.: Boundary-element method yields external vector potentials in complex industrial applications. Comput. Phys. **11**(2), 145–150 (1997)
148. Rappaz, J., Touzani, R.: Modelling of a two–dimensional magnetohydrodynamic problem. Eur. J. Mech. B/Fluids **10**(5), 482–507 (1991)
149. Rappaz, J., Touzani, R.: On a two–dimensional Magnetohydrodynamic problem, I: modelling and analysis. Model. Math. Anal. Numer. **26**(2), 347–364 (1992)
150. Rappaz, J., Touzani, R.: On a two–dimensional Magnetohydrodynamic problem, II: numerical analysis. Model. Math. Anal. Numer. **30**(2), 215–235 (1996)
151. Rappaz, M., Bellet, M., Deville, M.: Modélisation numérique en science des matériaux. Presses Polytechniques et Universitaires Romandes, Lausanne (1998)
152. Rappaz, J., Swierkosz, M., Trophime, C.: Un modèle mathématique et numérique pour un logiciel de simulation tridumensionnelle d'induction électromagnétique. Tech. Rep., École Polytechnique Fédérale de Lausanne (1999)
153. Raviart, P.A., Thomas, J.M.: A mixed finite element method for 2nd order elliptic problems. In: Bänsch, E., Dold, A. (eds.) Mathematical Aspects of Finite Element Methods. Lecture Notes in Mathematics, vol. 606, p. 503. Springer, New York/Rome (1977)
154. Reitz, J., Milford, F.: Foundations of Electromagnetic Theory. Addison–Wesley, Reading (1975)
155. Robinson, N.: Electromagnetism. Oxford Physics Series. Clarendon, Oxford (1973)
156. Rochette, D.: Contributions à la simulation d'écoulements de plasma haute pression appliquée aux appareillages de coupure et torches à plasma. Ph.D. thesis, Université Blaise Pascal (2012). Habilitation Thesis
157. Rodríguez, A.A., Valli, A.: Eddy Current Approximation of Maxwell Equations. Springer, Milan (2010)
158. Rodríguez, A.A., Valli, A., Hernández, R.V.: A formulation of the eddy current problem in the presence of electric ports. Numer. Math. **113**, 643–672 (2009)
159. Rogier, F.: Problèmes mathématiques et numériques liés à l'approximation de la géométrie d'un corps diffractant dans les équations de l'électromagnétisme. Ph.D. thesis, École Polytechnique (1989)
160. Roy, S.S., Cramb, A.W., Hoburg, J.F.: Magnetic shaping of columns of liquid sodium. Metall. Trans. B **26**(1), 1191–1197 (1995)
161. Sakane, J., Li, B., Evans, J.: Mathematical modeling of meniscus profile and melt flow in electromagnetic casters. Metall. Trans. B **19**(2), 397–408 (1988)
162. Schercliff, J.A.: Magnetic shaping of molten metal columns. Proc. R. Soc. Lond. A Math. Phys. Sci. **275**(1763), 455–473 (1981)
163. Schmidlin, G., Fischer, U., Andjelic, Z., Schwab, C.: Preconditioning the second-kind boundary integral equations for 3-D eddy current problems. Int. J. Numer. Meth. Eng. **10**(5), 482–507 (1991)
164. Sethian, J.: Level Set Methods and Fast Marching Methods: Evolving Interfaces in Computational Geometry, Fluid Mechanics, Computer Vision and Materials Science. Cambridge University Press, Cambridge (1999)
165. Sneyd, A., Moffat, H.: Fluid dynamical aspects of the levitation melting process. J. Fluid Mech. **117**, 45–70 (1982)
166. Steiner, G.: Simulation numérique de phénomènes MHD: application à l'électrolyse de l'aluminium. Ph.D. thesis, École Polytechnique Fédérale de Lausanne (2009)

167. Stephan, E.: Coupling of boundary element methods and finite element methods. In: Encyclopedia of Computational Mechanics, vol. 1, chap. 13, pp. 375–412. Wiley, Chichester (2005)
168. Szabó, B., Babuška, I.: Finite Element Analysis. Wiley-Interscience, New York (1991)
169. Touzani, R.: Un problème de courant de Foucault avec inducteur filiforme. C. R. Acad. Sci. tome 319, Série **I**, 771–776 (1994)
170. Touzani, R.: Analysis of an eddy current problem involving a thin inductor. Comput. Methods Appl. Mech. Eng. **131**, 233–240 (1996)
171. Vérité, J.C.: Trifou : un code de calcul tridimensionnel des courants de foucault. EDF Bulletin de la direction des études et recherches, Série C, Mathématiques et Informatique **2**, 79–92 (1983)
172. Vérité, J.C.: Traitement du potentiel scalaire magnétique extérieur dans le cas d'un domaine multiplement connexe. application au code TRIFOU. EDF Bulletin de la direction des études et recherches, Série C, Mathématiques et Informatique **1**, 61–75 (1986)
173. Wang, J., Xie, D., Yao, Y., Mohammed, O.: A modified solution for large sparse symmetric linear systems in electromagnetic field analysis. IEEE Trans. Magn. **37**(5), 3494–3497 (2001)
174. Wanser, S., Krähenbühl, L., Nicolas, A.: A computation of 3D induction hardening problems by combined finite and boundary element methods. IEEE Trans. Magn. **30**(5), 3320–3323 (1994)
175. Wendland, W.L.: On the asymptotic convergence of boundary integral methods. In: Brebbia, C.A. (ed.) Boundary Element Methods, pp. 412–430. Springer, Berlin (1981)
176. Xue, S., Proulx, P., Boulos, M.I.: Extended-field electromagnetic model for inductively coupled plasma. J. Phys. D Appl. Phys. **34**(4), 1897–1906 (2001)
177. Yamazaki, K.: Transient eddy current analysis for moving conductors using adaptive moving coordinate systems. IEEE Trans. Magn. **36**(4), 785–789 (2000)

Index

R. Touzani and J. Rappaz, *Mathematical Models for Eddy Currents and Magnetostatics: With Selected Applications*, Scientific Computation, DOI 10.1007/978-94-007-0202-8,

Printed in the United States
By Bookmasters